Praise for
Patriot Battles

"Spirited interpretation. . . . These observations set in motion a sharply focused view of the Revolution. . . . Even readers familiar with the military history of the Revolution will find fresh insights in Stephenson's accounts of the major battles."
—*Boston Globe*

"A fascinating story of how America's most important war was fought, and won."
—*Seattle Times*

"Engrossing. . . . *Patriot Battles* lets us feel what it was like to be in the heat and smoke, to realize, in all its material detail, what those men must have felt advancing on the enemy or fleeing from them as the surgeons waited, scalpel and musket ball extractor in hand, in the rear of the ragtag army."
—*Providence Journal*

"This excellent popular history should attract a wide audience with its fresh perspective."
—*Publishers Weekly*

"Comprehensive. . . . An interesting and easily digestible study appealing to both military-history buffs and general readers."
—*Booklist*

"An iconoclastic, provocative study of the [War of Independence]. . . . Contrarian and well written."
—*Kirkus Reviews*

"[An] engrossing and often revealing new look at our struggle for independence. . . . Stephenson artfully raises the harsh realities that played such a pivotal role in the outcome of the lofty strategies of the commanders. The book is a timely antidote that forces us to reassess the legends handed down through the generations and the factors that governed key decisions. . . . Sharp and different survey of our revolution. . . . The intimate knowledge of day-to-day life in the ranks that Stephenson has given us . . . refreshes our grasp of each battle. The maps are lucid and the analyses of strategy and tactics consistently sensible."
—*Philadelphia Inquirer*

About the Author

Michael Stephenson is the former editor of the Military Book Club, the largest of its kind in the world. He was the editor of and a contributor to National Geographic's *Battlegrounds: Geography and the History of Warfare* and the coauthor of *The Nuclear Casebook*. He lives in New York City.

PATRIOT BATTLES

How the War of Independence Was Fought

MICHAEL STEPHENSON

HARPER ● PERENNIAL

NEW YORK ● LONDON ● TORONTO ● SYDNEY ● NEW DELHI ● AUCKLAND

For my father,
Joseph Stephenson,
a soldier of the Second World War

HARPER ● PERENNIAL

A hardcover edition of this book was published in 2007 by HarperCollins Publishers.

FIRST HARPER PERENNIAL EDITION PUBLISHED 2008.

Map illustrations by Casebourne Rose Design Associates

Designed by Mary Austin Speaker

The Library of Congress has catalogued the hardcover edition as follows:
Patriot battles : how the War of Independence was fought / Michael Stephenson.
    p. cm.
Includes bibliographical references and index.
ISBN: 978-0-06-073261-5
ISBN-10: 0-06-073261-X
1. Battles—United States—History—18th century. 2. United States—History—Revolution, 1775–1783—Campaigns. I. Title.
973.3'3—dc22
E230.S75   2007   2006041317

ISBN 978-0-06-073262-2 (pbk.)

08 09 10 11 12 ID/RRD 10 9 8 7 6 5 4 3 2

CONTENTS

MAPS

ACKNOWLEDGMENTS

IT WAS MY GREAT good fortune to have been commissioned to write this book by one of the most experienced and revered editors of American publishing: Hugh Van Dusen of HarperCollins. Throughout he has been a model of kindness and constructive criticism. The associate editor, Marie Estrada, was a paragon of care, courtesy, and professionalism; and Vicki Haire, who copy-edited the manuscript, was wonderfully diligent and hawkeyed. I hasten to add, however, that any errors which may remain are, of course, entirely mine. Alex Hoyt has been a stalwart adviser and occasionally has had to endure the bleatings an author will invariably direct at his long-suffering agent.

Clay Smith, journeyman gunsmith at Colonial Williamsburg, was enormously helpful in educating me about eighteenth-century gunsmithing; and the ballistics expert Martin L. Fackler, M.D., an eminent battlefield surgeon and a retired colonel in the U.S. Army's Medical Corps, helped me understand the nature of gunshot wounds. I owe thanks to my friends Mike and Sue Rose of Casebourne Rose Design Associates for producing the maps. Peter Johnson, a friend of many years, kept me good company in tramping some of the major battlefields, and my dear friend LuAnn Walther tracked down an elusive (for me) Tolstoyan reference to explosive shells!

If it had not been for the truly heroic forbearance of my wife, Kathryn Court, this book would not have been written, for its author would have been found swinging from a beam in the barn—if we had had a barn.

INTRODUCTION

In Blenheim Palace, that lumpy and unlovely McMansion, there is a series of tapestries commissioned to glorify the military career of John Churchill, first duke of Marlborough. In one, the victorious general is, of course, front and center. Dressed in his great brocaded frock coat and full bagwig, he sits confidently astride his magnificent steed while pointing magisterially, if a little vaguely, to his battalions battling it out below on the Flemish plain. Tucked away in the bottom right-hand corner lies a man as dead as a doorknob but done nicely in perspective, his head to the viewer. He was a cavalryman or perhaps a mounted officer—whether French or English, whether friend or foe, hardly matters now. His wig, that essential mark of the gentleman, has fallen off to uncover, in shocking revelation, his shaved head. Under the coiffure lay the skull. It seems to me the most poignant detail of the whole heroic schema. It is warfare stripped of grandeur and grandiloquence. The image of that fallen soldier has guided not only the motivation but also the method of this book.

Patriot Battles is dedicated to a deceptively simple objective: to take off the wig and other accumulated finery that a couple of centuries and more of historiography has piled on, and look the war in the eye. Who fought? Why did they fight? How did they sustain themselves? With what did they fight? How did they fight? These are questions I have tried to address by deconstructing the traditional narrative battle histories as well as drawing on a range of specialist studies. Field Marshal Lord

Wavell, writing to the great military theoretician and historian Basil Liddell Hart, put his finger on it: "I think I should concentrate almost entirely on the 'actualities of war'—the effects of tiredness, hunger, fear, lack of sleep, weather. . . . The principles of strategy and tactics, and the logistics of war are really absurdly simple: it is the actualities that make war so complicated and so difficult, and are usually so neglected by historians." As John Keegan has so brilliantly demonstrated: God is in the details and the physical realities illuminate the larger picture.

My interest in the War of Independence started with a gentle paddling in the warm shallows of popular history. It was cozy and reassuring. For example, British soldiers were often characterized as criminals who had been beaten into submission and were led by sluggard, doltish, and venal commanders. The Americans were lean and freethinking, their commanders, by comparison with their sclerotic British counterparts, wonderfully gifted amateurs, fresh, imaginative, and free of petty squabbles. Washington emulated the noble Roman Fabius, a patriot soldier who chose to run away in order to "live to fight another day." Nimble partisan tactics rather than the tedious formality of an old European style of combat had won the war. The conflict had been a great rallying of popular will and determination: it was the first People's War. But as one got into the weft and weave of the war, it became clear that it had become trapped in amber: embalmed over the centuries by the slow accretion of national mythology and popular history.

Eighteenth-century warfare seems exotically formalized, strangely balletic, and "unreal" by modern standards. Certainly, the limitations of the primary weapons—musket and cannon—imposed on the soldiers a complicated ritual of arms drill. These "evolutions," as they were called, seem laughable or weird to those of us who have been fed on the wham-bam action of special-forces computer simulations and the techno-pornography of Hollywood movies. But there was nothing laughable about the realities of the combat of those times. The very limitations of weaponry (a common musket was fairly useless at ranges of more than sixty yards or so) imposed a shape on battle that demanded the highest degree of determination and courage on the part of the foot soldiers at the sharp end. To march to within, say, forty yards of an enemy, receive his fire, and then close in for the kill took a prodigious amount of nerve, as one of the great French military theoreticians of the eighteenth century, Jacques-Antoine-Hippolyte de Guibert, emphasized:

> *The kind of soldier who acts only under pressure will be frightened to see the enemy come so near, and he will often seek safety in flight without attempting to defend himself. The closer you approach the enemy the more fearsome you become, and a coward who will fire on a brave man at one hundred paces, will not dare so much as aim at him at close range.*[1]

The terrifying proximity so characteristic of eighteenth-century battle was described by Sergeant Roger Lamb of the 23rd Foot (Royal Welch Fusiliers) at the battle of Guilford Courthouse on 15 March 1781. When the British infantry approached within forty yards of the ranks of the North Carolina militia, who were resolutely ensconced behind a fence, "it was perceived," reported the understandably dismayed Lamb, "that their whole line had their arms presented, and resting on a rail fence. . . . They were taking aim with the nicest ["nice" meant "exact" in the eighteenth century] precision. . . . At this awful moment a general pause took place; both parties surveyed each other with the most anxious suspense."[2] Urged on by 23rd's Lieutenant Colonel James Webster, the British broke the spell and charged into the militia's fire. "Dreadful was the havoc on both sides," says Lamb.

Dreadful indeed, but nothing compared with the Civil War. Including the mortally wounded (but excluding deaths from disease), about 7,000 Americans were killed in battle during the War of Independence. The British lost approximately 4,000 killed in action; the Hessians 1,200. By comparison, over the course of the Civil War the Union suffered about 110,000 battle deaths; the Confederacy 94,000. At Antietam alone (generally reckoned to be the single bloodiest day in American military history) 26,000 men became casualties whereas the butcher's bill for the bloodiest battles of Washington's war rarely exceeded 1,500 killed and wounded for both sides combined. For example, at the battle of Bunker's Hill, one of the most sanguine of the whole war, 140 Americans were killed and 271 wounded. Britain lost 226 dead and 828 wounded, for a combined total of 1,465.

The Cinderella relationship of the War of Independence to the Civil War is a reflection, to some extent, of our taste for the red meat of military history. Big body counts may sell books, movies, and TV documentaries, but they should not obscure the often brutal realities of eighteenth-century warfare. For individual units the casualty levels could be fearful. At the battle of Brooklyn on 27 August 1776, for example, the 400-strong Maryland Brigade left 256 dead on the field after their heroic forlorn-hope counterattack against overwhelming odds. Lieutenant Frederick MacKenzie, a British officer at Bunker's Hill, reported that "most of our Grenadiers and Light-infantry, the moment of presenting themselves, lost three-fourths, and many nine-tenths, of their men. Some had only eight or nine a company left [a company would have had approximately sixty officers and men]; some only three, four, and five."[3] The 62nd Foot, one of the regiments in the center of the British line at the battle of Freeman's Farm (the first battle of Saratoga) on 19 September 1777, ". . . had scarce 10 men a company left," recorded Lieutenant William Digby. The regiment went from 350 officers and men to 60—a staggering loss of 83 percent.

A fortuitous collision of past and present added an extra dimension and relevance to the research for *Patriot Battles* and made me think about the "uses of history." The invasion of Iraq unrolled in March 2003. The two wars, separated by centuries, immediately set up a dialogue of comparison and cross reference: the past illuminated by the present; the present made more comprehensible by the past. For some the comparison was both good and bad, depending on what benefit they sought to extract. For example, the estimable David McCullough (who had published *1776* in 2005, a study of one of the most difficult years of the war as far as the patriot cause was concerned), politely but perhaps a tad testily rejected the notion put to him by a left-leaning radio talk-show host that there might be some connection between, say, the partisan tactics of 1776 involving assassination of loyalist opposition leaders, the destruction of loyalist property, and the general suppression of pro-Crown sympathizers and the similar tactics of "insurrectionists" in Iraq.

Not only was the specific analogy incorrect, insisted Mr. McCullough, but it was also intellectually inadmissible, even dangerous, to apply the lessons of history to contemporary events.

President Bush, however, an admirer of Mr. McCullough's book, seemed to suggest by association that he was in the same tight spot as George Washington had been in 1776 and, by emulating the perseverance and determination of the first commander in chief, would pull the nation out of the black hole of Iraq into which he had led it. What President Bush failed to see was that he had much more in common with George III than with George Washington.[4]

One of the leading historians of the War of Independence has called the comparison with the Vietnam war, for example, "overwrought,"[5] and there is an understandable instinct to insulate the sanctity of the great war of national liberation from any association with some of the more "awkward" periods of American history. But the comparisons are illuminating because colonial wars share a basic architecture that arises when an occupying power far from the mother country tries to suppress a popular uprising. Also, viewing the War of Independence through the lens of other imperialist wars, particularly America's involvement in Vietnam and Iraq, helps rescue it from the Disney World of history to which it has been consigned. By looking in the mirror of its own history, perhaps the nation can see its face more clearly, warts and all, and its own recent history may help it understand the dilemma in which Britain found herself in America almost two and a half centuries ago.

With the war in Iraq unfolding as I researched the long-ago conflict, it was impossible not to be struck by similarities. For example, the occupying power routinely constructs the necessary rhetoric of justification. It is a combination of "we are here to safeguard your best interests/we are here to protect the rule of law/we are here to protect the majority from the bullying of the insurgent minority." These are the fig leaves of moral respectability intended to mask the strategic and economic benefits to the occupiers. Flowing from this was an over-optimistic expectation from "friends" (that is, those presumed loyal to the occupying power). For example, Britain's strategy, particularly in the South, was built on what proved to be an unrealistic assumption of loyalist support. As with America in Vietnam and Iraq, the occupying power could neither mobilize loyalists effectively nor protect adequately those who did commit themselves. The issue is further complicated by the disdain the occupiers often have for their loyalist supporters. British regulars in North America were haughty about "provincial" troops whom they considered second-rate and undependable in battle—an attitude not a million miles away from that of the American military for the ARVN or the Iraqi army.

The armies of George Washington and George Bush also share some characteristics. They were both technically volunteer forces, but in reality economic hardship and the chance of some betterment, be it a $10 bounty and a suit of clothes for one, or the chance of a $40,000 enlistment bonus and a grant of up to $70,000 for a college education for another, swelled the ranks with those from the less privileged members of their society. After the first heady year of the war against the British, during which there was something approaching popular participation, the *rage militaire* subsided and the burden of the fighting fell on a small cadre of young men. It was they who endured the appalling privations of, for example, the bitter winters of 1777–78 at Valley Forge and 1779–80 at Morristown, where they were scandalously abandoned by the broad swath of Americans for whom they fought and died. Similarly in modern America. The Humvee in the shopping mall is a safer option than the under-armored Humvee in Iraq, and flag-waving in the gated community does not require a bulletproof vest. Certainly, any idea of sharing the burden more equitably was as politically unacceptable in eighteenth-century as in twenty-first-century America.

For the British army in America there was a massive logistical burden that constricted its strategic options. Unable to secure sufficient supplies locally (because of either scarcity or the relatively efficient denial of access by the patriots) it was dependent on the United Kingdom for almost all of its food, equipment, clothing, and reinforcements. Foraging, nonetheless, was a constant and pressing necessity. For example, each of the draft horses on which the army depended for transportation needed twenty pounds of hay and nine of oats each day, and the bulkiness of forage made it prohibitively expensive to ship transatlantically. The army "was a ship; where it moved in power it commanded, but around it was the hostile sea, parting in front but closing in behind, and always probing for signs of weakness. Whereas a defeated American army could melt back into the countryside from whence it came, a British force so circumscribed was likely to be totally lost. Its only hope was to fall back on a fortified port"[6]—a description that could just as well have been applied to American troops in Vietnam and Iraq.

It was an isolation not only of physical but also psychological space. The occupiers were aliens in the culture of the occupied. They often could not tell the difference between friend and foe. Their blunderings alienated potential allies and fortified their enemies. The fear of their isolation and vulnerability sapped their strength and set the hair trigger of overreaction. And as the fear grew it emboldened their adversaries: as true for the British in America as it was for Americans in Vietnam and Iraq.

As antidote, British commanders in insurgent America constantly demanded more manpower, but numbers alone could not solve the problem. The French foreign minister, Charles Gravier, comte de Vergennes, saw it clearly: "It will be in vain for the English to multiply their forces there, no longer can they bring that vast continent back to dependence by force of arms." Even within the loyalist press it found an echo: ". . . at more than 3,000 miles' distance, against an enemy we now find active, able, and resolute . . . in a country where fastness grows upon fastness, and labyrinth upon labyrinth; where a check is a defeat, and defeat is ruin. It is a war of absurdity and madness."[7] For Lord George Germain in eighteenth-century Britain, as for Donald Rumsfeld in twenty-first-century America, requests for more men and resources were often denied. Each had to juggle local logistical demands with the myriad others pressing in on an imperial world power. The manager could not also be the magician.

Tactically the War of Independence was a little schizophrenic. In the South, particularly, it was characterized by classic partisan warfare: hit-and-run, ambush, retreat-and-counterattack, isolate-and-overwhelm, interdict supplies (tactics, incidentally and ironically, informed by the Indians' "skulking way of war"—ironically because the victorious patriots saw off the Indians in double-quick time and gobbled up their lands) together with very effective political warfare that robbed the occupier of support. Like the picador, it goaded the enraged bull into suicidal attacks like that at Guilford Courthouse which weakened it sufficiently for the ritualized coup de grâce (Yorktown). All of this falls fairly neatly into the traditional strategy and tactics of colonial war.

What does not fit quite so easily is the fact that Washington did everything in his power to fight the war on European lines. He was uncomfortable with and dismissive of the partisan tradition. Those Pennsylvanian and Virginian riflemen were, in his estimation (and he was certainly not alone in his view), a liability. They were unreliable, uncouth, ill disciplined, and tactically fragile. They offended his patrician sensibilities and his military instincts. What he wanted, what he pressed Congress for, was a *proper* army, a good foursquare, stand-up-and-blast-away army like the British. In this he was a traditionalist and military conservative.

But conservative does not always equate with caution (as the George W. Bush administration can testify). Although history tends to depict Washington as an overwhelmingly defensive commander, avoiding battle at all costs, a good case can be made for the opposite. He was an instinctive fighter. On the battlefield he was tactically highly aggressive, sometimes recklessly so. One thinks of his hair-raising gambles at the battle of Brooklyn and at the second battle of Trenton, where his army could have been destroyed comprehensively (whether that would have ended the rebellion is another matter), or his eagerness to give battle at the Brandywine, Germantown, and Monmouth Courthouse.

It was not the skulking marksman in the fringed hunting shirt that did for the British. The battles that gutted them were formal affairs: Freeman's Farm and Bemis Heights (the two battles of Saratoga), Cowpens, Guilford Courthouse, and the siege of Yorktown. The naval battle of the Chesapeake Capes, which proved fatal to the British cause, was also fought à la mode.

Washington had, from his early days as colonel of Virginia's state troops, and as a respected member of the American oligarchy, a deep respect for European, and specifically British, military tradition. He wanted to fight a war in a style that would do credit to his class (with an army whose organization mirrored European social distinctions) and his country, and he wanted to fight it in a way he thought best suited to delivering grievous body blows to the British army. Washington despised the idea of sneaking in the back door and kicking the British out of the front. He preferred to come boldly in at the front and kick them out the back.

In general, the war was not revolutionary in any military sense or, one could argue, in any social one, either. On the one hand there were no technological or tactical innovations on the battlefield; and on the other, no restructuring of wealth or power within American society. An analogy might be that in a hostile corporate takeover an American management group replaced a British one that had become redundant and expensive and no longer added corporate value. Both regimes were oligarchic, but homegrown was preferable. It seemed to me, therefore, that the phrase "War of Independence" is a more accurate description of events than the "Revolutionary War."

The conservatism goes further. At some basic tactical level, eighteenth-century warfare shared a structure and dynamic with ancient and medieval warfare. All three were based on the phalanx: a compact body of men acting with strict discipline to deliver a heavy blow at close quarters. Battles in the different periods shared a shape. There was usually a standoff preliminary softening-up missile barrage (spears and arrows in earlier times; muskets and cannon later). This was galling but could be minimized by charging across the final 150 yards or so to come to grips with the enemy.[8] In the ancient world the close work was done with thrusting spear and short stabbing sword (80 percent of deaths in Greek battles came from thrusting spear wounds).[9] In the eighteenth century it was either close-up volleying of musketry or the bayonet charge. Whatever the differences in weaponry, some basics were shared. The body of men had to be compact to deliver the maximum weight of lethality, and it had to cover ground fast in order to get through the killing zone of incoming missiles as quickly as possible. Whether it was a phalanx or a regiment, it depended on what can be characterized as a "crystalline" formation: tight, coordinated, interdependent. If that molecular structure was breached, penetrated, or otherwise thrown off kilter, the whole entity could shatter. The Greeks called it *pararrexis:* the breaking of the line and the subsequent loss of cohesion. Frederick the Great, for example, saw the possibility of such a fracture and sought to exploit it with oblique strikes against the flanks: a tactic that became a staple of the great battles of the American war.

In one important sense the American War of Independence was extra- rather than intra-revolutionary. A new player was announced in the imperial game. The emulation of the Old World that Washington sought would lead to a more profound and long-lasting transformation for America than even he could have envisioned. To put an inflection on the adage "you are what you eat," America became what she beat. John Adams expressed it in a slightly different way when he wrote to an English friend in 1767: "We talk the language we have always heard you speak." Adams may have been alluding to such matters as the checks and balances of parliamentarian government and the principle of the supremacy of secular law, but there was another language, the language of imperialism, that America adopted. As the patriot general Nathanael Greene wrote with refreshing candor on 4 January 1776: "Heaven hath decreed the Tottering Empire Britain to irretrievable ruin—thanks to God since Providence hath so determined America must raise an Empire of permanent duration."[10] Prescient must have been his middle name. The westward expansion of American colonists Britain had sought to curb would be reversed. Similar to England's Highland clearances of the mid-eighteenth century, patriot America first dealt with its indigenous population. The Indians were pushed out and their lands expropriated, whether they had supported the cause or not. It was the first step to much, much bigger things.

In time the great wheel of history came sweetly full circle. As America rose to enjoy a world hegemony unparalleled in history, Britain fell from its imperial pinnacle to become an almost Ruritanian client nation; patronized as she had once patronized; condescended to as she had once condescended. The claustrophobic smugness and complacent assumption of superiority that had been the hallmarks of Britain's imperial ascendancy (and that had driven good republican Americans mad) have now been adopted by America. The War of Independence laid the foundation of the nation and subsequent historiography for the cult of the nation. America is at the far end of an arc that began with the victory at Yorktown on 19 October 1781. The nation, once a meetinghouse alive and energized by debate, is now a megalithic cathedral, reverberating with the endlessly rebounding boom of self-referential and self-reverential echoes.

PART ONE

THE NUTS AND BOLTS OF WAR

THEY FACED EACH OTHER, these men of the American and British armies, across a killing ground of perhaps forty or fifty yards. At that distance they would have clearly made out individual faces, perhaps not exactly looking into the "whites of the eyes" but certainly close enough to recognize their shared predicament and their soon to be shared fates. It was like gazing into a mirror. Apart from the fact that they were on opposite sides, more united than divided them.

In many, indeed most, ways the common soldiers of the War of Independence, British, American, German, could look across the gulf that divided them and see true brothers of the battlefield. They were each overwhelmingly drawn from the poorer and often the poorest strata of their respective societies. They were predominantly young and almost entirely without prospects (either before they entered the service, or during, or after), and few of them, given a more liberal choice of destiny, would have elected to be where they now were.

In fact they could also have been looking across the gulf of centuries and seen their reflections in the faces of many modern American and British servicemen of our "volunteer" forces for whom the promise of a steady job, training, and even college education provides a foot on the ladder denied them in civilian life. Then, as now, those with influence, moxie, and/or money afforded themselves a few more (less dangerous) options, as illustrated by at least two recent presidents and two vice presidents. To adapt a classic Pattonism, "It's not your job to die for your country, but to get the other poor bastard to die for your country." In some important ways war does not change very much.

Like their British counterparts, patriot men were drawn into the regular army through a variety of means, many of which, in one way or another, could be considered "coercive" rather than the result of an act of free choice in the fullest sense. The majority were technically volunteers (both sides resorted to the "press"—the forced enlistment of vagrants and those undesirables who were in receipt of parish aid; both sides offered criminals the chance to redeem themselves with their muskets), yet the single most important factor in their decision to enlist was economic necessity. In stark counterpoint to the "heroic" myth of sturdy, independent yeoman laying down the plow to follow the flag out of pure patriotic zeal—so beloved in popular histories and Hollywood movies, and still so stubbornly pervasive, despite an overwhelming amount of scholarship to the contrary—men joined the Continental army and the militia from the lowest ranks of society because it offered enlistment bounties, the promise (usually broken) of regular pay, the enticement (invariably disappointed) of regular food and clothing, and the chance (mainly illusory) of a roof over their head.

Eighteenth-century warfare demanded from the foot soldier an acceptance of subordination, a fortitude and obedience, and, above all, a willingness to accept suffering. On the battlefield this was suffering of an acute and, for us, almost unimaginable level of stress. For example, standing in compacted ranks while enduring hours of close-range bombardment as Lord Stirling's Americans did during the battle of Brooklyn or waiting to receive a massed volley at almost point-blank range, as the British did at Bunker's Hill or Guilford Courthouse. The poor, on whom both armies relied to fill the ranks, had a strength drawn from lives that were hard and invariably short. Although revolts among the soldiery would erupt during the war, particularly in the American lines, and desertion was a constant problem, what is remarkable was their capacity to endure.

1

"A Choaky Mouthful"

THE AMERICAN SOLDIER

After the first heady flush of enthusiasm following the spectacular successes over the British at Lexington, Concord, and Bunker's Hill during that glorious spring and summer of 1775 when close to 20,000 American patriots of all stations of society from the New England states had snatched up their motley collection of arms to support the insurrection, worthy patriots refused to join the ranks in impressive numbers. From that time on, the war, far from being a populist, "democratic" affair, became a military burden shouldered almost exclusively by the poorest segments of American society. No matter how persuasive the rhetoric of freedom, the siren call of self-interest and the urgent demands of survival were often more compelling. John Adams, writing on 1 February 1776, saw it clearly, if ruefully.

> The service was too new; they had not yet become attached to it by habit. Was it credible that men who could get at home better living, more comfortable lodgings, more than double the wages, in safety, not exposed to the sicknesses of the camp, would bind themselves during the war? I knew it to be impossible.[1]

And it would drive George Washington into regular conniptions throughout his tenure as commander in chief.

How did some and not other Americans end up looking down the business end of the barrel of a musket across that fateful fifty yards of killing ground? There were essentially three organizations in which they could volunteer or be forced to "volunteer." The first was the states' militias; the second, the states' troops who were normally drafted or "levied" from the militia for short terms of service and for specific tasks, such as guarding strategic points within the state; the third, after Congress "adopted" the "motley Crew" of citizen-soldiers on 14 June 1775 who were besieging the British at Boston, the Continental army—the regulars. All three types might appear on the monthly returns of regular army strength if militia and state troops had been co-opted to serve with the Continentals.[2]

⇥ THE MILITIA ⇥

The institution of the militia had been built into the fabric of the earliest colonies. The necessity not only to protect their settlements but also, where expedient, to expand their holdings, meant that technically every able-bodied man from the ages of sixteen to sixty was required to turn up, armed, for regular training and, if necessary, make himself available for longer periods of service. For example, the Patriot Committee of Frederick County, Virginia, proclaimed in the spring of 1775: "Every Member of this County between sixteen & sixty years of Age, shall appear once every Month, at least, in the Field under Arms; & it is recommended to all to muster weekly for their Improvement."[3]

In the beginning it had been a decidedly convenient arrangement for Britain to set up what were essentially trading satellites charged with the responsibility of defending themselves with little financial drag on the mother country. It was only after the French and Indian War (1754–63) that the cost/profit ratio of Britain's American empire shifted in an uncomfortable direction. Britain's national debt rose from £75 million to a whopping £130 million.[4] And in part, it was Britain's attempt to balance the increasingly wayward ledger books of its colonial investment that drove the colonies into insurrection.

Within each state not all were equally bound by the militia contract. Some, the lowest of the low in colonial society—slaves, Indians, white indentured servants and apprentices, and itinerant laborers—were exempt, not from some humanitarian impulse on the part of the white oligarchy but because it would too dangerous to arm groups that might at some future time turn their military experience in the wrong direction. In any event these people were property, someone else's property, and the rules and rights of property were at the sacrosanct heart of colonial society. It would be only during the severest pressures of the war that these rules would be bent or broken. At the other end of the social scale the more powerful could escape the inconveniencies of militia service by paying a fine or hiring a substitute, an avoidance long established in the colonial tradition: "No Man of an Estate is under any Obligation to Muster, and even the Servants or Overseers of the Rich are likewise exempted; the whole Burthen lyes upon the poorest sort of people," wrote Governor Alexander Spotswood of Virginia to the Board of Trade in 1716.[5]

A comparison of the original 1669 militia ordinance and the 1774 Militia Act for North Carolina shows how wide a gap had opened between the generally inclusive demands of the original ("all inhabitants and freemen . . . above 17 years of age and under 60") and the much more lenient expectations of the latter which excluded many categories of freeholders, including clergymen, lawyers, judges, millers, overseers, and constables.[6] The hierarchy of Virginia was acutely aware of the political fallout if too many militia obligations were placed on what we would now call its "core constituency," and the General Assembly regularly restricted militia service to those who were "not free-holders or house-keepers qualified to vote at the election of Burgesses."

Even back in the 1750s when George Washington was colonel of Virginia's state troops, he would get a taste of the problems that would gall him throughout the War of Independence. With the exemption of what he would have described as the "right sort of people," Washington was forced to draw "upon the lowest orders of society, whom he once portrayed as 'loose, Idle Persons that are quite destitute of House and Home.'"[7] And it would be just such as these who were to carry the main burden of the patriot cause whether in the militia battalions or Continental army. Most of the time, in those days of his colonelcy, the militia simply did not turn up (like trying to "raize the Dead," he wailed), and when they did turn up they were aggravatingly "bolshie": "Every *mean* individual has his own crude notion of things, and must undertake to direct. If his advice is neglected, he thinks himself slighted, abased, and injured; and, to redress his wrongs, will depart for his home."[8]

Within the intricately structured and close-knit societies of colonial America membership in the militia was something more than just an obligation; it was a part of being an acceptable member of the alpha group—white, male, property-owning—that held the largest stake and stood to benefit most from the self-protection the militia afforded. It was held together by "intricate networks of personal loyalties, obligations, and quasi-dependencies."[9] When the Concord, Massachusetts, militia assembled in March 1775, its colonel, Thomas Barrett, transmitted his orders through "a son and son-in-law, both captains, to a second son and a brother, both ensigns, down to yet another son and a nephew, both corporals, and ultimately to several other nephews in the ranks."[10] In societies so closely interlinked by marriage, property arrangements, local politics, and business, where everyone pretty much knew everyone else and had dealings with each other over a whole raft of activities, lies an explanation why, during the war, the militia were so notoriously disinclined (much to Washington's chagrin) to serve under any but their own officers or indeed showed little sympathy for fighting away from their home base.

The militia bands themselves reveled in their independence, especially among the New England colonies with their long tradition of "leveling." (More than half of the company officers in the Massachusetts militia who were mobilized during the French and Indian War identified themselves with manual occupations.)[11] Officers were often elected by the men, but it would be a sentimentalization to see this as some sort of happy band of equals. Officers, as could be expected from a hierarchical society, tended to come from the upper echelons: "We consider that our officers generally are chosen out of the best yeomanry of this colony, who live on their own lands in peace and plenty," declared the Connecticut Assembly. In the southern states (all those west of the Delaware) things would have been quite different. No leveling tendencies here. The more starkly stratified nature of planter society was reflected in militia and Continental army alike.

The militia caused the plutocrats, whether of the southern planter caste like Washington or the northern patroon class like Philip Schuyler, the greatest aggravation. Washington abhorred "familiarity between men of high and low position." On 20 August 1775 he vented in a letter to his cousin, Lund Washington, against both men and officers of the New England militia besieging Boston: "In short they are by no means such troops, in any respect, as you are led to believe of them from the accounts which are published, but I need not make myself enemies among them by this declaration, although it is consistent with the truth . . . they are exceedingly dirty and nasty people." (It was a sentiment of long standing. General James Wolfe had written during the French and Indian War of his American militia troops: "[They are] the dirtiest most contemptible cowardly dogs that you could conceive. There is no depending on them in action. They fall down dead in their own dirt [excrement] and desert by battalions, officers and all.")[12]

Nine days later Washington was writing to Richard Henry Lee (a fellow Virginian grandee, a prime mover in the Independence debate, and a signer of the Declaration of Independence) in similar vein: "[There is] an unaccountable kind of stupidity in the lower classes of these people which, believe me, prevails but too generally among the officers of the Massachusetts *part* of the army who are *nearly* all of the same kidney with the privates!"[13] It was an old complaint. Cadwallader Colden, lieutenant governor of the province of New York, writing to Lord Halifax in 1754, had steamed: "Our Militia is under no kind of discipline . . . The inhabitants of the Northern Colonies are all so nearly on a level, and a licentiousness, under the notion of liberty, so generally prevails, that they are impatient under all kind of superiority and authority."[14]

On 24 September 1776 Washington railed, as he did so often, against this leveling among the common soldiery and officers of the New

England militia: "While those men consider and treat him [an officer] as an equal, and, in the character of an officer regard him no more than a broomstick, being mixed together as one common herd, no order or discipline can prevail."[15] Such letters, and there were many, together with his general animosity toward the democratizing tendencies of the New England states earned him a good deal of animosity in the North, especially, and ironically, from John Adams, who had been Washington's principal sponsor as commander in chief.

Washington was not alone in his disregard for the "little people." (He was, after all, an eighteenth-century grandee; maybe not a grand grandee by European standards, but he shared something of their hauteur.) Where Washington condescended to yeoman-farmers as "the grazing multitude," John Adams referred to the "Common Herd of Mankind," and General Nathanael Greene complained that "the great body of the People [are] contracted, selfish and illiberal."[16] It is an interesting footnote that while Greene was lacerating the American people for being "contracted, selfish, and illiberal," he was contracting enthusiastically and liberally as a war profiteer while one of the foremost general officers in the Continental army.[12]

America in 1775 was a collection of fiercely independent colonies— that is, independent from each other—and, despite the enormous efforts of those who, during the war, sought to create a nation with the concomitant centralizing infrastructure, they fought hard to maintain their independence. It was precisely this dedication to their own, as they saw it, entirely healthy, self-interest that had led the colonies to reject Benjamin Franklin's Albany Plan of Union of 1754, and it was their fear of a centralized state that made Washington and Congress's work to create a standing army—the Continental army—so frustrating. Nathanael Greene, early in the war, put his finger on it: "It is next to impossible to unhinge the prejudices that people have for places and things they have long been connected with."[18]

But not everyone lamented, as did Washington and Greene, the resistance to a standing army. The militia model was just fine, they felt, both as a perfectly adequate military force and, equally important, as a safeguard of good republican virtues. "Caractacus," an anonymous contributor to a Philadelphia newspaper, wrote in August 1775 that standing armies corrupted soldiers and at the same time created a potential monster: "The military spirit, by being transferred from the bulk of the country to a few mercenaries [that is, any soldier who accepted pay] is gradually monopolized by them, so that in a few years, from being our servants, we furnish them the means of becoming our masters."[19] Dr. Benjamin Rush, that doughty and irascible patriot, though certainly no admirer of Washington, could write: "The militia began, and I sincerely hope the militia will end, the present war. I should despair of our cause if our country contained 60,000 men abandoned enough to enlist for 3 years or during the war."[20] Written in December 1777, this was an extraordinary dedication to republican virtue at a time when the militia had performed so disappointingly at the battles for New York.

The independent spirit that so aggravated Washington was rooted in something fundamental in colonial society: the idea of a contract between those who served and the political body that demanded they serve. After all, the establishment of each colony had itself been based on a contract implicit in the colonial charters. The obligations of each colony to Britain were to be balanced, in theory, by the responsibilities and obligations of the mother country. Militiamen (as well as Continental soldiers) were keenly aware of this contractual relationship—made between free men, and not coerced—and often invoked it during the war, much to the discomfiture of the military and political hierarchy.

General Richard Montgomery complained in 1775 of the militia fighting with him in the northern theater that "the privates are all generals . . . they carry the spirit of freedom into the field and think for themselves."[21] Washington also may not have liked it, but he was too perceptive and realistic a commander not to recognize that the inherent independency of the militia required different handling: "People unused to restraint must be led, they will not be drove, even those who are ingaged for the War must be disciplined by degrees."[22]

It was usually the length of service that proved the greatest stumbling block. Service in the militia was, as George Washington described it to John Hancock in September 1776, of variable lengths, but generally, there were "two sorts [of militiamen], the Six Month Men and those

sent in as temporary aid."[23] Men who were very often at the mercy of
the demands of the farming cycle could not make commitments that
took them from their farms for long periods. Men involved in the trades
who had families to support in an age with practically no welfare safety
net did not have the luxury to dedicate themselves to the cause for long
periods, no matter how impelling that cause might be. The history of
the war is punctuated with what might seem to us the most egregious
betrayals of the patriot cause by militia who, on the expiration of their
term of service, simply picked up sticks and headed home. (This, of
course, was a factor not only in the militia. The mutinies in several
Continental regiments in 1781 were predicated on the interpretation of
the soldiers' term of service.)

The leave-takings could be precipitous, and catastrophic. For
example, the timing of the disastrous attack on Quebec by Richard
Montgomery and Benedict Arnold on Sunday 31 December 1775 in part
may have been forced on the commanders by the imminent expiration
of the term of Arnold's New England militia volunteers at midnight of
the thirty-first.[24] Silas Deane, America's agent in France, wrote, "The
behavior of our soldiers had made me sick, but little better could be
expected from men trained up with notions of their right of saying how,
and when, and under whom, they will serve; and who have, for certain
dirty political purposes been tampered with by their officers."[25]

Nathanael Greene's dramatic retreat to the Dan River on the North
Carolina/Virginia border in February and March 1781 was a very close
run thing indeed, with the pursuing Charles Cornwallis breathing
down his neck. These were dire times, but when Edward Stevens
appealed to his militia "whose times were out" not to abandon the cause
until General Greene could get his troops up, he got short shrift: "To
my great mortification and astonishment scarce a man would agree
to it. And gave for an answer he was a good Soldier that Served his
time out. If the salvation of the Country had depended on their staying
Ten or Fifteen days, I don't believe they would have done it. Militia
won't do."[26]

Nathanael Greene was to experience an even more dramatic
demonstration of militia feistiness in April 1781 as he was preparing

to do battle with Lord Rawdon at "the fierce and sanguinary Battle of Hobkirk's Hill" (25 April 1781). Just five days before the battle, Major Guilford Dudley reported:

> *A most unpleasant and disgusting circumstance occurred which seemed for a moment to disturb the equanimity of the general himself. Lieutenant Colonel Webb's battalion of militia [North Carolina], which with my own constituted the command of Colonel [Jesse] Read, insisted on their discharge, alleging that their term of service had expired. This was at first refused and the allegation denied, when they evinced a spirit of mutiny, encouraged and heightened by Captain R. of that battalion, who was their chief spokesman. Persuasion and even entreaty was used by the field officers of the regiment, pointing to the enemy's works [at Camden] staring us in the face at a short distance and telling them not to desert their general but have patience and wait only a few days longer, when their services might be all important to him in the plain before us. But all this only made them more eager and determined upon being discharged, and finding our treaties unavailing, one of us went to the general and gave him the unpleasing information, when he, with great condescension, mounted his horse . . . and used all his persuasion and eloquence to detain them for a few days longer . . . but all to no purpose. Captain R. and the others became more clamorous, and General Greene, mortified and disgusted, directed Colonel [Otho Holland] Williams to write their discharge, which done they were instantly off.*[27]

A persistent myth attached to the militia was that every American was well versed in small arms and above every hearth hung a musket or rifle. Thomas Jefferson wrote to a friend in England, "Every soldier in our army [had] been intimate with his gun from his infancy."[28] Not so in Britain; Charles Lee, the ex-Britich Army officer turned patriot-general, pointed out that "the lower and middle people of England . . . [are] almost as ignorant in the use of a musket, as they are of the ancient Catapulta." And it is true that, compared with Britain, arms-bearing in the colonies was much more widespread. The militia tradition in England had lost much of the vigor of earlier times, although, technically, men between the ages of eighteen and forty-five were eligible for five-year stints which they could commute, as in America, by hiring a substitute or paying a fine. As the militia was funded by the land tax, it was controlled by the landed gentry, "and indeed the militia lists of the period are simply a catalogue of the leading county families . . . hence in many cases a regiment of militia became a very exclusive country club."[29]

However, the popular idea of the patriot citizen-sharpshooter born to the gun needs a little adjustment. During his colonelcy of Virginia state troops in 1756 Washington reported that the militia turned up "invariably minus weapons." He makes the point of having to drill men in basic firearms use—not something one would have thought necessary with men well versed in firearms. "One contingent of two hundred Culpeper County men reported with a total of only eighty firelocks."[30] The militia at Boston in 1775 were short of 2,000 muskets, and in the later phase of the war Robert Gray of Carolina reported in 1780 that "the people at that time [were] not much accustomed to arms."[31] The multitude of militia shooters during the retreat from Lexington is also persuasive testament to the fact that they were indifferent shots when one considers the number of militia involved, the time available to them, the vulnerability of the target, and the relatively low ratio of British casualties to the shots fired. About 3,700 militia spent the best part of a day expending about 75,000 rounds. The British had 273 killed and wounded, which meant that only one patriot ball out of 270 found its mark: "The entire proceeding should prove the fallaciousness of the belief so often expressed that the Yankees were superior marksmen, dead shots in fact."[32] But the militia of Concord and Bunker's Hill were not necessarily militarily unskilled. Colonial and Revolutionary-era militia muster rolls indicate many had fought in the French and Indian War, and many of their weapons were relics of that period. Perhaps the truth is that with the increasing settlement, cultivation, even urbanization of the older colonies, men had become much less reliant on the firearm and more on the ledger book.

The battles of 1775 in Massachusetts were the militia's glory days—the Norman Rockwell moments of the war. But the militia seems to have evinced only disappointment thereafter. George Washington was not a supporter. The militia around Boston revolted him. Their lack of deference infuriated him; their slapdash organization troubled him; and as far as their usefulness as a fighting force was concerned he had, throughout the war, the profoundest misgivings. On 24 September 1776, during one of the periodic droughts in enlistments to the Continental army, he wrote to the president of Congress, John Hancock, making his extended case for a small, permanent, and professional army.

> *To place any dependance upon Militia, is, resting upon a broken staff. Men just dragged from the tender Scenes of domestick life; unaccustomed to the din of Arms; totally unacquainted with every kind of Military skill, which followed by a want of confidence in themselves, when opposed to Troops regularly train'd, disciplined, and appointed superior in knowledge, and superior in Arms, makes them timid and ready to fly from their own shadow.*[33]

Nathanael Greene mirrored Washington's concern that tender civilians simply could not deal with the realities of battle: "[The militia were] not sufficiently fortified with natural courage to stand the shocking scenes of war. To march over dead men, to hear without concern the groans of the wounded, I say few men can stand such scenes unless steeled by habit and fortified by military pride."[34] After the near catastrophe of the battle of Brooklyn on 27 August 1776 he had come as close as a hair's breadth to losing the whole army and had been deeply humiliated when the militia ran away at Kips Bay in the face of the British and German amphibious assault on Manhattan. He knew very well the militia's strengths and weaknesses: "Place them behind a parapet, a breast-work, stone wall, or anything that will afford them shelter, and from their knowledge of the firelock, they will give a good account of their enemy; but I am well convinced, as if I had seen it, that they will not march boldly up to a work nor stand exposed in a plain."[35]

The British general Thomas Gage concurred with the first part of Washington's opinion, writing on 25 June 1775, "Whenever they find good cover they make a good stand, and the country naturally strong, affords it them."[36] General William Moultrie, one of the unsung heroes of the war, saw it clearly: "The militia are brave men, and will fight if you let them come to action in their own way."[37] In October 1780, following the disastrous patriot defeat at Camden, Washington reiterated his opinion: "I solemnly declare I never was witness to a single instance that can countenance an opinion of Militia or raw troops being fit for the real business of fighting. I have found them useful as light parties skirmishing the Woods, but incapable of making or sustaining a serious attack.[38]

And on the whole, the history of the war bore him out. Skirmishing along the road from Concord they had done well; firing from the redoubt on Bunker's Hill and from behind the beach fence they had inflicted crippling casualties; and the fighting in the woods at Bennington on 15 August 1777 under John Stark and Seth Warner had resulted in decimation of General John Burgoyne's Brunswick detachment—the dark foreshadowing of the British army's fate at Freeman's Farm and Bemis Heights.

They did fearsome work as partisan fighters in the South under Daniel Morgan, Andrew Pickens, and Francis Marion. When faced by Loyalist forces that were equally untutored in the formalities of European-style warfare, they performed well, as attested by the victories over the Highland Scots Loyalists at Moore's Creek Bridge, North Carolina, on 27 February 1776 (in a sort of miniature Culloden, the Scots attacked with the broadsword, only to be cut down by patriot musketry), and the smashing defeat of Major Patrick Ferguson's Loyalists atop Kings Mountain, South Carolina, on 7 October 1780. In fact the militia (actually more like tribal partisan bands) of the Carolina backcountry were the best in the country. Even Nathanael Greene, no admirer of militias, as we have seen, conceded that "there is a great spirit of enterprise among the back people; and those that come out as Volunteers are not a little formidable to the enemy . . . the rest of the Militia are better calculated to destroy provisions than the Enimy."[39] They were truly self-supporting and highly motivated. As James

Collins, a member of Captain Moffitt's band, described it: "[We were a] set of men acting entirely on our own footing, without the promise or expectation of any pay. There was nothing furnished us from the public; we furnished our own clothes . . . We furnished our own horses, saddles, bridles, guns, swords, butcher knives."[40] They were a hardscrabble bunch but true partisan warriors, who, in a later incarnation, would form the backbone of the Confederate army.

When it came to the terrifying realities of formal warfare, the militia often quailed and ran. Washington reported to Congress on the state of the militia after the battle of Brooklyn: "Instead of calling forth their utmost efforts . . . [they] are dismayed, intractable, and impatient to return. Great numbers of them have gone off, in some instances almost by whole Regiments, by half ones, and by companies at a time."[41] At Kips Bay on Manhattan in 1776 their failure reduced Washington to as black a despair as he was to feel in the whole war ("God, have I got such troops as these!")—so black that he sat his horse in a kind of trance until his aides led him away from the converging redcoats. Benjamin Trumbull, a clergyman and historian, saw that the failure of the militia was in fact a failure of leadership, of management: "The men were blamed for retreating and even flying in these Circumstances, but I image the Fault was principally in the General Officers in not disposing of things so as to give the men a rational prospect of Defence and a Safe retreat."[42]

At the battle of Camden (16 August 1780) Colonel Otho Holland Williams described the effect on militia of British troops rushing on, "*firing* and *huzzaing* . . . [which] threw the whole body of the militia into such a panic that they generally threw down their *loaded* arms and fled in the utmost consternation. The unworthy example of the Virginians was almost instantly followed by the North Carolinians."[43] General Horatio Gates was completely unable to stop their panicked flight, and into the gap opened by their defection rode the British cavalry to attack the American rear. It was, perhaps, the most comprehensive and humiliating defeat of American arms in the whole war, and it brought Washington back to his old lament: "I never was witness to a single instance, that can countenance an opinion of Militia or raw Troops being fit for the real business of fighting. I have found them useful as light

Parties to skirmish in the woods, but incapable of making or sustaining a serious attack . . . The late battle of Camden is a melancholy comment upon the doctrine. The militia fled at the first fire."[44]

But it did not have to be so. In the hands of a truly gifted commander the militia could perform well. General Daniel Morgan knew his men—their strengths *and* weakness—and, in response, devised the carrot-and-stick method of militia command that resulted in his crushing defeat of Banastre Tarleton at Cowpens on 17 January 1781. It was as unorthodox as it was imaginative. He wanted to keep his militias' feet to the fire and so took a position with his back to a river that made retreat impossible.

> *I would have thanked Tarleton had he surrounded me with his cavalry. It would have been better than placing my own men in the rear to shoot down those [militia] who broke from the ranks. When men are forced to fight they will sell their lives dearly and I knew that the dread of Tarleton's cavalry would give due weight to the protection of my bayonets [that is, his Continentals] and keep my troops from breaking . . . Had I crossed the river, one half of my militia would immediately have abandoned me.*[45]

He would place his weakest troops, the Virginia, Georgia, and North Carolina militias—which constituted about 600 of his available 800 men—in the front line. That was the stick. The carrot was a recognition that the militias would probably collapse under the British onslaught and so they were ordered to give two controlled volleys at about fifty yards and then retire in good order before the British got among them. They were then to take reserve positions behind Morgan's elite Maryland and Delaware Continentals, which constituted the main American battle line and the rock on which the British and Loyalist infantry would wreck itself. It worked to perfection, and the militias, reassured by their preplanned escape route (to echo Trumbell's observation, "a rational prospect of Defence and a Safe retreat"), performed their part to the letter.

Before the battle of Guilford Courthouse (15 March 1781) old Morgan, now almost completely disabled by arthritis, wrote to General

Nathanael Greene suggesting the wisdom of the tactics he had used at Cowpens. "If the militia fights you will beat Cornwallis; if not he will beat you and perhaps cut your regulars to pieces." Put the militia in the center, he advised, "with some picked troops in their rear with orders to shoot down the first man that runs. If anything will succeed, a disposition of this kind will."[46] Plenty of stick, but not much carrot. Greene certainly applied the stick but neglected to dangle the carrot of a guaranteed escape route, and the militia crumpled at Guilford. As Colonel Henry Lee wrote in his *Memoirs*, "To our infinite distress and mortification, the North Carolina militia took to flight . . . Every effort was made by . . . the officers of every grade to stop this unaccountable panic, for not a man of the corps had been killed or even wounded . . . All was in vain; so thoroughly confounded were these unhappy men, that, throwing away arms, knapsacks, and even canteens, they rushed like a torrent through the woods."[47]

Apart from their checkered career in battle, the militias performed a critical task for the patriot cause: to root out and ruthlessly suppress Loyalist opposition. To understand the nature of this activity, it is useful to look at the War of Independence through the lens of much more recent history. For example, the Communist political cadres of the Vietnam War, the IRA, and the "insurgents" of the second Iraq War were heavily invested in intimidating and killing those they perceived to be sympathetic to the occupying power. This is a constant element of an anticolonial war. Some American historians are squeamish about making such long-range connections between, say, Vietnam (or Iraq) and the American War of Independence (and right-wing talk-show hosts froth at the mouth when any comparison is made between the techniques of eighteenth-century American patriots and modern Iraqi "insurgents"), but the connections are numerous and, indeed, inevitable. Colonial wars share a certain geometry.

⊰ THE CONTINENTALS ⊱

Jesse Lukens, a patriot rifleman from the South, viewed the New England army besieging Boston in September 1775: "Such Sermons, such Negroes, such Colonels, & such Great, Great Grandfathers."[48] And exactly one year later we can hear its echo in the journal entry for Monday, 2 September 1776 of Ambrose Serle, Admiral Lord Richard Howe's private secretary: "Their army is the strangest that was ever Collected: Old men of 60, Boys of 14, and Blacks of all ages, and ragged for the most part, compose the motley Crew."[49]

Who made up that "motley Crew" and why did they join the army? Membership in this valiant band was small, select, and exclusive. Not because of any rigorous winnowing of recruits—quite the opposite— but because most sensible men, mindful of their families, their farms, their livelihoods, went to considerable lengths to avoid service in an army that could promise them only lousy conditions, low wages (if any at all), and no small chance of being put in harm's way, by either battle or disease. James Collins, for instance, a youngster from the Carolina backcountry, toyed with the idea of volunteering for the Continentals until his father, Daniel, a veteran of the French and Indian War, gave him some advice. James recalled his father saying: "The time was at hand when volunteers would be called for, and by joining them [the militia rather than the regulars] . . . I would be less exposed, less fatigued, and if there should be any chance of resting, I could come home and enjoy it; he said he had some experience and learned a lesson from it."[50]

After the initial ardor of 1775 the enthusiasm of the American people for its army was muted. For example, it is one of the more piquant ironies of American history that the army's suffering at Valley Forge during the winter of 1777–78 has been raised in the popular imagination to a quasi-religious expression of nationhood. In fact, the army was abandoned by the American people at Valley Forge. While the army starved and shivered and died, the local populace traded for hard cash—not the useless Continental paper money of the army—with the nearby British. Stung by the whining of the substantial burghers of Pennsylvania who demanded that the army at Valley Forge protect their property from British depredations, Washington let them have it with both barrels.

I can assure those Gentlemen that it is a much easier and less distressing thing to draw remonstrances in a comfortable room by a

good fireside, than to occupy a cold, bleak hill and sleep under frost
and Snow without Cloaths or Blankets . . . I feel superabundantly
for them [the soldiers] and, from my Soul, I pity those miseries.[51]

Albegence Waldo, a surgeon serving in the Connecticut Line,
echoed Washington's anger when he wrote in his journal at Valley
Forge: "People who live at home in Luxury and Ease . . . have but a very
faint Idea of the unpleasing sensations, and continual Anxiety the Man
endues who is in Camp . . . These same People are willing we should
suffer everything for their Benefit & advantage, and yet are the first to
Condemn us for not doing more!"[52]

The Continental army had been formally instituted on 4 July
1775, when Congress "adopted" the "troops of the United Provinces of
North America" then encircling the besieged British army in Boston.
(George Washington had been appointed commander in chief of "all
the Continental forces" about three weeks earlier, on 15 June.) In its
adoption Congress also inherited the terms of service of these former
militia recruits, many of them members of their respective states'
"provincial regiments" and thus part of a longer service commitment
than the plain militias. Their service was due to expire at the end of
the year. From the very beginning Washington was discouraged by
the failure to induce men to reenlist for any militarily useful length of
time—one year, three years, the duration of the war, anything other
than the short terms the majority of the men would accept.

Short-term enlistments were counterproductive. The men were not
in the army long enough to make them effective, yet in that short time
they consumed resources and clogged the administrative machinery.
"It takes you two or three months to bring new men in any tolerable
degree acquainted with their duty," said Washington to his secretary,
Joseph Reed, on 1 February 1776. "It takes a longer time to bring a
people of the temper and genius of these into such a subordinate way
of thinking as is necessary for a soldier. Before this is accomplished, the
time approaches for their dismissal."[53]

This constant chewing on the problem of getting men to enlist,
of getting the "nation" to fight its war of independence, was to be a

recurrent theme of his throughout the war and a constant source of anxiety and frustration that often wobbled his belief in the public's attachment to the cause.

> *Such a dearth of public spirit, and want of virtue, and stock-jobbing, and fertility in all the low arts to obtain advantages of one kind and another, I never saw before and pray God I may never be witness to again . . . Could I have foreseen what I have, and am likely to experience, no consideration upon Earth should have induced me to accept this command.*[54]

John Adams had recognized the problem when he had referred to the service as being "too new" and not yet "attached . . . by habit." George Washington was equally clearheaded about what would or would not induce men to leave their hearths and fight.

> *I do not mean to exclude altogether the Idea of Patriotism. I know it exists . . . But I venture to assert, that a great and lasting War can never be supported on this principle alone. It must be aided by a prospect of Interest or some reward. For a time, it may, of itself push Men to Action; to bear much, to encounter difficulties; but it will not endure unassisted by Interest.*[55]

So what were the "prospects of interest" that could entice a man to join the Continentals? In large measure they were no different from those offered to the British army recruit. First, there was the one-off up-front lump payment, the bounty; next, the monthly pay; then, the promise of regular food, drink (a not inconsiderable inducement as we shall see later), and shelter. There was sometimes the promise of land. For example, in 1778 the General Assembly of South Carolina offered 100 acres of former Cherokee territory in addition to the 100 acres promised by Congress.[56] Finally, but by no means the least important, something which is far harder to quantify but was, and always has been, a major factor in military recruitment, a cluster of "social" benefits for men who had been denied them in ordinary society: the chance to join a group, to share comradeship, to feel enveloped and involved, to escape the suffocating tedium of a life without prospects, to be lifted from the ordinary, to acquire a little bit of specialness in a life devoid of distinction. This last posed a strange paradox in eighteenth-century society. The professional soldier of the ranks, as compared with the officer, was seen as both a loser and an adventurer. He may have been denigrated by society at large as a rogue, but he had about him something of the flash and romance of the buccaneer. The uniform, the stories, a few drinks— *that* could get him some respect, and it was certainly no hindrance to bowling over the ladies! Joseph Plumb Martin, a foot soldier in the Continental army, is a good example. He enlisted for adventure, to escape his restrictive grandparents (who had taken him under their wing) and his circumscribed life as a farm laborer: "I wanted to recreate myself, to keep the blood from stagnating."[57]

Money, though, was highest on the list of inducements for regular army recruits. First there was the chance to acquire a lump sum—the bounty—which must have been highly attractive to men living either in poverty or on the tight rein of subsistence. Bounties were nothing new in the American military; individual colonies had been forced to resort to them during the French and Indian War, for example. In June 1776 Congress drew up its "plan to enlist militia" and offered privates a $10 bounty and monthly pay of $6⅔ (forty shillings) for a three-year service or the duration of the war. (On 27 May 1778 a differential pay scale was introduced that gave specialists a slight increase over ordinary infantrymen. For example, artillerymen and dragoons got $8⅓; fifers and drummers received an extra two-thirds of a dollar; saddlers and farriers received $10.)

The result, however, was anything but satisfactory as far as Washington and Congress were concerned. First, bounties simply did not attract sufficient men. (Of the 6,000–7,000 men of the "old" army who were issued reenlistment papers on 12 November 1776, only 966 took the offer. After a further week or so, no more than 3,500 had agreed to stay only until their term expired—sweetened by "a drink of rum, and were promised another on the morrow.") Second, a bidding war broke out between individual colonies and the Continental army. By the spring of 1777 Congress was forced not only to double its original $10 bounty for three years or the duration but to add a further inducement of 100 acres per man to be awarded at the war's end. But individual colonies were also competing for men, some to fill the quota for the Continental army that Congress had allocated to them on 16 September 1776 and some to fill their own provincial regiments.

The spiraling cost of bounties was confusing for potential recruits (who had every incentive to hold off making a commitment in order to benefit from a higher bounty) and often counterproductive for the Continental army. For example, Connecticut and Massachusetts added twenty shillings a month to the Continental base pay for privates in addition to a $33⅓ increase on the Continental bounty. New Jersey upped its bounty by $53⅓, which prompted Massachusetts and New Hampshire to increase their bounties to $86⅓. Such profligacy and self-serving on the part of individual colonies enraged John Adams: "I wish there was a laudable Spirit to give Bounties in Money, and Land, to Men, who would enlist during the war. But there is not. Congress offers Ten Dollars Bounty to inlist [sic] for three years when New Jersey, New York, Connecticut, Mass. Bay and New Hampshire are voting six, Eight, or Ten Pounds a Man to serve for six months. This economy at the Spigot, and Profusion at the Bung will ruin us."[58] In a grand show of humbug, John Hancock, one of the wealthiest Americans of his day, pontificated that he was "Pained . . . at this want of Public Spirit and Backwardness in the Soldiers." By May 1778 the Congress approved another hike in the bounty to $80 and on 23 January 1779 increased it to $200 plus a clothing provision if men would serve for the duration. And so it went on, in a pinball ricochet of competing inducements. Nevertheless, those who could afford to pass them up did.

But not all Americans could. The poor—the backbone of the army—needed the money and the ancillary benefits: "The hard-core of Continental soldiers . . . the men who shouldered the heaviest military burden, were something *less* than the average colonial Americans. As a group they were poorer, more marginal, less well anchored in the society. Perhaps we should not be surprised; it is easy to imagine men like these actually being attracted by the relative affluence, comfort, security, prestige, and even the chance of satisfying human relationships offered by the Continental army."[59]

The failure of volunteerism was obviously a problem, but it was a problem with a couple of dimensions. If fewer than needed would volunteer, some kind of coercion would have to be instituted: the draft. But the draft set off alarm bells among the radical Whigs—the prospect of a standing army. With a standing army came, so it was feared, the tyranny of the state. The American revolt had drawn inspiration from Britain's Great Revolution of the 1640s when a king's head had been lopped and parliament had staked out its proprietorial turf. But there was a problem with this heroic paradigm: tyranny of king had given way to tyranny of the regicide. Whichever way you carved it up, it was tyranny, and the army was the horse on which it rode. The battle throughout the American War of Independence would be between the nationalists like Washington who desperately wanted a professional army based on the British model and those who felt republican virtue could be safeguarded only by volunteerism and short-term commitments to military service.

Volunteerism was the preferred method of enlistment, as it was in Britain despite the still popular myth that the British soldier was either a released convict or a victim of the coercive "press." The problem was that volunteerism simply did not work "despite," as Washington wrote, the "allurements of the most exorbitant bounties and every other inducement that could be thought of," which served only to "increase the rapacity and raise the demands of those to whom they were held out."[60] In response, he advocated a draft or conscription of militiamen into the Continental army for twelve months. (Massachusetts and New Hampshire had already instituted drafts by 1777, and the rest of the colonies had them in place by 1778.) On 15 August 1776 Philip Vickers Fithian, a chaplain with the New Jersey militia in New York, noted, "Press warrants are issued to draft three Men out of every five of the Militia, till the several Brigades be filled.")[61] America followed Britain in using impressment (forced conscription): "Not uncommonly, as in Maryland, vagrants were inducted into the army—in that state for a nine-month period; and so were convicted felons, who received pardons for satisfactory military duty."[62] In South Carolina not even doubling Congress's promise of 100 acres, cash bonuses, or 10 percent promissory notes could induce men to volunteer, so "military duty in the Continental forces became the punishment for vagrants, anyone who harbored deserters, and those who hunted deer at night with fire."[63] As in the British army, criminals were released into the Continental ranks, particularly by Virginia, whose governor, Patrick Henry, became the target of Washington's protest. Nathanael Greene, likewise, complained about Virginia convicts serving with him in Pennsylvania in 1778, and the southern states "lightheartedly relieved themselves of criminals by compelling them to serve in the Continentals."[64]

The draft was carried out as a kind of lottery. Each state apportioned among its militia regiments the quota it was meant to furnish the Continentals. Militia colonels would then choose the men to go, sometimes by simply drawing their names from a hat. There were, however, two loopholes that favored the wealthier citizen: The draftee could simply pay a fine of £20 within twenty-four hours and free himself of his obligation; or he could pay a substitute to go in his place (20 percent of the New Jersey Line, for example, were substitutes).[65] Sometimes whole communities like Epping, New Hampshire, in 1777, escaped the draft by hiring substitutes from neighboring villages and towns. Charles Royster, in his milestone book *A Revolutionary People at War,* presents a somewhat tortured justification of the substitute system: "Both can claim to have done a virtuous act . . . one man has kept his personal freedom and given up money, while the other has gained money and given up some of his personal freedom."[66] The problem with Royster's argument is that the man who paid the money avoided the risk of death, while the man who took the money often did so out of economic necessity rather than free choice. The substitute did not merely surrender "some of his personal freedom," as though he were doing nothing more important than giving up a little free time, but ran a very real chance of death from battle or disease. In fact the man who paid for a substitute made a shrewd investment.

Also tapped for military service in America were "Redemptioners"—indentured servants from Britain (mainly Scotch-Irish) who were working to pay off the cost of their passage and were, essentially, the property of their masters. Rhode Island, New Jersey, and Maryland allowed them to enlist without first having to secure their master's approval.[67] Nor were criminals exempt from service: "Not uncommonly, as in Maryland, vagrants were inducted into the army . . . and so were convicted felons, who received pardons for satisfactory military duty."[68]

As in Britain, recruiting officers were sent out into the communities to "beat up" for enlistees. In July 1777 Congress established a system of district recruiters who received $8 for each recruit and $5 for every deserter apprehended. Such juicy incentives led to widespread graft.

> By all accounts, 1777 was the high point of graft among Continental Army recruiting officers. Officers drew bounty money, reported enlistments and disbursements, then reported the desertion of the recruit. No doubt, such reports were often true, but many of the reported recruits had never exsisted. Officers gambled away public money or spent it on their own pleasure . . . The officers' graft was crude and unimaginative compared to the inventiveness with which towns and citizens cheated quotas and conscription in the war years. They sent British deserters, prisoners of war, bounty jumpers, and the physically unfit; towns claimed credit for enlisting men who were already in the army . . . Enterprising businessmen could pay a willing recruit the official bounty and then offer him to draftees as a substitute. The higher bidder among the draftees sent the recruit to the army, and the entrepreneur pocketed the difference.[69]

Whatever the ordinary Continental soldier could get his hands on in terms of bounty and pay was quickly negated by the devastating drop in the value of the dollar as the war progressed. By the fall of 1777 hyperinflation had destroyed the value of Continental paper money. By September 1779 one dollar was worth thirteen cents. In May 1780 General Johann De Kalb paid $850 Continental for "indifferent" food and lodging for four for one night.[70] Private Martin wrote in 1779, "I had it is true, two or three shillings of old continental money, worth about as much as its weight in rags." The Virginia Continentals fighting with Nathanael Greene in the Carolinas in 1781 had not received a penny's pay for two years.[71]

As in the British army, the soldier also had "stoppages" deducted to replace uniforms, arms, or equipment that may have been lost or destroyed. So out of the roughly $75 a year, it was not unusual for a Continental private to receive only about $12 after deductions. Private Elijah Fisher complained bitterly: "If I had anone [known] of [stoppages] before I had Engaged I ever would have gone . . . They will promise them so and so and after they have got them to Enlist they are Cheated out of one-half they ought to have by one or another of the officers."[72]

It is interesting to compare the soldiers' pay with the civilians'. For example, an agricultural laborer averaged two shillings a day or £30 per year, assuming he worked a solid six-day week throughout the year, which was highly unlikely given the inevitable seasonal layoffs and sickness. If he was hired by the year as live-in help (food and lodging provided), he could expect about £12 per year (which is about what a common soldier could expect after stoppages). An artisan (a carpenter or bricklayer, for instance) could make a little more: £40–45, assuming he did not "live-in." Small farmers, who comprised about 30 percent of the total population but 40 percent of the white population, could expect to generate £30. At the other end of the spectrum, large landowners could anticipate an income of £1,000–2,000 per year. Judges were paid about £300, while a royal governor might make over £3,000.

A single man needed £10–13 a year for food, and another £13 for clothing and lodging. About £2½ went for clothes (a pair of "coarse laborer's shoes" cost seven shillings and sixpence; a pair of breeches £1), while the cheapest urban dwelling cost around £10 a year. The family man had a tougher time. If he was an artisan, he might make £45 a year, of which £20 would go for food, another £4 or so for clothes, and £4–6 per year for each child. So an artisan with three children would be living within a very tight margin of financial viability. The poor,

Not unlike Today!

those with zero assets, represented 40 percent of the total population, while at the other end of the scale the rich—those with assets of £5,000 or more—constituted 3 percent of the population. In between lay the middle class, mainly composed of farmers and artisans. "Some of them lived little above the subsistence level, spending nearly all their incomes on necessities . . . A majority, however, earned enough to buy a few luxuries and accumulate some property. Half of the farmers and two-fifths of the artisans owned between £100 and £500 in personal estate."[73]

The previous occupations disclosed by 273 common soldiers of the Pennsylvania Line are an almost exact reflection of recruits to the British army: the largest group (34 percent) were agricultural laborers; shoemakers (23 percent); weavers (19 percent); blacksmiths and carpenters (12 percent apiece). The 4th and 9th Massachusetts regiments were composed largely (over 50 percent) of agricultural laborers, almost entirely American-born, and overwhelmingly poor. (Fewer than 10 percent of the men who listed farmwork as their previous occupation had any taxable land.) A study of General William Smallwood's Marylanders (one of the crack regiments of the Continental army) shows the typical pattern: they had very little taxable wealth, the majority had been unskilled laborers, and their median age was twenty-one years. The New Jersey Brigade, in which half the men were under twenty-one at enlistment, was also made up of overwhelmingly poor young men. Sixty-one percent came from the poorest one-third of the population, while, on the other hand, relatively wealthy New Jersey taxpayers who owned 100 acres or more made up only 9 percent of the soldiery.[74] The stalwart yeoman farmers, sturdily independent artisans, and businessmen of patriot mythology were canny enough to avoid duty at the sharp end.

Smallwood's Marylanders were interesting in another way: 40 percent were foreign-born. Of 1,068 recruits of the Pennsylvania Line, 75 percent said they were not American-born; about 46 percent were Irish: "The middle states contributed nearly half of the total force of the main Continental army in 1777 and 1778. The Irish presence [was] around 45 percent of their entire strength."[75] One of every four Continental soldiers was Irish. The "American" army was highly dependent on poor non-American immigrants, mainly Scotch-Irish and Germans. By the start of the war there were 300,000 Irish in America.[76]

The British lieutenant William Feilding wrote to Basil Feilding, earl of Denbigh, on 18 July 1775 that the American army "is said to be above half Irish & Scotch, but far more the former than the latter."[77] In a similar vein, the Hessian captain Johann Heinrichs of the Jäger Corps wrote to a friend in 1776 stating that the war was "not an American rebellion [for] it is nothing more than an Irish-Scotch Presbyterian Rebellion." ("Nothing more than a Mac-ocracy," said the Continental army general Charles Lee.)[78]

German immigration had reached about 225,000 by 1775 (58,000 in Philadelphia alone) and was heavily concentrated in the backcountry of New York, Pennsylvania, Maryland, and North Carolina. By 1776 one-third of the population of Pennsylvania was German.[79] Congress acknowledged the Germans' value as a pool of potential recruits, seeing them almost as "internal mercenaries," men who, John Adams recognized, could free "the Natives [American-born citizens] of the country who were needed for Agriculture, Manufactures, and Commerce."[80] On 25 May 1776 Congress ordered the formation of the German Battalion. Its manpower was to be drawn from more than one colony, which was unusual, as Congress normally allowed individual states to raise and equip their own Continental Lines. Five companies were to come from eastern Pennsylvania, and four from Maryland, all to be commanded by Major Nicholas Hausegger of the 4th Pennsylvania.

In a somewhat more sinister example of the role of Germans in the Continental forces, the Marechaussee Corps, under the Prussian veteran Captain Bartholomew Von Heer, was formed on 27 May 1778. It was to act as Washington's mounted military police, sometimes being placed behind troops going into battle to prevent desertion, and also to assist the provost marshal in the grisly business of military executions.

How many Americans fought in the war? The population of the colonies in 1775 was about 2.5 million (eighteenth-century America experienced something of a population explosion, growing from 250,000 in 1700).[81] Roughly 560,000 were African slaves and therefore prevented from fighting in the militia or Continental army. To calculate the number of men of eligible fighting age, we would have to deduct women; the very old and children; Loyalists, Quakers, and other pacifist groups; and so forth. Men also enlisted multiple times (often to get the extra bounty), so it is difficult to arrive at a precise total. Benjamin Franklin estimated 250,000 men of fighting age in the late 1760s,[82] and Howard H. Peckham in *The Toll of Independence: Engagements and Battle Casualties of the American Revolution* (1974) estimated "200,000 men in service at one time or another."[83]

If, for argument's sake, we accept Franklin's and Peckham's numbers, we have a very high percentage of the men available for war actually participating, which seems to run counter to a great deal of evidence suggesting that the fighting was done primarily by the young and most economically disadvantaged. However, the historian Don Higginbotham[84] cites Peckham as asserting that "no more than 100,000 *different* [my emphasis] men actually bore arms," which would mesh with what we know about widespread avoidance of service among many Americans.

Frederick the Great's calculations on the size of his army in proportion to the population of Prussia provides an interesting eighteenth-century perspective. In *The History of My Own Times* he estimated the Prussian population to be not more than 4.5 million souls, from which he deducted 3.5 million women, minors, and the aged. Of the remaining 1 million (22 percent of the total population) "capable of bearing arms," he calculated that not more than 70,000 of his soldiers were native Prussians (1.5 percent of the total population and 7 percent of the potential soldiery), "and even this is a high number to endure." If the same percentages are applied to the American population in 1775, the numbers would be 550,000 capable of bearing arms and 38,500 who fought.[85]

As might be expected, the number of Continental rank and file recorded in the official returns (a duty of each regimental commander as mandated by the Articles of War) fluctuated throughout the war. For example, the first six months of 1775 for which records exist (July–December) show a strong and steady establishment with a monthly average of 18,950 rank and file, with only 14 percent on average out sick. The first six months of 1776 were a little bumpier. For example, a high of 18,887 in February was immediately followed by 7,720 in March—the lowest for the whole year. After Congress's reorganization of the army in September 1776 (known as the "eighty-eight battalion resolve"), the numbers jumped up to the highest they would reach in the whole war: 39,892 in September and 40,962 in November. But the toll of the crushing defeats at Brooklyn, White Plains, and Forts Washington and Lee between August and November 1776 followed by the main Continental army's bedraggled retreat through New Jersey reduced the number catastrophically: the return for December 1776 was 9,125 rank and file, of whom almost 35 percent (3,180) were sick, leaving Washington with just under 6,000 effectives. Following the defeat at Brooklyn in August, the percentage of sick escalated dramatically, and, tellingly, that proportion marked as "sick absent" (as against "sick present") grew significantly. For example in July, the month before the battle, 2,313 men were recorded sick present and 259 sick absent for a total of 2,572 (16.6 percent of all rank and file). By the time Washington's bone-weary men reached the Delaware in December, the reports showed 600 men sick present and 2,580 sick absent—a complete reversal of the previous patterns which is understandable in light of their demoralization.

Although the victories at Trenton on 26 December 1776 and Princeton in January 1777 were crucial to keeping the flame alight, the returns for spring 1777 mark the absolute rock bottom for the rebellion. In March Washington reckoned he had fewer than 3,000 men, of whom about 2,000 were militia whose term of service was about to expire.[86] The returns for April 1777 show that Washington's total rank and file numbered a paltry 2,188, of whom 474 were sick. Admittedly there was the Northern Army under Schuyler, and men in the New York Highlands, but the commander in chief could muster only 1,714 muskets.

By the time Washington faced off against Howe at the battle of Brandywine in September 1777, his army had recovered, hitting a high in October of 33,021. From December 1777 through to and including April 1778—the Valley Forge months—the average monthly count of rank and file was just short of 20,000, with sickness rates averaging about 30 percent. The interesting and quite understandable pattern, however, is that the ratio of sick present to sick absent moves strongly in favor of those present—a clear sign of rising morale—as those months progress.

Defeats were reflected in the sick present/sick absent ratio. The case of Brooklyn has already been noted, and the pattern was repeated after Brandywine and Germantown in September and October 1777 respectively. After the battle of Monmouth Courthouse on 28 June 1778 (a "technical" victory for Washington as the British quit the field, but in reality a savaging of the Continentals by the British rear guard), the ratio of sick present to sick absent was 951:5,116. No doubt some of those who were absent had had the stuffing knocked out of them, but the crudity of the army's medical provisions must have played some part. It would have been perfectly understandable and prudent for sick or wounded men to have returned home, if at all possible, to be cared for.

In 1778 Washington's main army averaged 24,000 a month with a steadily dropping rate of sickness. In 1779 it was 21,500 but with dramatically low sickness rates (just over 10 percent as a monthly average). During these two years one has the sense that Washington's military machine was now robust, inured, and inoculated. Annealed by battle and hardship, it had become the army Washington would eulogize so movingly after the war, praising the "unparalleled perseverance . . . through almost every possible suffering and discouragement." He saluted it as his "standing miracle."

The number of rank and file started to drop in 1780 (14,000 in an average month but still with low sickness levels), and as war weariness really began to take hold it plummeted in 1781 to around 6,000 per month. With his army almost drained of lifeblood, Washington knew how desperately it needed the 12,000 soldiers and sailors that the comte de Rochambeau brought with him from France in July 1780. So pitiable was the size of the Continental army that it caused the American high command considerable embarrassment; so much so that Alexander Hamilton thought it "good policy to keep out of sight [of the French] the disappointments we have met with in the number of men &c."[87]

The 12,000 men of the initial French *expédition particulière* (of which about 4,000 were infantry) were, ironically for an army come to aid a people in revolt against a monarchy, a microcosm of the ancien régime. The army officer corps was strictly noble and represented about 6 percent of the total force. Officer salaries, however, exceeded that for the whole of the other ranks combined.[88] As with the British and American high commands, the French indulged itself in the spirited animosity that grew out of close association at court. They drew together in their dislike of Lafayette—who had gone native (in this they shared with their British counterparts a disdain for officers who, in their estimation, had forsaken the European tradition) and was also considered "pretentious."

Rochambeau was given a free hand in choosing the men and officers he wanted to take to North America. He chose steady but not stellar regiments that were pretty much at full strength (67 officers to 1,148 men each): Bourbonnais, Saintonge, Soissonnais, and the "German" regiment Royal Deux-Ponts which, he reasoned, would be doubly useful because it might be a convenient repository of deserters from Britain's German auxiliaries. Three thousand more men, from the Touraine (Lafayette's father had been killed leading the regiment at the battle of Minden in 1759), Agenais, and Gâtinais, would be brought by Admiral the comte de Grasse in August 1781.

The ordinary French foot soldiers were drawn from the artisanal class, certainly not the dregs of French society, and came from every part of France, although the departments of Alsace and Lorraine were the seedbeds of the French army. They were professional soldiers but not veterans; hardly any had seen battle. Most were in their early twenties, starting out on their first eight-year hitch, but with a leavening of older soldiers the median age was twenty-seven. The oldest was eighty, the youngest twelve.[89]

They were certainly stylish in their tight-fitting white uniforms (the Deux-Ponts, however, wore "celestial blue" in line with other German regiments), gaiters with twenty-four buttons, and shoes with one-inch-high heels, and their hair was drawn back and tied with a black ribbon. A single curl descended in front of each ear. On their tricorne hats was a white cockade which after a while was joined by a black one, the preferred color of the Continentals, to symbolize the alliance. Armed with the dependable 1777 model Charleville musket, the French expeditionary force of 1780 "was probably the most sophisticated military instrument ever dispatched to the New World."[90]

The benefits of the alliance, however, were not always universally appreciated. The French, after all, had always been the traditional enemy in North America, and papists to boot. John Adams, for example, had written himself a note on 1 March 1776: "What connexion may we safely make with her? No military connexion; receive no troops from her."[91] Charles Lee was characteristically blunt when he wrote in an undated letter to James Monroe, "The remedy is worse than the disease,"[92] and went on to speculate, with his trademark sarcasm, that before long America would be welcoming the Russian army to its shores. By September 1779 nobody gave a fig for Lee or his opinions, and the time had passed for Adams's punctilio. Washington's Continental army was as thinned out as breakfast gruel, and he made sure Lafayette passed the message back to Paris that French soldiers would be most welcome. In fact one of the very few occasions on which the commander in chief seems to have broken through that famously icy reserve and shown downright dance-a-jig joy was when he learned that Admiral de Grasse's fleet had defeated the Royal Navy on 5 September 1781 at the battle of the Chesapeake Capes and had arrived safely back in Chesapeake Bay. "I caught sight of General Washington," wrote Rochambeau, "waving his hat at me with demonstrative gestures of the greatest joy." They embraced, reported Baron Closen with italic emphasis, "*warmly.*"[93]

Lobsterbacks

THE BRITISH SOLDIER

They were the best of men; they were the worst of men. Many historians of the war want it both ways. On the one hand, the men of the British army are invariably described as "veterans," highly trained and experienced: tough nuts to crack. "The British army was the best in the world." On the other hand, British soldiers are invariably described as "scum," the refuse of society, jailbirds, "their ranks replenished with vagabonds and criminals."[1] Another American historian makes the astonishing assertion that "three British regiments in the American Revolution were composed entirely of reprieved criminals."[2] In this version men were snatched from the streets or disgorged from the prisons and then brutalized into mechanical obedience. "This hackneyed and hostile overview of the British Army has enjoyed a particular resonance in North America . . . it fits snugly with the comforting knowledge that such 'professional' British redcoats were subsequently worsted by 'amateur' American patriots."[3] Of course there were criminals in the British army, as there were in the Continental army. If recruitment comes from the least privileged, it will certainly scoop up a good few unsavory characters. And this is probably as true today as it was then.

If the assessment of some popular historians is to be believed, the reprieved criminals in the British army had all committed murder or other violent acts; but in eighteenth-century Britain it did not take much—poaching, the theft of a loaf of bread, rent arrears, trespassing, literally hundreds of petty crimes—to land the hapless culprit in the slammer. For example, a justice of the peace in Surrey wrote ingratiatingly to Lord Barrington, secretary of war, on 10 September 1777 that "John Quinn an Irish American 29 Years of Age near six feet high very dirty and ragged seemingly of slow understanding was this morning convicted before me of Orchard Robbing. He is willing to serve as a soldier. I have therefore committed him to the House of Correction in Guildford to await your Orders."[4] No doubt the secretary of war was reassured.

As far as attracting men to the colors, the experience of Britain during the war was not much different from that of America. Not many, or rather not as many as necessary, found it an attractive proposition. If one had work, and pretty regular work, there was little financial incentive to enlist. But many were not so lucky. Several factors were in play in eighteenth-century Britain that drove men into the ranks from what a contemporary called "the compulsion of destitution." As in America, population rates were rising sharply: 5.7 million in 1750, 8.6 million by 1800,[5] with large surges after 1750 when the annual rate of increase rose from 3 to 10 percent. This, together with the impact of industrialization on what had been labor-intensive trades (such as shoemaking, weaving, and spinning), as well as on those who were highly dependent on seasonal work (agricultural laborers, carpenters, bricklayers, masons; and small subsistence farmers who were forced off their lands by rising property prices), created a highly competitive low-wage job market. Poverty was seen both as a just reward for personal shortcomings (laziness, stupidity, "creature urges") and as necessary to the maintenance of the fabric of society. One of the most influential of British economists and agricultural reformers, Arthur Young, wrote cynically or sarcastically, "Everyone but an idiot knows that the lower classes must be kept poor or they will never be industrious."[6]

The economic tectonic plates of depressed wages and rising prices (the cost of living doubled between 1770 and 1795)[7] ground against each other, and those trapped between would be fodder for the army: "Military service [was] preferable to starvation."[8] Only the lowliest stratum of the British proletariat—agricultural laborers—was considered beneath a soldier. Which was saying something when one considers how contemptuously soldiers were regarded. A Royal Navy saying of the eighteenth century described a pecking order of disdain: "A messmate before a shipmate, a shipmate before a stranger, a stranger before a dog, a dog before a soldier." The navy always seemed to win the hearts of the people; the army their contempt. One reason was that if you were an unlucky citizen with spare room in your home, an unwashed and uncouth soldier might come to live with you.

Billeting of soldiers in private houses and inns was commonplace in eighteenth-century Britain. There were no barracks or government funds for purpose-built military housing. The soldiery was spread around the country in various policing roles (chasing down smugglers; putting down civil uprisings) and had to lay their not-too-fragrant heads somewhere. Billeting was widely resented in Britain and also became one of the bones of contention in America before the war. The parish constable, the official usually responsible for allocating billets, made sure he did not antagonize the bigwigs (the term *bigwig* derives from the large bagwigs worn by the powerful in the late seventeenth and early eighteenth centuries) of his community, "Squire, Parson, Gentleman or able Farmer, by sending so disagreeable a Thing to their Houses as a Red Coat."[9]

Enforced conscription was not an option in Britain. First, there was widespread mistrust of a standing army—as indeed there was in the colonies—as a potential instrument of absolutism. Second, the landowning class, which formed the bulk of the members of parliament, would not pay for a conscripted army through taxation or make the political concessions it might involve: "A whole nation could not be called to arms without at least the illusion of political partnership; and the ruling classes of the eighteenth century had no intention of paying the price of total war."[10] Third, the people simply would not accept it, and the machinery of rule, through monarch and parliament, was not as absolute as in, say, Prussia, Russia, or France, where more draconian methods could be employed to coerce men into the ranks.

When enforced drafting into the militia was tried in 1757, it resulted in widespread rioting and had to be abandoned. The British army was, until the introduction of conscription in 1916, technically a volunteer force—in theory if not always in fact.

Broadly speaking, there were two principal ways in which the government could get soldiers. The first, and most widely used, was *beating up* (a term alluding to beating the drum to attract would-be recruits rather than physical coercion). The second was impressment. Beating up was usually conducted by a small team of regulars: a recruiting officer who was often an NCO, perhaps a drummer, and one or two others, all dressed in their finery, who would visit the towns and villages within the catchment area of the regiment involved. Winter was often a fruitful time, particularly in agricultural areas where seasonal dips in employment made laborers particularly vulnerable, as were carpenters, bricklayers, and stonemasons. The recruiter's job was to sell the adventure, the camaraderie, the glorious traditions of the regiment, the sterling qualities of its colonel, as well as the financial "benefits" of army life, all washed down with liberal amounts of alcohol (and in this American recruiters were no different from their British counterparts). Alcohol played an important role at all levels of eighteenth-century society and was a major factor in army life.

In times of peace, service was usually a lifelong commitment, but when the pill had to be sweetened to induce men to sign up for America, the commitment for volunteers was three years. Enlistees were paid a bounty, and, as in recruiting for the Continental army, the difficulty of attracting sufficient numbers of men led to increases in the bounty during the period of the American war. It started at a guinea (£1 and one shilling) but rose to £3 in 1778, and three guineas in 1790. In addition, the recruit got the "king's shilling" (actually a crown), which was invariably spent immediately on quaffables. Out of the bounty (which the recruit did not get in his sticky fingers), there were deductions for kit and clothing (not until 1856 would British soldiers be issued free kit and uniforms on enlistment), which reduced the net bounty considerably. Occasionally, patriotic towns raised subscriptions among the citizenry to offer bounties to stimulate enlistment. Such, for example, was the case with Bath, Bristol, Birmingham, Warwick, Coventry, and Leeds in 1778 when 15,000 men were added to the rolls.[11] Cavalrymen received a higher bounty of £4 fourteen shillings at the start of the war, and in general the cavalry attracted a "better" class of recruit, which was reflected in the lower desertion rates compared to the infantry.

The pay of a British private has been described as "a racket in which everybody received money but the soldier himself."[12] He was paid eight pence per day (there were twelve pence to the shilling; twenty shillings to one pound sterling), a rate that that had been established in 1660 by parliament. (It would rise to the princely sum of one shilling per day in 1797 and remain there for a further seventy years.) Of the fifty-six pence per week gross basic pay, forty-two pence was earmarked for "subsistence," supposedly for food, but other items were also included. For example, six pence a week went for shoes, stockings, and repair of arms; one penny per week was a fee split between the regimental surgeon and paymaster, and fourteen pence a week for the "gross off-reckoning" which included what we might call overhead: payments to the paymaster general, to Chelsea Hospital (the home for military pensioners that had been founded in 1690), and to the regimental agent. The arithmetic has a Micawberesque neatness to it: fifty-six pence in; fifty-six pence out.

To supplement his pay, the soldier was often forced to look for moonlighting work. Redcoats at Boston prior to the outbreak of hostilities hired themselves out as laborers, which caused a good deal of resentment among the local workforce and led to violence. But what was true in the British army was also true among many armies in the eighteenth century. For example, the *Connecticut Journal* of 14 June 1775 reported that men of the American Continental army hired themselves out in the "husbandry business" for eighteen pence per day.[13] In Prussia, short on funds but large on army, moonlighting was built into the system, and French soldiers hoped for garrison duty at Brest, where "everyone could find a job."[14]

The Press Acts, the legal mechanism that enabled the British government to forcibly draft men, were resorted to at several critical points for British arms in the eighteenth century, but two fell within the period of the war against America: 28 May 1778 and 9 February

1779. All "able-bodied, idle, and disorderly Persons, who could not, upon Examination, prove themselves to exercise and industriously follow some lawful Trade or Employment, or to have some Substance sufficient for their Support and Maintenance" were eligible. They were essentially the poorest of the poor, those who were "on the parish," that is, subsisting on the meager provisions of the local poor law. The assumption was that the state had a claim on those who were a burden on society. Impressed men did not receive a bounty; instead, the parish was reimbursed to the tune of twenty shillings for a single man and forty shillings if the man left a family dependent on parish relief. All voters (men of property) were exempt from the press, as were farm laborers during the harvest season and apprentices whose masters demanded their release. There was no escape from the press through substitutes because, in the words of Sir Charles Jenkinson, secretary of war, men who procured a substitute would "be left at liberty to return to the very neighborhood where they had proved obnoxious."[15] In any case, not many could have afforded to procure a substitute.

Some men mutilated themselves by cutting off the thumb and forefinger of the right hand rather than suffer impressment, while others joined the militia which, as Jenkinson pointed out, was no bad thing "for the Men being troublesome and obnoxious, became at once of use to their Country, to whose defence they contribute as much by voluntary Service in the Militia, as they had been impressed into the regular Army."[16] In fact impressment was not a substantial source of recruits. For example, during the whole of the 1779 press 2,200 men were procured, and there was even a backhanded benefit for volunteerism. During the same period the rate of volunteerism rose by 33 percent over normal with an intake of 7,000.[17] Many, it seemed, would rather jump (and get the bounty) than be pushed.

Recruiting officers were restricted to enlisting men between the ages of seventeen and forty-five who were sound in body (they had to undergo a rudimentary medical exam) and at least five feet six and one-half inches tall. Although recruits were hard to come by as the war progressed, the height criterion was adhered to. At the end of the war the average height of a British foot soldier was five feet seven inches (five feet nine inches in the cavalry), with an average age of thirty years. The men had joined up when they were, on average, twenty years old, and by 1782 had an average of nine years' service under their belt.[18]

The men who would be shipped out to America, far from being criminal "scum," were representative of a wide swath of the British laboring poor; men who had fallen on hard enough times to make service in the army, if not the most appealing prospect, at least a viable one. Twenty percent of the recruits in the Coldstream Guards and the 58th Foot, for which good records survive, were from the textile industry. (Weavers had been particularly hard hit by the mechanization of the Industrial Revolution.) Forty percent of the Guards and 15 percent of the 58th were displaced agricultural laborers.[19] As in the Continental army, they tended to be young (45 percent of the population of Britain was under twenty years old), although not quite as young as their American counterparts. Sixty percent of them were English and Welsh; 24 percent Scots; 16 percent Irish. (Perhaps due to some genetic Celtic wizardry, the artillery had a substantial majority of Welsh and Scots—63 percent.)[20] These percentages had shifted as the eighteenth century progressed. At the time of the French and Indian War, as revealed by the 1757 returns, 30 percent of recruits to the British army had come from England and Wales, while over 50 percent were from Scotland. (Another 15 percent were made up of American-born and European, particularly German, enlistees.)[21] Although Scotland still accounted for a sizable chunk of the total (especially in proportion to its population, which of course was much smaller than England's), it was, ironically, the drain of Scots immigrants to America earlier in the century that accounts for a lower percentage of Scots soldiers.

The victory of the Hanoverian army over the Jacobites at Culloden in 1746—a victory of disciplined musketry over the broadsword and pike—had destroyed the old clan system. In its aftermath, the wearing of the tartan and the carrying of arms were forbidden to clansmen. Destitution also followed in its wake. The Highland clearances—large-scale appropriations of land for cattle and sheep farming by landowners who had proved their loyalty to the Hanoverian cause—had driven great numbers of clansmen from their livelihoods. And in one of those ironies of history, thousands of dispossessed Scots joined the British army; they not only joined but became, and have ever since been, a fighting elite, with an esprit de corps that was, and is, remarkable. They were to become perhaps the finest infantry in the war against America. For example, the 71st Foot (Fraser's Highlanders) would see more action in America than any other British regiment, from the battle of New York in 1776; Brandywine in 1777; Stony Point, Briar Creek, and the defense of Savannah in 1779; Camden in 1780; and the great battles of 1781: Cowpens, Guilford Courthouse, and Yorktown.

The British army, far from the being "the greatest in the world," was one of the least significant among the armies of Europe. When Frederick the Great contemplated the balance of military power in Europe, this is how he saw the comparative strengths of the armies.

France:	180,000
Austria:	160,000
Prussia:	150,000
Sardinia:	50,000
Spain:	40,000
England:	20,000[22]

ENGLAND: Not a Great Army but a great Navy

Admittedly the armies of France, Austria, and Prussia were national troops with a very significant contingent of mercenaries (all eighteenth-century nations used them liberally), and Frederick's numbers represented only those forces that could be committed to European warfare. Britain, more than any of the others listed, had extensive commitments to its colonial possessions. Nevertheless, it is a salutary correction to the "greatest army" myth. Britain, as seen at least by Frederick the Great, was only a bit player in the military picture of eighteenth-century Europe as far as its land forces were concerned. The "greatness" to which many historians of the War of Independence allude—Britain's success against, primarily, France during the Seven Years' War—was built largely on its navy, which certainly was the most powerful in the world. Ironically, Britain's failure in America might quite plausibly be laid at the Royal Navy's feet.

In 1760, at the height of the Seven Years' War, Britain had mustered about 203,000 men (including German mercenaries).[23] But by the eve of the American war fifteen years of peace had reduced that number to about 48,500,[24] of whom approximately 36,000 were infantry,[25] 7,000 cavalry, and 2,500 artillery. Of this total, 7,000 men were already in or en route to America and Canada, while another 10,000 were in the West Indies, Minorca, and Gibraltar. In April 1775 Britain would have 7,000 men in North America, and its army would reach its peak of 50,000 in February 1778.[26] These were relatively small numbers (as British commanders in chief continually pointed out). For example, Frederick the Great had 77,000 and his French adversary approximately 85,000 at the battle of Hohenfriedberg in 1745, numbers that would be dwarfed during Napoleon's wars. Bonaparte had more than 100,000 men each in seven of his battles. At Wagram, for example, 167,000 Frenchmen were pitted against Archduke Charles's 130,000.[27] At Waterloo Wellington had 107 regiments of infantry and 40 regiments of cavalry, which, with gunners and his German auxiliaries, gave him a total of 60,000 combatants. By comparison, the numerically largest battle of the American war was at Brooklyn (on 22–29 August 1776), where 15,000 British faced 9,500 Americans.

Were the redcoats battle-hardened veterans as so many historians claim? As far as the mass of the British soldiery was concerned, it would have been doubtful if it could boast of many who had seen action. In 1775 the British army had not fought a major land battle since 1759: at Minden on 1 August under George II (the last British monarch to personally lead his army in battle) and at Quebec under James Wolfe on 13 September. In the fifteen years that had since elapsed there would have been only a sprinkling of rank and file who had seen action back then.[28] A small number of NCOs would also have had battle experience from that earlier period, but most privates would have passed out of the service. In fact the American militia probably mustered a much higher proportion of men who had seen action during the French and Indian War because age did not prevent them from serving the patriot cause. On the other hand, it was true that many of the British senior command—Thomas Gage, William Howe, Henry Clinton, John

Burgoyne, and Guy Carleton, for example—had been blooded either in Europe or in America, as indeed had patriot commanders like Charles Lee, Frederick von Steuben, and Johann de Kalb in Europe; and Horatio Gates, Philip Schuyler, Richard Montgomery, Daniel Morgan, and, of course, George Washington in the French and Indian War.

In wartime, during periods of intensive recruiting, the ratio of raw recruits to seasoned soldiers rose. In peacetime, the percentage of recruits in the infantry represented about 16 percent of the establishment (9 percent in the cavalry), but during the war with America it rose to 27 percent (19 percent in the cavalry), and by the end of the American war the average infantry regiment could muster only 59 percent of its paper establishment, and the percentage of recruits had risen to 24 percent— an arc that is perfectly consistent with most wartime armies.

The young British lieutenant William Feilding, who was in action at Bunker's Hill, reported on the performance of this largely inexperienced force: "I can only say from the oldest soldiers here that it was the hottest fire they ever saw not even the battle of Minden which was reckoned one of the greatest actions ever gained by British Troops . . . many who had never seen a Ball fired . . . behaved with the greatest courage."[29] Lord Rawdon, who was also in the battle, saw things a little differently: "Our confidence in our own troops is much lessened since the 17th of June. Some of them did, indeed, behave with infinite courage, but others behaved remarkably ill. We have great want of discipline among officers and men."[30] Whether they did well or ill, there is a sense that these men were inexperienced, and that years of peacetime inactivity had dulled their war-fighting skills. Nor could their German allies claim to be veterans. Mostly young farm laborers and impoverished artisans, they were lambasted by a British officer: "Government has been Cheated by their sending one half Militia, and the greatest part of the others are Recruits, very few Viterons [veterans] amongst them."[31]

<hr />

A typical British infantry regiment (or battalion—the terms were pretty much interchangeable) consisted of eight "battalion" companies and two "flank" companies (one of grenadiers, the other of light infantry), so called because when the battalion was drawn up in its battle lines, they were positioned on each flank or wing: the grenadiers on the right, the position of honor, and the "light bobs" on the left. On paper, though rarely in reality, each company had fifty-six privates. (In actuality there were fifty-three because three were fictitious "contingency" men whose pay was collected by the company captain to defray, theoretically at least, the cost of weapons' repair and other company expenses.) In addition, each company would have a captain, a lieutenant, an ensign (the most junior officer rank), three sergeants, three corporals, a drummer, and a fifer. A colonel, a lieutenant colonel, and a major formed the headquarters staff, and a chaplain, surgeon, surgeon's mate, and quartermaster completed the establishment.

Technically, a British battalion complete with its flank companies had 642 officers and men and could muster 560 muskets (privates plus corporals). But the difference between paper strength and battle strength was enormous, due to losses through illness, casualties, desertion, or shortfall of recruitment. For example, of the eleven British line regiments at New York in 1778, the average number of effectives per regiment was 411.[32] Lieutenant Frederick MacKenzie reported the rank-and-file strength of the 23rd Foot (Royal Welch Fusiliers) after the Lexington-Concord escapade, in which the regiment had taken four killed, twenty-six wounded, and six missing, as:

Grenadier company	29
Light infantry company	35
Eight battalion companies	218 *
Total	282[33]

* an average of 27 per company

Even before the action the regimental total would have been only 318.

The first grenadier companies had been raised in 1667 in the French army by selecting the tallest and strongest men (those physically best suited to toss grenades) of the Régiment du Roi and by the late 1670s had been adopted in the British army.[34] Although by the time of the American war they had long ceased to be grenade throwers, they were still distinguished from the ordinary foot soldier by their height and build, and their role as shock troops conferred an elite status. Britain, unlike many European countries, did not create specific regiments of grenadiers, and British grenadiers in America are frequently confused with the Grenadier Guards (known in this period as the 1st Foot Guards and who were given their name only in 1815 in recognition of their achievement at Waterloo fighting as guardsmen, not because they were grenadiers).[35] The confusion is fueled perhaps by the fact that there was a brigade of Guards in the war made up of fifteen men seconded from each of the sixty-four companies of the three Guards regiments: 1st Foot Guards, 2nd Foot Guards (Coldstream), and 3rd Foot Guards (which was named the Scots Guards in 1877), and further muddled by the fact that grenadier companies from a wide assortment of regiments and at various times in the war were taken from their regiments and brigaded together to fight as independent corps: the 1st and 2nd Grenadiers (as indeed were the light companies).

The creation of these separate grenadier and light infantry corps has been identified as an organizational weakness that put British regiments at a significant disadvantage when it came to exchanges of volleys with their patriot counterparts. The argument goes that American regiments—at least according to the theoretical reorganization Congress approved on 4 November 1776—would have had eight companies each containing seventy-six privates who (when added to the four corporals per company) would have totaled 640 muskets per regiment, compared with the 536 muskets of a British battalion shorn of its flank companies— a 16 percent advantage to the Americans.[36] However, what appears on paper rarely appears on the battlefield. The Continental army, with the best will in the world, would never come close to the theoretical strength mandated by Congress. In September 1776, for instance, the main army under Washington stood at 31,000. Over half of its men—17,670—were militia, and of the 13,283 Continentals 11,590 were rank-and-file musket men. Sickness and detached postings further reduced the number of muskets available to 6,081 spread over twenty-five regiments, an average of only 243 muskets per regiment.[37] Likewise, under the reorganization of the Continental army of 4 November 1776 North Carolina was mandated by Congress to raise nine regiments for a total of 7,000 men. By July 1777 those nine regiments totaled only 1,094, of whom 963 were musketeers—an average of only 107 muskets per regiment.[38]

⚜ BRITAIN'S GERMAN AUXILIARIES ⚜

There were over 300 states in "Germany" in the eighteenth century: a rag-bag patchwork of small electorates, princedoms, duchies, margravates, landgravates, bishoprics, and free cities, many of them not much larger than New England townships. Each had its court to maintain, and each was constantly in need of money. Their armies were assets that could, to use a modernism, be "monetized." The use of foreign mercenaries was part and parcel of eighteenth-century warfare. Frederick the Great's army, for example, was heavily dependent. In 1751 his army consisted of 83,000 foreigners and 50,000 native Prussians:[39] "Mercenary troops are absolutely necessary," he had declared. Twenty percent of the French wartime army was composed of mercenaries: Germans, Swiss, Italian, Irish, and Scots.[40] Given the German origins of the eighteenth-century British monarchy, it is not surprising that German mercenaries had been a staple of Hanoverian warfare. Hessians had fought for the Crown in Europe in 1739, 1740, and 1742. At Culloden in 1746 they had been a contingent of the duke of Cumberland's army. They had even been hired out to contending armies. In 1743 Hessian fought Hessian: 6,000 under the banner of George II and an equal number fighting against them for the Austrian emperor, Charles VII.

The accession of the House of Hanover and Brunswick to the British throne in 1714 brought with it an intricate network of connections with several German principalities. For example, George I married Sophia Dorothea of Brunswick-Lüneberg; George II married Caroline of Brandenburg-Ansbach. Their son Frederick married Augusta of Saxe-Coburg-Gotha, and their daughter Mary married Frederick II of Hesse-Cassel. George III was married, particularly happily, to Charlotte Sophia of Mecklenburg-Strelitz. In fact, this German connection would be carried right through into the modern British monarchy. (Under pressure from anti-German sentiment during World War I, Saxe-Coburg-Gotha changed its name to Windsor.) It is not surprising, therefore, that Britain should turn to a tried and trusted source of mercenary auxiliaries. The greatest of the German states, Frederick the Great's Prussia, refused to supply troops (nor would Catherine of Russia), but there were still those German princelings, ever needy, ever greedy.

The first deal was struck with Charles I, duke of Brunswick, in January 1776. He agreed to supply 3,964 fully equipped infantry and 336 mounted dragoons (minus their mounts) at the rate of £17 four shillings four and a half pence a head per year. In addition, the British government would pay an annual "fee" of £11,517 seventeen shillings one and a half pence starting from the day of signature of the treaty. For each of the two years following the return of the Brunswickers to Germany, the fee would be doubled. As a bonus to be paid directly to Charles, Britain also agreed to cover the costs of replacing any Brunswicker killed in British service (at £17 four shillings four and a half pence a head), with a macabre footnote that three wounded men were to count as one dead man. (The War Office paid the sum "off the books" to avoid public censure—an eighteenth-century Iran-contra arrangement.) For his part, Charles would have to replace any deserters or any soldier who fell sick with anything other than an "uncommon contagious malady."[41] By war's end Brunswick would send a total of 5,723 troops.

The end of January 1776 saw a treaty with the largest supplier of men: Hesse-Cassel (hence the umbrella word *Hessian* to describe all Germans fighting in America). The per-head rate was the same (as it would be with the smaller states who participated), but the landgrave, Frederick II, cannily managed to wring a substantially better deal than Charles: £108,281 five shillings per year (paid up to a year after the troops' return) together with a British defense commitment should Hesse-Cassel be attacked. Frederick was to supply 12,000 men, a figure that grew to almost 17,000 by war's end. Treaties with the smaller states of Hesse-Hanau, Anspach-Bayreuth, Waldeck, and Anhalt-Zerbst completed the haul. In all, the total throughout the war was just short of

price for mercenaries

30,000, about 30 percent of the British military presence in America, and General Wilhelm von Knyphausen would rise to be deputy commander in chief of the allied army in America.

The Hessians—"of a moderate stature, rather broad shoulders, their limbs not of equal proportion, light complexion with a b[l]ueish tinge, hair cued as tight to the head as possible, sticking straight back like the handle of a skillet,"[42] with their extravagant mustachios (the British and American armies of the time were clean-shaven)—were considered exotic and a little savage by British and Americans alike. But in fact, the common Hessian soldier was not unlike the common British or American soldier. They had been agricultural laborers, textile workers, shoemakers, all of whom had fallen on hard times, or fallen for the sucker punch of an unscrupulous recruiting agent. They were, to use the wonderful phrase of the poet Johann Gottfried Seume, himself an unwilling recruit, a "true pell-mell of human souls." Seume listed among the occupations of men in his group: "a runaway student from Jena, a bankrupt merchant from Vienna, a haberdasher from Hannover . . . a Prussian Hussar guard, a cashiered major from the fortress," and even a monk from Würzburg.[43]

Not even their allies always appreciated them. Ambrose Serle, Admiral Lord Howe's secretary, thought they would "tend to irritate and inflame the Americans infinitely more than two or three British armies."[44] Captain John Bowater, writing from New York in May 1777, called them "the worst troops [he] ever saw . . . They are exceedingly Slow, their mode of discipline is not in the least Calculated for this Country and they are strictly enjoined by the Landgrave, not to alter it. They are so very dirty that they always have one half of their People in the Hospitals . . . [The Americans] have now got a very Considerable Army together, and you may depend upon it, they will beat the Hessians every time they meet (the Granadiers & Chassures only excepted.)"[45] Serle and Bowater's attitude is explained to some extent by good old British chauvinism, for given the circumstances that had led many of the German troops into the ranks, and given that they had no interest in the ideological basis of the war, they fought with surprising tenacity. At Long Island, at the attack on Fort Washington in Manhattan, at

White Plains, at Brandywine, even during the Saratoga battles in which they took tremendous punishment, and later in the South, particularly at Guilford Courthouse, they gave good accounts of themselves, as evidenced by a short list of the senior German officers killed in battle: Johann Rall at Trenton, Friedrich Baum at Bennington, Heinrich Breymann at Bemis Heights, and Carl von Donop at Fort Mercer.

"The Idea which we at first conceived of the Hessian riflemen was truly ridiculous," wrote an American officer of the battle of Brooklyn, "but sad experience convinces our people that they are an Enemy not to be despised."[46] A Hessian officer explained why: "[The rebels] have some very good marksmen [and] they are clever at hunter's wiles. They climb trees, they crawl forward on their bellies . . . But today they are much put out by our greencoats [jägers] for we don't let our fellows fire unless they can get good aim, so that they dare not undertake anything more against us."[47] In fact, from the first they put the fear of God into the Americans, as Colonel Samuel Atlee, commanding a Pennsylvania battalion, attested at the battle of Brooklyn: "The opinion we had formed of these [Hessian] troops determined us to run any risk rather than fall into their hands."[48] Forced to surrender, Atlee much preferred to take his chances with a battalion of Highlanders which, given their reputation for less-than-gentle treatment, seems like a choice between the devil and the deep blue sea. The Hessians were noted for their use of the bayonet, and it is probably this that put the wind up the unfortunate Atlee.

Hessian desertion rates, certainly in the early ears of the war, were surprisingly low. Even though Congress and Washington mounted a concerted campaign to woo Germans, including the offer of fifty acres for each deserter, only 66 men deserted in 1776 and just over 100 in 1777. As the war progressed, and as more Germans became prisoners, desertion rates inevitably rose. The big jump was in 1778 (approximately 480) and can be explained by the large number of Germans taken prisoner after Burgoyne's surrender at Saratoga in October 1777. The rate fell off during 1779 and 1780 but began to climb again as war weariness ground down morale. In 1782 it was about 500, and in 1783 approximately 750.[49] There were those who repaid the cynicism of their rulers with a little of their own. An exasperated German officer wrote: "Most of the recruits, mainly foreigners, behave very badly and desert at the first opportunity . . . Many of them have intended to take the chance of free passage to this country, and finally to quit Europe. They would have had to work about four years to pay the cost of their crossing."[50] Of the 30,000 who came to fight in America, 7,500 would lay their bones there: 1,200 in battle and 6,300 by disease. Another 5,000 simply stayed to make a new life in the new nation.

⊰ LOYALISTS ⊱

They are the forgotten ones; sharing the gloomy penumbra of history with impoverished White Russian aristos working as waiters in Paris in the 1920s after the Bolshevik revolution, or once-wealthy, cultivated South Vietnamese government officials opening delis in Los Angeles and New York after their country fell. They have about them a sepia sadness, the corners curled and the colors faded. Who cares? They were the losers.

The traditional estimate of Loyalist strength (first proposed by John Adams) is that one-third of Americans were patriot, one-third pro-British, and the remaining third neutral or persuadable one way or the other. Some historians think this three-way split is "too simple, and too high for the king's adherents";[51] while others consider it too low.[52] Somewhere between the two lay a very considerable body of people for whom the patriot cause was the destruction of the known, true, ordered, and trusted world, from which, at war's end, over 80,000 of their persuasion would be banished.

There were two kinds of Loyalist. Claude Van Tyne in his classic, *Loyalists in the American Revolution* (1902), one of the first books devoted to their history, had no doubt who one kind was: "They were the prosperous and contented men, the men without a grievance . . . Men do not rebel to rid themselves of prosperity." And for him they *were* Loyalism. But his description is by no means true of many, if not most. The other kind, particularly in the Carolinas and Georgia, were backcountrymen, mainly Celtic Scots immigrants (as against the predominantly Protestant Scotch-Irish), and mainly poor. They nursed bitter grievances (particularly against the Whiggish plantation society of the tidewater) that would, as the war progressed, work themselves out in a bloody civil war of startling savagery.

The wealthier Loyalist came predominantly from the middle colonies, particularly New York and New Jersey, which were the strongholds of Loyalism during the revolution. Those two colonies alone represented as much as 50 percent of American Loyalism,[53] a fact reflected by Tom Paine's outburst in the first issue of his broadsheet "The American Crisis" of 10 December 1776: "Why is it that the enemy hath left the New-England provinces, and made these middle ones the seat of war? The answer is easy: New-England is not infested with Tories, and we are."[54]

This particular strain is usually described as passive, overdependent on its British protectors, lacking in initiative, hesitant, crippled by its own gentility, and always a few fatal steps behind its better organized, more energized, more determined patriot enemy. They may have been relatively wealthier, but they were not effete grandees. When Massachusetts passed its Act of Banishment in 1778, the occupations of the 300 or so Loyalists slated spanned a broad gamut. A third were merchants, lawyers and other professionals, and "gentlemen"; another third were farmers; and the rest were artisans, laborers, and small shopkeepers.[55] When over 1,000 Loyalists fled Boston with General Howe, the majority were middle- and lower-class. Two hundred and twenty-five were government officials, Anglican clergy, and large farmers; but 382 were small farmers, traders, and "mechanics." Another 213 were merchants, and 200 unspecified.[56]

Nor should we assume that Loyalists were somehow more "English" in their background. In fact, the crucible of the revolution was in the most "English" of the colonies (Massachusetts and Virginia), and it was here that Loyalism was relatively weak (only about one-tenth of Tories were in the Northeast). In fact, perhaps the most militarily active strain was far removed from genteel "Englishness." The southern strand (approximately one-third of all Loyalists), with its Celtic impetuosity, its clannishness, its yearning for the heroic to the point of foolhardiness, rose up early in the war and was cut down almost as precipitately. On 27 February 1776 a Loyalist force of about 1,600, of whom 1,300 were Highland Scots under the command of Brigadier General Donald McDonald, their vanguard swordsmen led by Captains John Campbell and Alexander McCloud, charged across the bridge at Moore's Creek, North Carolina, to attack 1,000 patriots dug in on the opposite bank. The Highland broadswords were met by concentrated musketry and artillery. Campbell, McCloud, and 40 of their men were killed; 850 were captured. The patriots lost one man killed and one wounded. A tribal style of war fighting, no matter how passionate, could not alter the cool mathematical logic of volleyed muskets and artillery.

Loyalism was a siren call for the British. They were constantly bending their strategy to conform to the chimera of Loyalist support that was assumed to be there but somehow never materialized. Howe's Philadelphia campaign and Burgoyne's invasion from Canada, as well as the British strategy in the South, were based on the assumption that large numbers of Loyalists would rise in support, if only sufficiently encouraged and protected, as many people—influential and knowledgeable people—whispered seductively. Ambrose Serle was among them. He had been an undersecretary to Lord Dartmouth when Dartmouth was secretary of state for the colonies between 1772 and 1775, and had come to America as Admiral Lord Howe's secretary. Reporting on the situation in Pennsylvania in May 1778, he confidently reported Loyalist hopes: "Had some Conversation with Mr. Andw Allen. His most material Remark was . . . that five Sixths of the Province were against the Rebels, our Army had only to drive off Washington & to put arms into the Hands of the well-affected, and the Chain of Rebellion would be broken; especially if we restored the Province to the King's Peace . . . That this wd affect all the Southern Colonies; & that then the Northern Colonies could not long stand out."[57]

In the South, James Simpson, the attorney general of South Carolina, reported in August 1779 to General Clinton and Secretary of State for the Colonies Lord George Germain: "I am of opinion, whenever the King's Troops move to Carolina, they will be assisted, by very considerable numbers of the inhabitants, that if the respectable force proposed, moves thither early in the fall, the reduction of the country without risk or much opposition, will be the consequence."[58] Perhaps the most powerful voice in the British ear was that of William Knox, a wealthy Georgia landowner and sometime member of the Georgia Council, who had come to England in 1761 as agent for the state and had become undersecretary in the American Department. In his opinion Britain should cast off the insurgent northern colonies completely and re-form its empire in the South.

Broadly speaking, there were two main ways in which a Loyalist could serve the Crown in arms: in provincial regiments or in the militia. During the first years of the war Britain preferred, primarily for reasons of economy, to encourage Loyalists to raise their own regiments through a system known as "raising for rank." It was, in essence, an entrepreneurial activity by which a prominent citizen set himself up as a recruiting agent (bearing most of the recruiting expenses) and prospective colonel of his regiment. For the British the system had some distinct advantages. It was cheap. Apart from shouldering the recruiting expenses, the recruiting officer was eligible for pay only when he had raised 75 percent of his regiment, and the officers of the regiment did not qualify for permanent rank with the British army or go on to half pay if the regiment was downsized or disbanded. At first, instead of a money bounty, each private soldier was given a royal grant of 50 acres while NCOs received 200 acres. Unlike the regular army, Loyalist regiments had no provision for medical care, and wounded officers did not receive invalid pay. And to underline their inferiority, if Loyalist field officers served with a regular British regiment, they were demoted to the most junior rank of the grade below their provincial grade.

As the pressure of the war bore down on resources, Britain was forced to pay more attention to the Loyalists as a military resource. In November 1776 Howe had asked Germain for 15,000 reinforcements, a number that horrified him. He might be able to muster perhaps half that, he thought. There would have to be more reliance on American manpower, and this requirement, to an important extent, dictated British strategy for the rest of the war. Howe's 1777 invasion of Pennsylvania and the capture of Philadelphia, a stronghold of Loyalism—dismissed by most historians as strategically pointless and simply another indication of British dim-wittedness—makes more sense when it is seen in the light of the growing importance of hitching the Loyalist draft horse to the royal wagon. The load was becoming too heavy for the regular army to haul alone. As Howe explained in a rationale for his Philadelphia strategy, "The opinions of people being much changed in Pennsylvania, and their minds in general, from the late progress of the army, disposed to peace, in which sentiment they would be confirmed, by our getting possession of Philadelphia."[59]

Most of the important provincial regiments were raised following the fall of New York in September 1776. They included Oliver De Lancey's brigade from New York; Robert Rogers's Queen's Rangers; Cortland Skinner's New Jersey Volunteers ("Skinner's Greens"); Sir John Johnson's Royal Greens; Beverly Robinson's Loyal American Regiment; and the New York Volunteers from Nova Scotia. In the winter of 1776 and spring of 1777, more important units were added, including Edmund Fanning's King's American Regiment from New York; John Coffin's King's Orange Rangers, a mounted rifle regiment from New York; Monteforte Brown's Prince of Wales American Regiment, composed largely of Connecticut men; William Allen's Pennsylvania Loyalists, who could boast Sir William Howe himself as their colonel; and James Chalmers's Maryland Loyalists. Although almost eighty Loyalist regiments were raised over the course of the war (including black corps like the Jamaica Rangers, which was established in the summer of 1779 and had about 800 men of all ranks) and an estimated 15,000 men served at some time, it was those of 1776–77 who formed the backbone of the provincial service and within whose ranks two-thirds of all Loyalist troops would serve.[60]

Three other units deserve special mention. Each was led by charismatic and maverick officers. Major Patrick Ferguson's American Volunteers was raised in New York in 1779. Ferguson, an Irishman, had been wounded at Brandywine in 1777 leading his specialized corps of sharpshooters armed with Ferguson's own invention, a unique breech-loading rifle. Appointed inspector of militia in May 1780 in South

Carolina, Ferguson quickly wearied of the dreariness of administration, preferring the hand's-on business of active soldiering. He would die at the head of his men in a defiant sword-in-hand last stand trapped on the summit of Kings Mountain, South Carolina, on 7 October 1780.

The famous, or infamous, British Legion was raised in New York in July 1778 under its commander, Lieutenant Colonel Banastre Tarleton, a young British cavalry officer with a swaggering, almost casual brutality that is reminiscent of the Confederate cavalry leader Nathan Bedford Forrest. (Both men also shared a taste for flamboyant hats.) Tarleton's defeat of Colonel Abraham Buford at Waxhaws, South Carolina, on 29 May 1780 was accomplished with an "astonishing cruelty" that is illustrated by the comparative casualties: 113 Americans killed and 150 wounded, while the Legion lost five killed and twelve wounded. ("Tarleton's quarter" would become a battle cry of revenging patriot troops thereafter.) Tarleton would meet his own comeuppance at the battle of Cowpens just over six months later at the hand of Brigadier General Daniel Morgan when the Legion was cut to pieces and its commander was lucky to escape with his life.

The name of the third regiment—the American Legion—has a decidedly ironic twist when one considers its commander: Brigadier General Benedict Arnold, the "Dark Eagle" of the war. It was raised on Long Island, New York, in October 1780 and from there was led by Arnold on its raiding campaigns into Virginia in June 1781, and New London, Connecticut, in September 1781.

Spurred by the entry of France into the war following Burgoyne's defeat at Saratoga, the British authorities tried to jump-start Loyalist enlistment by offering, in December 1778 and January 1779, a bounty (initially three guineas but later increased by twenty-two shillings and six pence) and £40 per year per regiment as a contribution toward medical services. Provincial officers would now be entitled to half pay if their regiments were reduced or disbanded, and, as George Germain put it, if they "happen to be wounded in action, so as to lose a limb, or be maimed, shall be entitled to the same gratuity of one year's advanced pay as officers of His [Majesty's] established army."[61] Unfortunately for the British, these plums did not whet the appetite of too many enlistees.

By December 1778–January 1779, when the reforms were announced, there were 7,400 men in the provincial line, and it rose only to 9,000 by December 1779 and 10,000 by December 1780, where it remained steady.[62]

The truth was, perhaps, that the British seemed never to have had much respect for the "Friends of Government." They neither understood their culture nor knew how to harness their military potential. Too often Loyalists were viewed as the country bumpkins, the "provincials." There were a few notable exceptions where British officers—Major Patrick Ferguson, Lieutenant Colonel Banastre Tarleton, and Colonel Francis Rawdon, for example—led Loyalist units with a passionate commitment. Ferguson described his men as "very fit for rough & irregular war, being all excellent woodsmen, unerring shots, careful to a degree to prevent waste or damage to their ammunition, patient of hunger & hardship & almost regardless of blankets, cloathing, rum, and other indulgences."[63] But on the whole the British military establishment was superior and dismissive: "I should be very sorry to trust any one of them out of my sight," said Captain John Bowater of the Royal Navy to Lord Denbigh on 5 June 1777. "They swallow the Oaths of Allegiance to the King, & Congress Alternately, with as much ease as your Lordship does poached Eggs."[64] The Carolinian Loyalist Robert Gray saw it clearly: "Almost every British officer regarded with contempt and indifference the establishment of a militia among the people differing so much in custom & manners from themselves."[65]

The "failure" of the Loyalists to flock to the royal standard cannot be understood without appreciating how effectively they had been locked down and neutralized by the patriots. Loyalists could not buy, sell, or bequeath property or other assets. They were barred from all legal recourse to recover debts or redress any other injury. They could not practice law or teach unless they had taken an oath of allegiance to the cause. They could not be executors of estates or be a guardian to a child. "Any person who wrote, or spoke, or by any overt act libeled or defamed Congress, or the acts of the Connecticut General Assembly, should be brought to trial." All known Loyalist sympathizers were disarmed, yet they were coerced to join the patriot militia; any refusal was met with heavy fines or punishments like tar-and-feathering or riding a rail ("grand Tory Rides," as a patriot called them). Although these punishments may now seem comic (electricity has done so much to improve our torture techniques!), they were in fact excruciatingly painful. A patriot described the procedure for tarring.

> First strip a Person naked, then heat the Tar until it is thin, and pour it upon the naked Flesh, or rub it over with a Tar Brush, quantum sufficit. After which, sprinkle decently with the Tar, whilst it is yet warm, as many Feathers as will stick to it. Then hold a lighted Candle to the Feathers, and try to set it all on Fire; if it will burn so much the better.[66]

Hot tar (pine was preferred) that has cooled will strip off a layer of skin when it is removed. Riding a rail could cause severe damage to the genitals and anus. The Committees of Safety and their enforcers brought a Jacobin-like enthusiasm to suppressing and neutralizing Loyalists. Cornwallis wrote to his superior, Sir Henry Clinton, on 29 August 1780: "We receive the strongest professions of Friendship from North Carolina; our Friends, however, do not seem inclined to rise until they see our Army in motion. The severity of the Rebel Government has so terrified and totally subdued the minds of the people, that it is very difficult to rouse them to any exertions."[67] The young Lord Rawdon wrote to Cornwallis, his commander in the South, on 5 December 1780 that those coming into his camp professing attachment to the Crown were often fifth columnists: "My conduct towards the inhabitants, and the extraordinary regularity of the troops under my command, I must assert to have been such as ought to have conciliated their firmest attachment; yet I had the firmest proofs that the people who daily visited my camp . . . used every artifice to debauch the minds of my soldiers and persuade them to desert . . . Several small detachments from me were attacked by persons who had the hour before been with them as friends in their camp."[68]

Throughout the war Britain disastrously mismanaged its Loyalist dependents. Time after time, starting with Howe's evacuation of

Boston in 1776, Loyalists who had exposed themselves by declaring for the Crown during British occupations were abandoned and left to patriot retribution—"the fury of their bitterest enemies," as an English officer put it. When Howe quit Boston the city was thrown into a wild panic. A contemporary diarist reported the desperation: "We are told that the Tories were thunder-struck when orders were issued for evacuating the town, after being many hundred times assured, that such reinforcements would be sent, as to enable the king's troops to ravage the country at pleasure . . . Many of them, it is said, considered themselves as undone . . . One or more of them, it is reported, have been left to end their lives by the unnatural act of suicide."[69] This pattern would be repeated when the British quit Philadelphia in 1778. Ambrose Serle was ashamed of his country's betrayal of its Loyal supporters in the city.

> This [the news of the army's evacuation] was soon circulated about the town, and filled all our friends with melancholy on the apprehension of being speedily deserted, now a rope was (as it were) about their necks, and all their property subject to confiscation. The information chilled me with horror, and with some indignation when I reflected upon the miserable circumstances . . . I now look upon the contest as at an end. No man can be expected to declare for us when he cannot be assured of a fortnight's protection. Every man, on the contrary, whatever might have been his primary inclinations, will find it his interest to oppose and drive us out of the country.[70]

Like many colonial regimes since, the British were too often ignorant and clumsy when it came to understanding the complex dynamic between friend and foe. To some extent they were caught in the classic colonial dilemma of trying to appease an enemy in an attempt to win "hearts and minds" even when it resulted in undermining Loyalist support. Howe's policy of issuing general pardons to rebels who would go through the motion of taking an oath of allegiance outraged Loyalists, who denounced his "sentimental manner of waging war," and even roused Germain's ire. He lashed Howe's policy as "poor encouragement for the Friends of Government who have been suffering under the tyranny of the rebels, to see their oppressors without distinction put upon the same footing as themselves."[71]

The British were trapped in this dilemma throughout the war. Later, in the South, they stumbled around with the same uncertainty. The Carolinian Loyalist Robert Gray detailed the mess.

The want of paying sufficient attention to our Militia produced daily at this time the most disagreeable consequences. In the first place, when the Rebel Militia were made prisoners, they were immediately delivered up to the Regular Officers, who, being entirely ignorant of the dispositions & manners of the people treated them with the utmost lenity & sent them home to their plantations upon parole . . . the general consequence of this, that they no sooner got out of our hands than they broke their paroles, took up arms, and made it a point to murder every Militia man of ours who had any concern in making them prisoners, on the other hand whenever a Militia Man of ours was made a prisoner he was delivered not to the Continentals but to the Rebel Militia, who looked upon him as a State prisoner, as a man who deserved a halter, & therefore treated him with the greatest cruelty. If he was not assassinated after being made a prisoner, he was instantly hurried into Virginia or North Carolina where he was kept a prisoner without friends, money, credit, or perhaps hopes of exchange. This line being once drawn betwixt their militia & ours, it was no longer safe to be a loyalist.[72]

The "collateral damage" that is inevitable in insurgency warfare hurt friend as well as foe. The Quaker merchant Robert Morgan watched as British troops in Philadelphia in November 1777 destroyed Loyalist houses because they were being used by patriot snipers: "This morning about 10 o'clock the British set fire to Fair Hill mansion House, Jon'a Mifflin's and many others . . . The reason they assign for this destruction of their friends' property is on acco. of the Americans firing from these houses and harassing their Picquets." But what astonished Morgan was the ignorant glee with which British troops burned the furniture of the Loyalists, something, he points out, "Gen'l Washington's Army cannot be accused of. There is not one instance to be produced where they have wantonly destroyed and burned their friends' property."

And in the end, for Britain, there was simply an implacable blank wall. When Cornwallis (perhaps the only British senior commander who felt a genuine bond with the Loyalists)[73] urged shifting the main theater of operations into Virginia following the failures—his failures—in the Carolinas, Sir Henry Clinton wrote to his pugnacious subordinate: "Your Lordship will, I hope, excuse me, if I dissent from your opinion . . . there is no possibility of re-establishing order in any rebellious province on this continent without the hearty assistance of numerous friends. These, my Lord, are not, I think, to be found in Virginia; nor dare I positively assert, that under our present circumstances they are to be found any where else, or their exertions when found will answer our expectations."[74] Still, Clinton could not help himself when he added, "But I believe there is a greater probability of finding them in Pennsylvania." He still had an ear cocked to the siren's call.

3

"Men of Character"

THE OFFICER CLASS

George Washington knew what he wanted. Just as he modeled his Continentals on the British army—"a respectable army" as he put it—rejecting a militia model as being wholly inadequate to fight his war, so too did he strive to mold his officer corps into an institution with decidedly European characteristics—a place for gentlemen.

> *You must have good officers, [and] there are, in my Judgement, no other possible means of obtaining them but by establishing your Army upon a permanent footing, and giving your Officers good pay, this will induce Gentlemen, and Men of Character to engage; and till the bulk of your Officers are composed of such persons are actuated by Principles of honour, and a spirit of enterprise, you have little to expect from them—They ought to have such allowances as will enable them to live like & support the Characters of Gentlemen.[1]*

When he wrote this, in the early stages of the war, Washington was struggling mightily with an army composed mainly of unruly militia regiments and an officer corps that seemed to him utterly unprofessional, ill-trained, and, above all, ill-bred: "Their officers generally speaking are the most indifferent kind of people I ever saw." Too many officers, in his opinion, were not leaders of men but led by their men. The diary of Aaron Wright, a New Jersey private, underlines Washington's point. Although Wright and his fellow common soldiers were "sworn to be true and faithful soldiers in the Continental army, under the direction of the Right Honorable Congress," Wright goes on to say:

> After this we chose our officers . . . When on parade, our 1st lieut. came and told us he would be glad if we would excuse him from going, which we refused; but on consideration we concluded it was better to consent; after which he would go; but we said 'You shall not command us, for he whose mind can change in an hour is not fit to command in the field where liberty is contended for.' In the evening we chose a private in his place.[2]

Washington, the patrician Virginia planter, and a man who always kept a fastidious distance from those of lower rank, was aghast at the easy familiarity, between officers and their men. This New England "leveling" spirit appalled him.

> To attempt to introduce discipline and subordination into a new army must always be a work of much difficulty, but where the principles of democracy so universally prevail, where so great an equality and so thorough a leveling spirit predominates [it is even harder]. . . . You may form some notion of it when I tell you that yesterday morning a captain of horse . . . was seen shaving one of his men on the parade near the house.[3]

The British general Sir Guy Carleton was equally shocked when he reviewed the American officers he had captured at Quebec in 1775: "You can have no conception what kind of men composed their officers. Of those we took, one major was a blacksmith, another a hatter. Of their captains there was a butcher . . . a tanner, a shoemaker, a tavern-keeper etc. Yet they all pretended to be gentlemen."[4] Henry Knox, a major general in the Continental artillery, put it a little more bluntly when he wrote to his brother on 23 September 1776: "There is a radical evil in our army—the lack of officers . . . the bulk of the officers of the army are a parcel of ignorant, stupid men, who might make tolerable soldiers, but are bad officers."[5]

The early American officer corps had to be purged, and within a few weeks after his arrival at Cambridge the commander in chief could report that he had made "a pretty good slam" at doing so. A regimental chaplain, William Emerson (grandfather to Ralph Waldo Emerson), saw the new broom sweeping clean: "Great overturning in the camp as to order and regularity. New lords, new laws. Now great distinction is made between officers and soldiers. Everyone is made to know his place and keep it."[6] But there remained huge problems. Although Congress reserved the right to appoint general officers, field officers were appointed by individual colonies. Reform was a knotty problem, made more difficult not only by the jealously guarded "rights" of each colony but also because of a euphoric confidence that attended those early successes of American arms: "The army of the united colonies are already superior in valour, and from the most amazingly rapid progress in discipline, we may conclude will shortly become the most formidable troops in the world," James Warren, a patriot leader from Massachusetts, could brag to Samuel Adams in August 1775. It was from this sticky treacle of self-congratulation that Washington would have to extricate his army.

As the war progressed the American officer corps increasingly reflected Washington's ideals. It became more homogeneous in its social caste and in its feeling of separateness from civilian society. Increasingly, it came to resemble its British counterpart. Shared suffering welded the Continental officer class into a tight-knit exclusivity, and at war's end this sense of brotherhood and shared grievances would almost tip America over the edge into military dictatorship.

The old militia officer corps, although not exactly a democratically elected body, was influenced by a certain egalitarianism. For example, in the French and Indian War more than half of the officers in the Massachusetts provincial regiments identified themselves with manual occupations.[7] Corporal John Adlum, a member of the Pennsylvania Flying Camp, was captured at New York in 1776 and during his interrogation gave an interesting insight into the social composition of the American officer corps during this early phase of the war.

> The four companies of the town were commanded by Capt. Charles Lukens, the sheriff of York County (and to which I belonged), Capt. William Bailey, a respectable man, a coppersmith, Capt. Rudolph Spangler, a silversmith, and Capt. Michael Hahn. I am not certain as to his occupation, but I believe it was a smith ... Lieutenant Sherriff, who was appointed brigade major and who had been a schoolmaster in Yorktown; Lieutenant Holzinger ... was a fellow prentice ... to Michael Doudle, a tanner ... of Captain Trett's company I only recollect Ensign Myers, a blacksmith, and who was the most uncouth-looking man in the army and one of the greatest dunces.[8]

A French officer was disconcerted to find that "there are shoemakers who are Colonels; and it often happens that the Americans ask the French officers what their trade is in France."[9] Colonial America, however, although much more socially dynamic than the Old World, was still hierarchical, and this balance was reflected in its officer cadre—a very mixed bag indeed. For example, the colonel of Ipswich, Massachusetts, militia regiment was the wealthiest man in town, but the fourth wealthiest was only a lieutenant, while a captain and two lieutenants were "poor."[10] Officers would tend to come from the wealthier segments of society, whether elected or not. Half of the militia officers on Concord, Massachusetts's tax list came from the town's most wealthy 10 percent; and in Goshen, New York, about one-third came from the top 10 percent and two-thirds from the richest 20 percent.[11] In the New Jersey Line, for example, 84 percent of the officers were drawn from the wealthiest one-third of society, many of them coming from the wealthiest 10 percent.[12] The Maryland Line was similar. Senior regimental officers came from the colony's political leadership. William Smallwood and Francis Ware, the colonel and lieutenant colonel respectively, had both been members of the Maryland Convention, as had four majors.[13] The less wealthy were not disbarred from becoming officers (as they were not in the British army, as we shall see), but the general picture is fairly clear. As in the British army, wealth and influence were important factors, particularly at the higher levels of command.

At the outbreak of the war there were 3,700 infantry officers and 400 cavalry officers in the British army. Sixty-six percent of them had bought their commissions, and it is this fact that seems the most difficult to reconcile with modern ideas of an effective officer corps. The conclusion that many popular historians jump to is that British officers must have been, ipso facto, effete amateurs. Certainly, there were deadbeat young twits who had bought their way into rank, but the overall picture is a little different. At the highest levels of military command there is no doubt that the wealthiest and the most powerfully connected dominated, but at every level in between the officer corps was much more diverse. About 25 percent of regimental officers and 50 percent of proprietary colonels ("proprietary" because the colonel owned the business and ran the regiment like a CEO) and general officers tended to come from the highest echelons of British society. But the greatest source of officers came from the broad swath of "good families"—the noninheriting younger sons of respectable stock.

The commission purchase system (abolished in the British army in 1871) is usually the single issue pounced upon by the "lions led by asses" school of historians: the victory of privilege over merit; tradition over efficiency. In fact, it proved to be more open to talent than might appear at first glance. All ranks below that of colonel could be purchased. (Colonels were appointed by the king.) The lower the rank, the less expensive the price. (Exceptions were the engineers and artillery, whose officers required technical expertise; they constituted a meritocracy whose ranks were filled by middle-class technocrats, much to the distaste of aristocratic officers. Wellington, for one, could not abide them.)

A vacancy occurred when an officer retired, sold out, transferred, was killed, died, or was cashiered. Assuming the reason did not fall in the last three categories, the seller usually had to offer his commission to the most senior officer of the grade immediately beneath his. If that officer declined, it was offered to the next senior, and so on, creating a chain reaction of promotions. A retiring officer had a choice to make. Either he could sell out and take the purchase price as a lump-sum pension fund, or he could elect to go on half pay. If he chose the former, the price he received was his original purchase price plus a supplement. Since he had bought his office from the government, it was the government that now paid back the purchase price. The purchaser repaid the government, which held the funds in bond as an assurance of good behavior; thus, if an officer was cashiered, he would lose his bond. The purchaser paid the supplement—the "profit" on the commission—directly to the retiring officer (a negotiation usually handled by the regimental agent). The negotiations could be complicated. Here, John Peebles, the senior lieutenant of the 42nd Foot, describes the Byzantine horse-trading that went on to move himself up a rung in 1777. His colonel had asked him to pay a little over the regulation price to the vacating Captain Lieutenant Valentine Chisholm, who was too ill to carry on in the service.

> I thought it was too much [an extra £50], and am of the opinion that Mr Chisholm should either sell or serve . . . if he sold the regulation price was as much as he could expect in the current state of affairs, however to facilitate the matter & make it as well for poor Chisholm as we could I agreed to give the £50, 20 of which Lts Rutherford and Potts agreed to make up equally betwixt them on the above conditions & Ensign Drummond gives £30, which with the regulation price from Ens. Campbell makes up the 600 guineas to Chisholm if I succeed to this captaincy.[14]

Death also brought its reward, if not for the demised, then certainly for his fellow officers. This was the opportunity for a free leg up—a windfall from the fallen—because the families of the dead man could not sell his commission. A vacancy created by death was filled strictly by seniority. So battle was not only an occasion of risk but also an opportunity for advancement and profit. And this was as true for the American officer corps as it was for the British. Lieutenant William Scott, an American captured at Bunker's Hill, gave his interrogator an honest insight into one of the major motivations for joining the army: "I offered to enlist upon having a Lieutenants Commission; which was granted. I imagined myself now in a way of Promotion: if I was killed in battle, there would be the end of me, but if my Captain was killed, I should rise in Rank, & should have a Chance to rise higher."[15]

Combat brought other fringe benefits. Pocketing the pay of men killed in battle was one of the more sanguinary perks, and not to be sniffed at. It was not unknown for a British officer in the War of Independence to make a profit of £800 (the annual salary of an average infantry colonel) in this way.[16] In the event of losing a limb or eye in combat, an officer could expect a full year's pay and all medical costs paid. And if he was killed, his widow was entitled to £16 to £50 per year, depending on the deceased's rank.

There were openings for British officer-candidates without money for purchase. In fact one-third of all entrants came through this door. For example, the king might appoint half-pay officers, or the sons of poor officers' widows, or those of serving officers in "straitened circumstances." Wartime was always a boost, especially to senior noncommissioned officers. (Over 200 gained their ensigncies during the Seven Years' War.) An interesting group is the "gentlemen volunteers," young men from respectable but impecunious families who started in the ranks with the hope that their performance in battle might bring them the recognition that would catapult them into the officer corps. First, they had to hack through basic training "with a squad composed of peasants from the plough and other raw recruits." But for these young men, ambitious and hungry for war, relishing the prospect of throwing the dice on the battlefield, it was worth it. The odds, however, could be fatally unfavorable. Thomas Brotherton, a light dragoon officer in the Peninsular War, recorded their particular brand of do-or-die bravery: "The *volunteers* we had with the army . . . always recklessly exposed themselves in order to render themselves conspicuous, as their object was to get commissions given to them without purchase. The largest proportion of these volunteers were killed, but those who escaped were well rewarded for their adventurous spirit."[17]

The British officer corps was socially more diverse than probably any other comparable European army. The French army, for example, was systematically cutting off access to its officer cadre for men of nonnoble rank. (By 1781, for instance, any officer-candidate would have to show proof of four generations of nobility.) In France lesser nobility proliferated (there were 110,000 to 120,000 by 1781), while in Britain, where primogeniture applied a more rigorous filter to nobility, there were only 220 peers by 1790. There were officers whose noble birth gave them a decided advantage. (For example, George Lennox, the second son of the duke of Richmond, was an ensign at thirteen, a lieutenant colonel at twenty, and a full colonel at twenty-four—the same age Wellington would attain that rank.) But they simply were not numerous enough to dominate the officer corps. Even so, for those without influence it was a long and hard slog up through the officer ranks of the British army.

As with the common soldier, there was a significant representation from Britain's Celtic fringe in the British officer class. Men from Scottish and Irish backgrounds were a very important segment throughout the eighteenth century, and even a cursory look during the War of Independence illustrates the point: Fraser, Murray, Mackay, Hamilton, O'Hara, Gordon, Stuart, MacKenzie, Campbell, McNab, MacDonnell, and MacLeod feature largely among the higher echelons of British regimental command.

The numbers tell a story. In 1777, for example, the average length of service from first commission for a lieutenant colonel in a foot regiment was thirty years; for a major, twenty-three and a half; for a captain, seventeen; and for a lieutenant, ten years.[18] They had to sweat out many years in order to climb the ladder. The impact of war can be seen clearly in the difference between the number of years an officer had spent in his present rank in peace and war. Just prior to the outbreak of the American war a lieutenant colonel would have spent an average of eight years in his present position; a major, three and a half; captains and lieutenants, six; and an ensign, three. War, however, speeded things up, particularly among the junior ranks, on which combat took a higher toll. A lieutenant colonel spent five years in his present rank; a major, three; a captain, four and a half. The greatest impact was among the lieutenants and ensigns, with three and a half and one and a half years respectively.[19]

Such statistics may be dry, but they were irrigated by the blood of the battlefield. Officers, whether British or American, were expected to lead from the front. Conspicuous bravery was a matter of honor—and honor was at the core of the officer code. The prestige of high rank bestowed no exemption from the price that had to be paid. General Sir William Howe, for example, leading the British attack on the American left wing at Bunker's Hill, had addressed his men before the battle and reassured them, "[I do not expect you] to go a step further than where I go myself at your head." A promise he kept. Every member of his staff was shot down. General Simon Fraser was killed in action at Bemis Heights; General Charles O'Hara was twice wounded at Guilford Courthouse, and his commanding officer, Cornwallis, had two horses shot from under him. Lieutenant Colonel James Webster was killed leading his regiment (the 23rd Foot) in the same battle "with a flamboyance typical of his profession." Brigadier General James Agnew, colonel of the 44th, was killed at Germantown, as was Lieutenant Colonel John Bird of the 15th. Lieutenant Colonel Henry Monckton of the 45th Foot died at Monmouth Courthouse. At Bunker's Hill nineteen officers were among the 226 British dead, and at Eutaw Springs on 8 September 1781 almost a quarter (20) of all the British killed (85) were officers.

Combat also claimed American officers of senior rank. General Richard Montgomery was shot down at the head of his men at Quebec in 1775; General Hugh Mercer was killed at Princeton, General Johann De Kalb at Camden, and General Casimir Pulaski at Savannah in 1779. The militia general Nathaniel Woodhull died of his wounds at Jamaica, Long Island, in August 1776. General Francis Nash was killed at Germantown. Benedict Arnold was wounded in combat twice (in the same leg), once at Quebec and again at Bemis Heights. Washington himself was conspicuously brave in battle. He was almost captured at Kips Bay when the British attacked Manhattan in 1776, and was saved only when aides physically pulled him away from danger. He came close to being killed at Germantown and was in the thick of the fighting at Princeton. At Yorktown the following exchange between Colonel Cobb and Washington, recorded by Dr. James Thacher, shows his sangfroid under fire.

> *During the assault, the British kept up an incessant firing of cannon and musketry from their whole line. His Excellency General Washington, Generals Lincoln and Knox, with their aids, having dismounted, were standing in an exposed situation waiting the result. Colonel Cobb, one of Washington's aids, solicitous for his safety, said to His Excellency, "Sir, you are too much exposed here. Had you not better step a little back?"*
>
> *"Colonel Cobb," replied His Excellency, "if you are afraid, you have liberty to step back."[20]*

Even in the face of imminent death an officer, if he was a man of honor, was expected to show indifference, a sanguine acceptance. In October 1780 the British major John André was caught spying. (He had been instrumental in abetting Arnold's defection to the British and his treacherous plan to surrender his command of West Point.) André was condemned to hang—a common criminal's death and shockingly insulting for a man of his class. Even though he had petitioned Washington twice in "concise, but persuasive terms" to allow him to die by firing squad, Washington and Congress were unmoved. Alexander Hamilton, who was there, described André's demeanor at the point of death.

> *In going to the place of execution, he bowed familiarly as he went along to all those with whom he had been acquainted in his confinement. A smile of complacency expressed the serene fortitude of his mind. Arrived at the fatal spot, he asked with some emotion, must I then die in this manner? He was told it*

had been unavoidable. "I am reconciled to my fate (said he) but not to the mode." Soon however recollecting himself, he added, "it will be but a momentary pang," and springing upon the cart performed the last offices to himself with a composure that excited the admiration and melted the hearts of the beholders. Upon being told the final moment was at hand, and asked if he had anything to say, he answered: "nothing, but to request you will witness to the world, that I die like a brave man."[21]

Honor was inextricably bound to social standing. In the world of the eighteenth century the common man was accorded no honor; indeed he had no need for it. Expediency circumscribed the narrow boundaries of his moral dimension. The reason Washington placed so much emphasis on "gentlemen" officers and fought so hard to win adequate benefits for his officers ("allowances as will enable them to live and support the Characters of Gentlemen") was that he wanted them to be able to rise above the considerations of mere expediency. The marquis de Crénolles wrote in 1764 that "if one of them [the nobility] had not been born with honor in his blood, the fear of losing caste, the shame which would result for his family, would act as a check on him and to some extent replace what he may lack in courage. Certainly it is only the nobility who are like that."[22] Honor, in the marquis's and Washington's understanding, was not simply a decorative notion. It had practical consequences in warfare. Men can be counted on to perform, to die if necessary, when the pressure of their caste's expectation is brought to bear. How this pressure worked can be seen in Colonel Otho Williams's anguished reaction to the inglorious retreat of the American army at Hobkirk's Hill on 25 April 1781: "Many of our officers are mortally mortified at our late inglorious retreat. I say mortally because I cannot doubt but some of us must fall in endeavoring the next opportunity to re-establish our reputation. Dear Reputation! What trouble do you occasion, what dangers do you expose us to!"[23]

The American officer cadre was less socially homogeneous than Britain's. There simply were not enough men of "refined" birth to provide the officer corps with the numbers it required, and so a wider social net had to be cast. This practice had something porous, shifting, about it, which to someone like Washington was unnerving. Two traditions were constantly rubbing up against each other. On the one hand, there were Washington's "European" elitist aspirations, and on the other, the egalitarianism that was deeply rooted in the American experience. The resulting defensiveness expressed itself in interesting ways. The American officer cadre was notably sensitive to slights and deeply aware of the subtle delineations of status. As John Trumbull, an aide-de-camp to Washington, put it in 1777, "A soldier's honor forbids the idea of giving up the least pretension to rank." When Colonel John Lamb was outranked in seniority by Colonel John Crane (they had the same date of seniority in the Continental artillery, but Congress decided in favor of Crane), Lamb took serious umbrage, writing to Washington: "It is impossible for a soldier, who is tenacious of his honor (the only jewel worth contending for) to suffer himself to be degraded by being superseded; and his right torn from him, and given to another, without resenting the injury . . . I must frankly acknowledge, that my sensibilities are deeply wounded by this event."[24] One of the roots of Benedict Arnold's treason taps into his disgust at being passed over for promotion.

British and French officers tended to pride themselves on "an easy social intercourse and fellow feeling among officers."[25] John Burgoyne underlined this convention among the British officer corps: "Any restraint upon conversation, off parade, unless when an offense against religion, morals, or good breeding is in question, is grating; and it ought to be the characteristic of every gentleman neither to impose, nor submit to, any distinction but such as propriety of conduct, or superiority of talent, naturally create."[26] It was just such a generosity of spirit that earned Burgoyne not only the regard of his subordinate officers but also the devotion of his men.

Continental officers seem to have been a fractious lot (not unlike their British confreres). An exasperated John Adams exclaimed that he was "wearied to Death with the Wrangles between military officers, high and low. They Quarrell like Cats and Dogs. They worry one another like Mastiffs Scrambling for Rank and Pay like Apes for Nuts."[27] Washington's genius had as much to do with his negotiating between his senior commanders as it did with his performance on the battlefield. Their shenanigans do not have to be gone into in detail, but even a précis gives some idea of the problems the commander in chief had to deal with. Schuyler detested Gates; Arnold detested Gates; Gates conspired against Washington; Lee felt superior to Washington; even Washington's adjutant general, Colonel Joseph Reed, undermined him with flagrant disloyalty; and so on in a depressing litany of backbiting.

The "Quarrell" could turn lethal, as American officers resorted to dueling with increasing frequency. In fact the duel became almost a cult in the Continental army. Gérard de Rayneval, the French minister to the United States, wrote in March 1779: "The rage for dueling here has reached an incredible and scandalous point. . . . This license is regarded as an appanage of liberty."[28] The social fragility felt by America's newly elevated officers seems to have fed into the rage. Dueling, with its well-bred European associations, seemed to them an eminently suitable way of settling intractable arguments while reinforcing social cachet. An American officer, however, saw something less flattering underlying the gallantry: "These new gentry expect a great deal of deference, their ideas are sublimed, and, fond of imitating their betters, they cannot abate an iota of this article."[29] Even at senior levels of command dueling was accepted, if not encouraged. For example, General John Cadwalader seriously wounded General Thomas Conway, an Irish-born soldier of fortune who had found a berth in the Continental army, where he indulged himself in extensive slander and backstabbing of Washington; John Laurens, Washington's devoted aide, wounded General Charles Lee in a duel, after Lee had disparaged Washington following Lee's dismissal for gross mismanagement of his troops at the battle of Monmouth Courthouse in June 1778.

Dueling may have been a breach of the British army's Articles of War, but it was an incontrovertible article of the gentlemanly code of honor, and in the British army dueling was endemic in the eighteenth century. An officer of junior rank could not challenge his senior, although a senior could agree to forego the niceties. (For instance, Colonel Hervey Aston of the 12th Foot was killed by his major.)[30] The paradoxical coexistence of legal constraint through the Articles and the strong support of dueling as an honorable act could be reconciled. More than a blind eye was turned. When Ensign Charles Shirreff of the 45th Foot was goaded beyond endurance by Ensign William Wetherhead over some trivial dispute, Shirreff called him out. Wetherhead refused to fight. At his court-martial, Shirreff claimed that "[he] should be unworthy to bear His Majesty's commission" if he had not made the challenge. The court had no option but to find him guilty and cashier him. But the court noted that if he had not been willing to duel, he "might have fallen into a breach of the 23rd Article of the 15th Section of the Articles of War"—conduct unbecoming an officer and a gentleman. Shirreff was reinstated, and it was Wetherhead who was kicked out— for conduct unbecoming an officer and a gentleman![31]

For those Continental officers who did not have sufficient financial means to support themselves in the manner of a gentleman, the pressures mounted as the war progressed. Their needs and aspirations increasingly ran afoul of the massive deflation in the Continental dollars in which they were paid. Unlike the men they led, who were going through the same pinch, officers were spared the ignominy of desertion. They could simply resign, as some hundreds did. In March 1778 Washington reported to the president of Congress, Henry Laurens, that since August of the previous year "between two and three hundred officers have resigned their Commissions and many others with difficulty disswaded [sic] from it."[32] Although the commander in chief was alarmed, and many good men did leave, some of those who quit were the poseurs, the speculators, the flotsam. As the heat of combat and hardship rose, it winnowed out the lightweights just as it did within the ranks. (Desertion rates tended to drop after the initial six months of enlistment.)

But the mounting hardships created a tinder-dry underbrush of resentment within the officer corps that threatened to burst into flame. Officers' disillusion with the Revolution could be profound. Lieutenant Colonel Ebenezer Huntington of Connecticut, for instance, wrote: "I despise my countrymen. I wish I could say I was not born in America . . . The insults and neglects which the army have met with from the country beggars all description."[33] Inflation and devaluation reduced the annual salary of a captain to little more than the price of a pair of shoes. In 1780 officers of the New Jersey Line petitioned their state legislature for relief. They wanted pay in coin, "Spanish milled dollars," not the now worthless Continental paper dollar. Nothing happened except "legislative windbaggery," which drove them to threaten resignation en masse. After a good deal of huffing and puffing about not being bullied, the lack of republican virtue in the army, and so forth, the state agreed to pay each officer £200 and provide new clothing. The revolt was defused, at least temporarily.

As Continental officers were responsible for providing for their clothing and food from their pay and whatever private means they could draw on, those who came from humbler backgrounds suffered along with their men. Small things mattered. For example, in 1782 officers complained that food contractors were sending them beef carcasses without the kidney tallow from which they made candles for their families.[34] Lieutenant Joseph Hodgkins, an Ipswich, Massachusetts, cobbler, wrote from Valley Forge in the winter of 1777–78 to his wife, on whom he depended entirely for his clothing, that without her supplies he "must go naked."[35] Nor were generals immune from neglect and want. General John Glover, who had led his Marblehead corps with such distinction, received no pay for two years.[36] General Daniel Morgan complained that his pay was months in arrears and that he was reduced to such shabbiness he felt embarrassed to be seen in public. If Morgan shared some of the ordinary soldiers' hardships, he also held a radical view that the greater wealth held by a few, and further enlarged by war profiteering, should be shared: "The War should not end till the Soldiery were provided out [of] the Estates made by it and of such as had too much Property to their Share."[37] It was a revolutionary idea, but far too radical for this revolution.

Ironically, it was the imminent peace that drove the Continental officer corps closest to outright revolt. Faced with the prospect of being pitched back into civilian society while still waiting for Congress to honor its 1780 promise to award half pay and to settle outstanding back pay, officers drew up a list of grievances in the fall of 1782. It was carried to Congress by General Alexander McDougall, Colonel Matthias Ogden, and Colonel John Brooks. As a concession, the officer corps would accept the commutation of half pay for life into a lump-sum pension. The prospect of an army in revolt terrified most members of Congress, though not all. Nationalists like Robert Morris, Alexander Hamilton, and Gouverneur Morris thought such a threat might scare the states into granting more power to central government: "The Army have swords in their Hands . . . I am glad to see Things in their present Train . . . When a few Men of Sense and Spirit get together & that they are the Authority such few as are of a different Opinion may easily be convinced of their Mistake by that powerful Argument the Halter."[38]

On 10 March 1783 two anonymous declarations of grievances—the Newburgh Addresses—were circulated among the officers at Newburgh, New York. In a striking way the addresses echo the language of patriot revolt against Britain itself, except America was now the oppressor: "A country that tramples upon your rights, disdains your cries and insults your distresses." In order to head off the insurgents, Washington preempted a meeting they had called and, with a brilliant coup de théâtre—reaching for his glasses, he said, "Gentlemen you will permit me to put on my spectacles, for I have not only grown gray, but almost blind in the service of my country"—he won them over. The officer corps returned to what it had always been during the war: loyal to the nation beyond all reasonable expectation and more steadfast to republican ideals than most of the people it served.

4

What Made Men Fight

We fight, get beat, rise, and fight again." It was General Nathanael Greene's phrase. He was looking back at his campaigns in the South but could just as well have been describing the whole of the American effort. Its pithy pugnacity encapsulates almost the entire history of the war and does more than most tomes to explain why the patriots won and why Britain was doomed to defeat. But what motivated soldiers, American and British, to go through the whole bloody cycle Greene described?

Motivation falls into two broad categories. The first is ideological. Patriots experienced it as a shared sense of national purpose that encompassed a passionate desire to defend the nation against the imagined and real oppressions of Britain in America. On the British side there was also an ideological commitment based on a strong sense of national pride built up over centuries. For many though by no means all Britons, the rebellion was wrong because it attempted to impose the will of a relatively few fanatics, as the British saw them, who sought to break a sacred contract. The second category of motivation is much less cerebral. When a soldier is terrified, or starving, or outnumbered, what makes him fight; what makes him run?

The ideological debate seems to be divided into two camps. On the one hand, there is the argument that it was self-interest rather than patriotic zeal that motivated the American soldiers. They were poor and needed the benefits in the form of pay, food, clothes, and shelter that service in the army would provide (even if, as it turned out, the promise most often overreached the reality). Given the lack of opportunity in civilian life, men from the underclass chose the lesser of two evils: destitution or soldiering. The other camp, which emphasizes the "heroic" commitment to revolutionary ideals, even among the least intellectual soldiers, was championed by Charles Royster in his landmark book, *A Revolutionary People at War* (1979). Royster conceded the socioeconomic description of the common American soldier. Yes, they were the poorest members of society. Yes, they had the least opportunities. But, Royster argued, if self-interest was the primary motivator, why did poor men fight when the benefits—pay, food, clothing, shelter—failed to turn up? He asks the question: if American soldiers were motivated only by the self-interest that characterized European armies (the assumption being that British soldiers, for example, fought only for money), why didn't they stay on in the army after the war? The answer may be that there was no army to stay in. After the war Congress all but disbanded it.

To counter the argument that most common soldiers joined up because they had little to lose, Royster points out that the mortality rate was very high, and therefore they indeed had a lot to lose. A counterargument might be that enlistees did not know this at the time, particularly early in the war, and when they did see the risks as the war developed, enlistment dropped. Royster's book seeks to rescue the common soldier from mere expediency; to accord him, no matter how ragged, how smelly, how inarticulate, the dignity and the honor of having fought for a great cause. The truth, as it often does, has an awkward way of lying somewhere between these two camps.

Men joined because it was the best deal on offer in a world that did not offer them any really good deals. They also joined because they believed in the cause. It may not have been a deeply-thought-through ideological commitment. It may have been inflamed by populist propaganda and wrapped-in-the-flag tub-thumpers. It may have been a crudely colored cartoon version of the issues, but in that the War of Independence is no different from the wars of our own time. None of these reasons are mutually exclusive. Although it is legitimate to state that it "is impossible to believe that Washington's little army was held together over eight grim winters merely by concern over economic issues,"[1] the corollary is not necessarily that that every soldier was a firebrand patriot. This was, after all, a revolution about which Louis Duportail, a French volunteer and chief engineer in the Continental army, could say, "There is a hundred times more enthusiasm for this Revolution in any Paris café than in all the colonies together."[2]

The high rates of desertion in the Continental army in late 1777 and early 1778—the winter of Valley Forge—indicate that revolutionary zeal was not a sufficient motivator on its own. Deprived of bare necessities, many chose to quit (officers as well as rank and file). Those that stayed displayed a gritted-teeth stoicism that deeply moved Washington. On 16 February 1778 he wrote:

> For some days past there has been little less than a famine in the camp. A part of the army has been a week without any kind of flesh and the rest three or four days. Naked and starving as they are, we cannot enough admire the incomparable patience and fidelity of the soldiery, that they have not been ere this excited by their suffering to a general mutiny and dispersion.[3]

A Continental staff officer described the men at Trenton in 1777: "Christmas 6 p.m. . . . It is fearfully cold and raw and a snow-storm setting in. The wind is northeast and beats in the faces of the men. It will be a terrible night for the soldiers who have no shoes. Some of them have tied old rags around their feet, but I have not heard a man complain."[4]

Comradeship was a warmer fire for the soul on those terrible nights than political philosophy could ever be. The Continental soldier Joseph Plumb Martin described the deep attachment to his comrades.

> I can assure the reader that there was as much sorrow as joy transfused on the occasion. We had lived together as a family of brothers for several years (setting aside some little family squabbles, like most other families,) had shared with each other the hardships, dangers and sufferings incident to a soldier's life, had sympathized

with each other in trouble and sickness; had assisted in bearing
each other's burdens, or strove to make them lighter by council and
advice; had endeavoured to conceal each other's faults, or make
them appear in as good a light as they would bear. In short, the
soldiery, each in his particular circle of acquaintance, were as strict
a bond of brotherhood as Masons, and, I believe, as faithful to each
other.[5]

His counterpart in the British army, Sergeant Roger Lamb, would have recognized immediately what Martin was talking about: "Attachments of persons in the army to each other terminate but with life . . . it is like friendship between school-boys, which increases in manhood, and ripens in old age."[6]

Martin, neither here nor in the rest of his narrative, makes much mention of patriotism. He does, however, rail regularly against the lack of support from civilians and government: lashing *their* lack of motivation. At the war's end he felt bitterly betrayed: "When the country had drained the last drop of service it could screw out of the poor soldiers, they were turned adrift like old worn out horses."[7]

"To shun the dangers of the field is to desert the banner of Christ," wrote the American cleric John Murray in 1779. Religion was an important psychological tool to shore up the soldier's commitment, to motivate him, and to hold his feet to the fire. The importance of clergymen in the Continental army was much greater than in the British. Britain, compared with the colonies, was skeptical and secular. "Enthusiasm"—that fundamentalist passion that had been a defining characteristic of colonial America—was considered in bad taste in late eighteenth-century Britain, certainly among the gentry. In the army religious observance was more in the omission than the commission. Thus the idea of a "holy war," which was promoted by the patriot clergy, particularly irked the British. To them it smacked of fanaticism: "It is your G-d damned Religion of this Country that ruins the country; damn your Religion," wrote an irate Major Harry Rooke at Bunker's Hill.[8]

The role of the chaplain in the American army was to weld together the trinity of the cause: godliness, patriotism, and discipline. On 27 May

1777 Congress appointed chaplains with the rank of colonel to each brigade. They were not only to attend to the men's morality but also to "strengthen the Officers hands by publick and private exhortations to obedience." Brigade commanders were encouraged to choose clergymen who were "zealously attached to our glorious Cause, who will not begrudge the exertion of every nerve in the service."[9] The Reverend Mr. John Gano is a good example. He informed his congregation of short-term enlistees that "our Lord and Saviour approved of all those who engaged in His service for the whole warfare."[10] The intervention of the French in 1778 was marked by the Americans with religious services of thanksgiving. Chaplains read a summary of the treaty, and at Valley Forge Chaplain John Hurt urged the soldiery, "[Let us] redouble our diligence, and endeavour to acquire the highest perfection in our several duties; for the most we do for ourselves, the more reason have we to expect the smiles of Providence."[11] At the funeral for General Francis Nash, killed at the battle of Germantown on 4 October 1777, the presiding chaplain could not stop himself from launching into a tactical lecture.

> Remember the mistake of that day, and never do you commit it. What I mean is—never, in pursuit of a flying enemy, never separate or break your ranks ... But though I recommend bravery, and applaud you for it, yet beware of rashness. Regularity and the strict attention to orders is the life of every action ... In retreat—(and the bravest have often retreated) good order is equally essential.[12]

The soldiery, however, was not always receptive. Sometimes sanctions had to be applied to encourage the men to submit themselves to the exhortations of the worthy: "A few hours spent digging out stumps in a New York woodland proved effective."[13] Their resistance to godly assistance was lamented by Philip Vickers Fithian, a chaplain with the army in New York in August 1776: "The Lords Day is come once more. But the Sabbath is scarcely known in the Army. Profaned is all religious Exercise. Dreadful is the thought that Men who expect an Engagement every Day with a obstinate, wise, & powerful Enemy, should dare be so ungodly."[14] This sense of their irrelevance as padres could put a fearful stress on deeply committed chaplains. Abiel Leonard was the pastor of First Congregational Church of Woodstock, Connecticut, and one of the most influential military chaplains, praised by Washington and Nathanael Greene. By the summer of 1777 Leonard was deeply depressed at the lack of respect accorded to chaplains and that they "were able to do no more good in their Place." On 27 July 1777 he went to Judge Coe's tavern in Kakiate, New York, and there cut his throat.

In the British army the role of religion had a similar function. By emphasizing the moral virtue of submission to authority, military effectiveness would be strengthened. An influential religious tract of the day, *The Soldiers Faithful Friend,* intoned: "The duty of a *peaceable* and quiet submission to commanders is more incumbent on you as a *soldier* than in any other contract for service . . . Whenever you act turbulently or unsoberly, you bring scandal on your corps, and possess people with prejudice against your profession. This ought to put you on your guard as a *good subject,* as well as a good *soldier* and a good *Christian.*"[15] Church attendance was compulsory, and twelve pence could be deducted for the first noncompliance and imprisonment for a second offense. Officers could be court-martialed. Nevertheless, the British soldier, like his American counterpart, seems to have been stubbornly resistant to the motivations of religion, and church attendance was desultory. In this the British soldier was not much different from many of his countrymen. An eighteenth-century bishop of Durham wailed, "The deplorable distinction of our age is an avowed scorn of religion in some and a growing disregard of it in the generality." The great Methodist preacher John Wesley went further: "Ungodliness is our universal, our constant, our peculiar character . . . High & low, cobblers, tinkers, hackney coachmen, men and maid servants, soldiers, tradesmen of all rank, lawyers, physicians, gentlemen, lords are as ignorant of the Creator of the World as Mohametans & Pagans."[16]

British army chaplaincies were purchased just like officer commissions. Absenteeism (sometimes for years on end) was widespread. The provision for regimental chaplains was laid down in the Articles of War, but many chaplains treated their positions as sinecures, hiring substitutes to do, or not do, the work. General Thomas Gage complained that he had only two chaplains with his army at Boston who "did duty occasionally . . . and none were properly fixed to any particular battalion, so that none knew who was their proper Chaplain."[17]

In striking contrast to British troops, the Germans were overtly and sometimes loudly pious (something which does not seem to have prevented them from gaining a reputation as the war's most energetic plunderers). Ambrose Serle ruefully noted the difference. Writing from Staten Island in the summer of 1776, he observed, "It was pleasing to hear the Hessians singing psalms in the evening, with great solemnity; while to our shame, the British navy and army in general are wasting their time in imprecations or idleness."[18] The British general Lord Rawdon was struck by the same comparison. During the amphibious assault on Kips Bay on 15 September 1776,

the Hessians, who were not used to this water business and who conceived that it must be exceedingly uncomfortable to be shot at whilst they were quite defenseless and jammed together so close, began to sing hymns immediately. Our men expressed their feelings as strongly, tho in a different manner, by damning themselves and the enemy indiscriminately with wonderful fervency.[19]

Not infrequently, histories of the War of Independence claim that, in contrast to the American army, the British lacked motivation. Statements like "Though patriotism was a more compelling force in America than across the Atlantic . . ."[20] and "The American soldier, unlike British derelicts and Hessian mercenaries, faced the invaders by an act of free choice and beat them"[21] are common. Apart from the fact that economic necessity significantly narrowed the "free choice" of American soldiers, it is worth looking at the other side of the equation. Just as it is difficult to reconcile the heroic efforts of ordinary American soldiers if their only motivation was self-interest, the same is true for British soldiers. If only fear or automatism motivated them, it is not possible to explain the performance of the ordinary British soldier in battle, which, time and again, was heroic.

For instance, at Germantown, Lieutenant Martin Hunter describes the difficulty officers had in getting their men to retreat: "This was the first time we had retreated from the Americans, and it was with great difficulty we could get our men to obey our orders."[22] At the point of surrender at Saratoga Lieutenant William Digby recorded: "All thoughts of a retreat were then given over, and a determination made to fall nobly together, rather than disgrace the name of British troops . . . We were called together and desired to tell our men that their own safety, as well as ours, depended on making a vigorous defence; but that I was sure was an unnecessary caution,—well knowing they would never forfeit the title of Soldiers."[23] The British lieutenant Thomas Anburey recorded the extraordinary esprit de corps among the defeated British troops at Saratoga.

> [We were] in this state of weakness, no possibility of retreat, our provisions nearly exhausted, and invested by an army of four times our number that almost encircled us . . . In this perilous situation the men lay continually upon their arms, the enemy incessantly cannonading us, and their rifle and cannon shot reaching every part of our camp . . . True courage submits with great difficulty to despair, and in the midst of all these dangers and arduous trials, the valour and constancy of the British troops was astonishing: they still retained their spirits.[24]

Brigadier General John Glover was also in the Saratoga battles and reported, "The enemy in their turn sometimes drove us. They were bold, intrepid and fought like heroes, and I do assure you Sirs, our men were equally bold and courageous & fought like men."[25]

The effects on ordinary British soldiers of the crushing defeat at Yorktown were recorded by Captain Samuel Graham of the 51st foot.

> It is a very sorry reminiscence, this. Yet the scene made a deep impression at the moment, for the mortification and unfeigned sorrow of the soldiers will never fade from my memory. Some went so far as to shed tears, while one man, a corporal, who stood

near me, embraced his firelock and then threw it on the ground,
exclaiming, "May you never get so good a master again!"[26]

None of this, nor the extraordinarily courageous frontal attacks
at Bunker's Hill or Guilford Courthouse, for example, were the acts of
"derelicts." In the smoke, noise, and unholy confusion of battle there
were always opportunities "to flinch and fumble, struggle or stumble,
and avoid coercion."[27] But as Sylvia Frey, an American historian of the
British army, says, "During the American Revolution British troops
fought with a revolutionary ardor." How was this possible?

The British sense of national virtue was as strong, and as complex,
as the newfound nationalism of the colonists. It had been built over
centuries, a potent brew of fact and mythology. "Albion," "Britannia,"
with its traditional freedoms, its constitution, its parliamentary
government, was inspirational not only for Britons but also for
Americans who constantly referred to it as a benchmark. The American
Revolution, in some interesting ways, was less a revolution and more
a conservative and nostalgic movement to restore in the colonies the
traditional freedoms of "old England." Britons were proud: "Crucially,
Albion was uniquely successful in war, conquest and colonization;
eupeptic patriotism and profit became the heads and tails of Britannia's
golden guinea."[28] In fact, they were often arrogantly proud and
dismissive of their American "cousins." And they paid a price for it.

The great muscle that powered the British army was attached to the
regiment.—"the ultimate and most important motivation in battle."[29]
The tribal bond of the regiment transcended notions of patriotism and
religion, or any other lofty but generalized ideas that were meant to
inspire. "Everything that one can make of the soldiers consists in giving
them an *esprit de corps* or, in other words, in teaching them to place their
regiment higher than all the troops in the world," wrote Frederick the
Great.[30]

The symbol of this immensely powerful bond was the colors.
There were two per regiment: the royal standard and the regiment's.
They were made of silk, about six feet by six feet, and carried on a
ten-foot staff. When a regiment was presented with its colors, they

were consecrated in a religious ceremony. Lieutenant Colonel Francis
MacLean addressed his regiment at a consecration of the colors of the
Royal Highland Emigrant Corps in 1777.

> *Though we do not worship the colours, yet the awful ceremony of
> this day sufficiently evinces, that they are with us, as in ancient times,
> the object of particular veneration; they hold forth to us the ideas
> of the prince whose service we have undertaken, of our country's
> cause which we are never to forsake, and of military honour which
> we are ever to preserve. The colours, in short, represent everything
> that is dear to a soldier; at the sight of them all the powers of his
> soul are to rouse, they are the post to which he must repair through
> fire and sword, and which he must defend while life remains . . . to
> desert them is the blackest perjury and eternal infamy . . . to lose
> them, no matter how, is to lose everything; and when they are
> in danger, or lost, officers and soldiers have nothing for it but to
> recover them or die.[31]*

Carried in the center of the front rank, they attracted the hottest fire,
and so to be allotted (*awarded* might be a better word) the task of
carrying them (usually given to the most junior officers—ensigns—
flanked by sergeants of the color guard, and passed on down through
the junior officer ranks if the bearers fell) was an honor. At Waterloo,
for example, Sergeant William Lawrence was finally called on to carry
the colors of the 40th Foot because "[earlier] that day fourteen sergeants
[had been] already killed and wounded while in charge of these colours,
with officers in proportion, and the staff and colours were almost cut to
pieces."[32]

To lose the colors in battle was not only a matter of the deepest
shame but also, in a purely practical sense, the signal of defeat. (To
"strike the colors" on a warship, for instance, was the sign of surrender.)
At Saratoga the baroness von Riedesel had the Brunswick colors sewn
inside a pillow to prevent them from falling into enemy hands.[33] At
Albuera, a particularly bloody battle in Spain during the Napoleonic
War, "Ensign Richard Vance, sixteen years old and just six weeks in

the service, carried the regimental colour of the 29th [during the War of Independence its flank companies had been captured at Saratoga] until there were so few soldiers of the regiment left on their feet that he thought its capture inevitable. He tore it from its staff and hid it in his jacket: it was found on his body that evening."[34] The charismatic Baron von Steuben, in his *Regulations for the Order and Discipline of the Troops of the United States*, the training manual of the Continental army, instilled in the ensigns under his tutelage the same reverence for the colors: "As there are only two colours to a regiment, the ensigns must carry them by turns . . . When on that duty, they should consider the importance of the trust reposed in them; and when in action, resolve not to part with the colours but with their lives."[35]

"Because there were only a very limited number of social roles open to them . . . heroism in battle provided at least an opportunity for recognition if not for glory."[36] The individual soldier in combat can be driven forward not only by training, the fear of retribution, and the pressures of his peers but also by a profound sense of the possibility of the heroic: what might be called "internal glory." It is that moment—an adrenaline flash—when a man picks himself off the ground and rushes a strongly held enemy position. It is the moment when an extraordinary feat of courage overcomes terrible fear. But that stupendous, transforming, adrenaline rush comes with a price, a debilitating drain of energy that follows close on its heels. The nervous system is burned out, like an electrical surge frazzling a computer's motherboard. The psychological and physical crash that follows battle to some extent explains the inability of generals (Howe after Brooklyn and Brandywine; Cornwallis after Camden, for example) to follow up victory with vigorous pursuit. The truth was that their men were spent.

— ◦◦◦ —

Private Joseph Martin called it "the good creature." And if adrenaline needed a little artificial boost now and then, alcohol was often called upon as liquid motivation. (Seventeenth-century British troops in the

Netherlands developed a taste for the local "genever" gin and gave us the term *Dutch courage*.) Frederick the Great recognized its crucial importance to the fighting man: "If you contemplate some enterprise against the enemy, the commissary must scrape together all of the beer and brandy that can be found en route so that the army does not lack either . . . all of the brewers and distillers, especially of brandy, must be seized so that the soldier does not lack a drink, which he cannot do without."[37] The tipple of choice in both armies in America was rum, as it remained in the British army up through World War I. (A British medical officer of the Black Watch reported to the Shell Shock Committee in 1922, "Had it not been for the rum ration I do not think we should have won the war.")[38]

Drinking was a constant thread that wove its way through most men's military careers, starting with recruitment. The British rifleman Benjamin Harris's experience as a member of a recruiting party in the early part of the nineteenth century would have been just as true fifty years earlier.

> *We started on our journey in tip-top spirits from the Royal Oak at Cashel: the whole lot of us (early as it was) being three sheets to the wind. When we paraded before the door of the Royal Oak, the landlord and landlady of the inn, who were quite as lively, came reeling forth with two decanters of whisky which they thrust into the hands of the sergeants . . . The piper then struck up, the sergeants flourished the decanters, and the whole commenced a terrific yell.[39]*

John Claspy was an American recruiter for the Continental army who, in his pension request of 1833, described the role of liquor as an inducement: "His [Claspy's] destination always was where there were the largest gathering of the people in their civil capacity and where whiskey was most likely to induce them to assume a military one."[40] Robert Morris, chief justice of the Supreme Court of New Jersey, lamented to George Washington on 6 March 1777 that unscrupulous sutlers kept soldiers "drunk while the money holds out; when it is gone, they encourage them to enlist for the sake of the bounty, then to drinking again. That bounty gone, and more money still wanted, they must enlist again with some other officer, receive a fresh bounty, and get more drink."[41]

Although Sylvia Frey contends that British "Regimental memoirs of the Revolution make no direct reference to the distribution of spirits before or during battle,"[42] rum was issued daily, starting in 1777, to the British army at the rate of about a half pint per man, and the army drank its way through more than 360,000 gallons per year—representing the single largest cost of all supplies. One of the reasons the British diverted troops from the conflict in North America to send them to the West Indies was to defend their massively profitable sugar plantations (sugar was the main ingredient of rum) from French incursions. In effect, the eighteenth century saw a battle between France, Britain, and, to some extent, Spain for control of a drug cartel. The drug was not opium or cocaine but sugar, which, in its way, proved to be as highly addictive.

Drunkenness was far and away the most common cause of disciplinary proceedings in both armies. In fact, one historian of this phenomenon contends that most British soldiers of the eighteenth century were habitual drunkards.[43] British officers, also, were certainly prodigious imbibers. Burgoyne invaded America from Canada with an impressive stock of champagne (a wine that was newly fashionable in the eighteenth century). General William Howe fought his battles and loved his bottle. The Continental general Adam Stephen was as tight as a drum at the battle of Germantown and had his men fire on their compatriots of Anthony Wayne's division. Found later, passed out, Stephen was cashiered. This was the world of the eighteenth century, high and low: William Hogarth's proletarian Gin Lane was next door to Mayfair's aristocratic port and champagne club land. Captain John Peebles of the 42nd Foot in America on 29 March 1777 recalled "getting foul with claret" when thirty-one officers drank seventy-two bottles of Bordeaux, eighteen of Madeira, and twelve of port.[44]

In battle alchohol was a necessary fortifier and prop to motivate men under extreme stress. At the height of the battle of Bunker's Hill the patriot drummer boy Robert Steele was summoned by a sergeant.

"You are young and spry, run in a moment to some of the stores and bring some rum. Major Moore is badly wounded. Go as quick as possible." . . . I seized a brown two-quart, earthen pitcher and drawed it partly full from a cask and found I got wine. I threw that out and filled my pitcher with rum from another cask. Ben took a pail and filled [it] with water, and we hastened back to the entrenchment on the hill . . . our rum and water went very quick.[45]

General Nathanael Greene wrote, with a touch of desperation, to the governor of North Carolina in 1782, "Without spirits the men cannot support the fatigues of the campaign," and begged him to send some double quick. The lack of spirits could have disastrous results, as the hapless Horatio Gates discovered prior to the battle of Camden in August 1780. One of his officers, Colonel Otho Holland Williams, explained: "As there were no spirits yet arrived in camp; and as, until lately, it was unusual for troops to make a forced march, or prepare to meet an enemy without some extraordinary allowance, it was unluckily conceived that molasses, would, for once, be an acceptable substitute."[46] The result, as Williams delicately put it, was that the men were "much debilitated" by this "cathartic."

If too little of "the creature" posed problems for the soldiery in battle, too much had the predictably opposite and equally disastrous effect. Before the battle of Eutaw Springs in South Carolina on 8 September 1781 General Nathanael Greene, needing time to allow his raw troops to fortify their spirits for battle, "halted his columns, and after distributing the contents of his rum casks, ordered his men to form in the order of battle,"[47] Colonel Otho Williams reported. "We . . . moved in order of battle about three miles," continued Williams, "when we halted and took a little of that liquid which is not unnecessary to exhilarate the animal spirits upon such occasions."[48] But, with the battle seemingly won, the Americans fell on the abandoned British camp and, "dispersing among the tents, fastened upon the liquors . . . and became utterly unmanageable."[49] A British counterattack drove them from the field and forced Greene to retreat with his whole army. Booze had snatched defeat from the jaws of victory.

At various times attempts were made to control the thirst. After his victory at Trenton, Washington ordered all the captured Hessian rum casks to be destroyed, which must have been a great sacrifice to men who had marched, some barefoot, through the foul and freezing weather of the night, and who would have relished a warming tot. But with Cornwallis lurking nearby, the commander in chief (who was not a prude about such matters) could not take the chance of his men being caught off-guard and sozzled. Ironically, much later in the war, Cornwallis would take a leaf from Washington's book and destroy all his army's rum ration before taking off in pursuit of General Greene in the frantic "dash to the Dan" in North Carolina.

Fear, like alcohol, is a double-edged sword of motivation. If harnessed and directed, it can be powerfully effective, as in the fear of punishment and the fear of defeat. If uncontrolled, it corrodes the soldier's ability to fight effectively or even fight at all. The raw militia army at Bunker's Hill had great difficulty dealing with fear that spread through several groups, infecting and affecting their willingness to fight. It was entirely understandable among inexperienced troops facing an enemy of frightening reputation. Captain John Chester set off from Cambridge in the company of three regiments of reinforcement for the fighters on Breed's Hill, but when he arrived, "there was not a company with us in any kind of order. They were scattered, some behind rocks and hay-cocks, and thirty men, perhaps, behind an apple tree." Some on Bunker's Hill (perhaps a mile or so behind the scene of the fiercest fighting taking place at the redoubt on Breed's Hill) were wounded, and a gaggle of frightened men would volunteer to carry them to the safety of the rear. John Chester again: "[There were] frequently twenty men round a wounded man, retreating, when not more than three or four could touch him with advantage. Others were retreating seemingly without excuse, and some said they had left the fort with leave of their officers, because they had been all night and day on fatigue."[50]

At moments like these it takes an inspired officer to stop the rot and remotivate terrified men. The narrow neck of land that attached Charlestown Peninsula to the mainland was heavily bombarded by British gunboats, and to cross it was a severe test of nerve. When Colonel

John Stark, leading his New Hampshires, approached the Neck, he found it blocked by troops too terrified to hazard the crossing. Stark ordered them to stand aside and let him take his men across, which he did with an agonizing slowness. Captain Dearborn, leading the front company, was understandably keen to quicken the pace. Stark was not impressed, and, as Dearborn later recalled, "he fixed his eye on me and observed with great composure—'Dearborn, one fresh man in action is worth ten fatigued men." It was one of those "Damn the torpedoes" moments. The outcome turns you into either a hero or a damned fool. A Farragut or a Custer.

The application and control of fear are two of the essential weapons in the arsenal of command and were exercised in eighteenth-century armies through a comprehensive system of punishments that ranged from fines, extra drill, and demotion at one end of the scale, up through draconian floggings and public execution at the other. Frederick the Great's famous, or infamous, dictum is taken as the classic statement: "Many soldiers can be governed only with sternness and occasionally with severity. If discipline fails to keep them in check they are apt to commit the crudest excesses, they can be held in check only through fear."[51] The British army, though certainly not the most brutal of the eighteenth century (a dubious distinction vied for between Prussia and Russia), was not loath to motivate through fear. General Howe's orderly books are filled with entries like: "Boston, 24th Nov. 1775. Thomas Bailey, Grenadier in His Majesty's Corps of Marines, tried by the General Court Martial . . . for Striking Lieut. Russell of the 4th King's Own Regiment . . . to receive Eight Hundred Lashes on his bare back with a Cat of nine Tails."[52] Burgoyne's orderly book is also dotted with punishments of as many as 1,000 lashes (and neither Howe nor Burgoyne was a notable martinet; in fact they were regarded by their men as humane commanders). Even though the lashes were not administered all at one time, and a regimental surgeon was on hand to monitor the victim, a cat (an eight-thonged whip with each of its "tails" knotted to maximize injury) inflicted appalling wounds. In his *Memoirs* Sergeant Roger Lamb relates, "I well remember the first man I saw flogged. During the infliction of his punishment, I cried like a child."

The historian Richard Holmes describes the ritual of flogging in the British army.

A soldier who was to be flogged in barracks was marched onto the parade ground or inside a large building like a riding school, with the men of his regiment formed up, in full dress, in hollow square. The adjutant read out the sentence and its confirmation, and then turned to the prisoner, ordering: "Strip, sir." The prisoner removed his shirt. He was then tied up, an infantryman to a large iron triangle, derived from the traditional pyramid of sergeants' halberds, and cavalryman to a short ladder made fast to a wall or tree. Short whips called cat o' nine tails were already on hand in green baize bags, in charge of the drum major (for infantry units) or the farrier major (for the cavalry). There was a bucket of water and a chair, a hospital orderly, and the regimental surgeon stood close by to monitor the prisoner's condition. When arrangements were complete, the adjutant reported to the colonel, who ordered: "Proceed." The first cat was removed from its bag, and a farrier or drummer struck the prisoner with it, the sergeant major calling out each stroke.

The punishment went on, with floggers being replaced as they grew tired, and the cats being exchanged for fresh ones as they became worn or clogged with blood and tissue. Onlookers frequently fainted or vomited.[53]

A soldier described the effects of being flogged: "I felt an astounding sensation between the shoulders, under my neck, which went to my toe nails in one direction, my finger nails in another, and stung me to the heart, as if a knife had gone through my body."[54] Flogging in the British army was abolished as a peacetime punishment in 1868, and in wartime in 1881; but it persisted in military prisons, limited to twenty-five strokes, until 1907.

Just as George Washington wanted a Continental army structure, officer corps, and training that approximated the British model, he also urged Congress in the fall of 1776 to give him disciplinary powers more in line with the British model. Until Washington became commander in chief the maximum number of lashes was thirty-nine (a strange number but one, apparently, sanctioned in the Bible). He lobbied Congress for a maximum of 500 but compromised at 100. He also wanted death for desertion and "bounty jumping" (enlisting multiple times). Even during the appalling conditions at Valley Forge, when starving men were forced to sustain themselves by stealing, Washington bit down hard on those unfortunates unlucky enough to be caught. Dr. James Thacher, in his capacity as military surgeon, witnessed a flogging.

> *The culprit being securely lashed to a tree or post receives on his naked back the number of lashes assigned to him, by a whip formed of several small knotted cords, which sometimes cut through his skin at every stroke. However strange it might appear, a soldier will often receive the severest stripes without uttering a groan or once shrieking from the lash, even while the blood flows freely from his lacerated wounds ... They have, however, adopted a method which they say mitigates the anguish in some measure. It is by putting between the teeth a leaden bullet, on which they chew while under the lash, till it is made quite flat and jagged.*[55]

"Running the gauntlet" was another punishment common in European armies that was adopted by the Continental army. Edward F. Patrick, a prison guard at Salisbury, North Carolina, recounted the punishment of a Tory who had been "inducted" into the Continental army but had attempted to escape.

> *Langham [the officiating officer] placed two hundred men in two direct lines consisting of one hundred each, leaving a space of six feet between, and the men passing each other all with hickreys [hickory rods] in their hands, and then calling for the before-named Elrod, who was soon brought, and I was placed at one end of the lines thus drawn up, right opposite to the space between the two lines, with orders to turn Elrod, who was placed in the space between*

the two lines . . . Those hickreys were nimbly used on Elrod, as the orders were if anyone favored him, that they should undergo the like discipline. Elrod was in one general gore of blood.[56]

The cruelest cat-and-mouse game involved execution. The deft use of the reprieve seems to have been a staple of eighteenth-century military and civil punishment. As the number of capital offenses in Britain grew from 50 in 1689 to 200 by the end of the eighteenth century, so too did the number of reprieves. Forty people a year were hanged in London and Middlesex in the final decades of the century, compared with 140 in the seventeenth century. One in three people condemned to death was reprieved,[57] a pattern followed in colonial America, where as many as 25–50 percent of all condemned men were pardoned.[58] In the British army the number of capital offenses listed in the rules of war actually ran counter to civil law and fell during the century. As a result, the death penalty as a percentage of all punishments dropped from just over 30 percent to just over 20 percent over the course of the century. Corporal punishment, however, increased sharply from about 15 percent of all punishments to 46 percent over the same period. The rate of acquittal also rose, from about 9 percent to 25 percent of all men charged.[59]

In the War of Independence there were many instances of the gruesome ritual of execution, whether by firing squad or hanging, being interrupted with last-minute pardons. Mock execution was used by both armies, and the often drawn-out ritual was quite deliberately manipulated by commanders. Israel Trask was a very young soldier (he volunteered at age ten) when he witnessed what can only be described as a grotesque piece of psychological theater.

The criminals were heavily ironed and strongly guarded and were by the sentence to be so kept until the day of execution. The door of the prison was left open during the daytime, with free permission to receive the visit of all, whether drawn by friendship or curiosity. Of the latter, nearly the whole army availed themselves of the liberty given. When I visited them, I learned they were both

natives of Marblehead and both married men, and their wives,
respectable-looking women, had taken up their temporary abode
in the same prison with their husbands, the ghastly countenances
of the latter on which the deepest contrition portrayed, the tears of
penitence coming down their rough cheeks, made impression on a
young mind not easily effaced. The stern purposes of Washington
were inflexible to the prayers and supplications of the friends of
the criminals. He continued to receive in silence all solicitations
in their favor until those purposes were attained. He then freely
granted the unexpected pardon.[60]

Ebenezer Wild, a Continental soldier, described in his journal how the condemned were marched to the place of execution with their coffins preceding them. With the whole brigade in attendance the sentence was read out, the graves dug, and the coffins placed beside them. Each condemned man knelt "beside his future resting place in mother earth while the executioners received their orders to load, take aim and . . ." At this moment a messenger galloped up with the reprieve.[61]

But not all were so lucky. Washington, and even more so subordinates like Anthony Wayne and Henry Lee, could be ruthless when the occasion demanded. On 1 January 1781 two brigades (about 1,500 men) of the Pennsylvania Line mutinied at Morristown. Several officers were killed, and Washington and Congress were forced to negotiate with the mutineers. It was a lesson in humiliation neither their commander, Anthony Wayne, nor Washington would forget. When the New Jersey Line mutinied, 20–27 January 1781, Washington ordered Major General Robert Howe to make an example of some of the leaders even though they had been given a general pardon by their commanding officer, Colonel Elias Dayton. Howe rounded up the men at dawn and executed two sergeants. He used the mutineers themselves to form the firing squad. (The first squad missed, even though the victim was kneeling only a few paces, away, and had to be replaced by a second squad.) In May 1781 the Pennsylvania Line again mutinied, and Wayne moved against the ringleaders swiftly and decisively: "Macaroney Jack" was the first to fall, followed by a soldier named Smith who had his head hideously blasted to bits by a volley of musket fire; the firing squad stood less than ten feet from their victims, so close that the handkerchiefs covering the mutineers' eyes caught fire."[62] Ensign Ebenezer Denny of the 7th Regiment recorded the, literally, awe-inspiring execution.

> *The regiments paraded in the evening earlier than usual; orders passed to the officers along the line to put to death instantly any man who stirred from his rank. In front of the parade the ground rose and and descended again, and at the distance of about three hundred yards over this rising ground, the prisoners were escorted by a captain's guard; heard the fire of one platoon and immediately a smaller one, when the regiments wheeled by companies and marched round by the place of execution. This was an awful exhibition. The seven objects were seen by the troops just as they had sunk or fell under the fire. The sight must have made an impression on the men; it was designed with that view.[63]*

Wayne wrote to Washington a few days later that "a liberal dose of niter [gunpowder]" had done the trick. Almost exactly one year later the Connecticut Line revolted, and Private Lud Gaylord was executed. In June 1783 the Pennsylvanians again mutinied, but by this time the war was practically over and Washington did not feel that the methods he had used previously were appropriate.

Henry "Light Horse" Lee (Robert E. Lee's father) had no doubt about "motivating" through fear. He wrote to Washington suggesting beheading as a suitable method for dealing with deserters, a suggestion so grotesque that it prompted Washington to respond:

> *The measure you propose of putting deserters from our Army immediately to death would probably tend to discourage the practice. But it ought to be executed with caution and only when the fact is very clear and unequivocal. I think that part of your proposal which respects cutting off their heads . . . had better be omitted. Examples however severe ought not to be attended with*

an appearance of inhumanity otherwise they give disgust and may
excite resentment rather than terror.[64]

The key word here is *terror*. Washington is less concerned with issues
of humanity in their own right than with the countereffectiveness of
Lee's suggestion. But, unknown to Washington, Lee had already acted.
Brigadier General William Irvine wrote to General Anthony Wayne
on 10 July 1779 that on 8 July a party of Lee's cavalry had captured
three deserters and, after some haggling among Lee's men, agreed to
decapitate one, "a corporal of the First Regiment": "[His] head was
immediately carried to camp on a pole by the two who had escaped
instant death . . . I hope in future Death will be the punishment for
all such. I plainly see less will not do."[65] Lee informed Washington of
his action, which elicited a sharp reply from the commander in chief,
who reminded him of their recent correspondence and ordered Lee to
bury the body quickly to prevent it from falling into enemy hands and
perhaps providing the enemy with some juicy propaganda.

Feeding the Beast

When Frederick the Great humbly declared, "It is not I who commands the army but flour and forage [that] are the masters," it was a salutary recognition of the formidable constraints faced by eighteenth-century commanders. Inspirational rhetoric, regimental pride, the sting of the lash, the lure of bounty, and all the other arts of persuasion and coercion played their part in motivating men to fight, but at the end of the day the army had to be supplied, fed, clothed, and sheltered if there was to be an army at all to face battle. The problems of supply and transport dictated much of the strategy of eighteenth-century warfare. For example, the traditional campaign season ran from approximately May to October when roads were usable and provisions more plentiful. In the winter eighteenth-century armies went into camp. (Washington's attack on Trenton began 25–26 December 1776 and Montgomery and Arnold's assault on Quebec on 31 December 1775 were highly unusual, and both, to a degree, driven by desperation.) General Howe's strategy immediately following his evacuation of Boston in 1776 was largely dictated by lack of supplies. He wrote to Lord George Germain from Halifax in May 1776, "I tremble when I think of our Present State of Provisions, having now Meat for no more than thirteen days in store." It is interesting to speculate that if Howe had amassed the necessary supplies to have started his New York campaign earlier in the season (he fought the battle of Brooklyn at the very end of August 1776), he may well have had enough of the season left to have mopped up Washington's defeated army as it retreated through New Jersey. As it turned out, it was early December when Cornwallis halted at the Delaware, his supplies practically exhausted: "We subsist only on the flour we found in the country." The British had run out of time.

Both the American and British armies struggled with their supply and transport problems throughout the war. If the organization of supply for the British was antediluvian, for the Americans it was prenatal. They both wrestled with the same boa constrictor of bureaucracy: how to establish a system that worked more or less (usually less) efficiently.

For the Americans it was a frantic game of catch-up. A centralized supply infrastructure had to be created within a political system almost tailor-made to frustrate it. The only centralizing agency, Congress, was a lion tamer without a whip. With almost no power to force individual states to conform to a central strategy, it could only appeal rather than enforce, beg rather than demand. After all, it had taken more than sixteen months of acrimonious debate before a "mission statement"— the Declaration of Independence—had been wrung from the fractious states, and throughout the war the battle between centralized and local control worked like square wheels on a cart. The states and Congress competed with each other in foreign markets to purchase armaments, gunpowder, and all manner of military stores. For example, in January 1778 there were three competing sets of commissary agents operating in Pennsylvania: congressional commissioners, Board of War procurement commissioners, as well as Pennsylvania's own supply commissioners. The competition between the three conspired to drive up prices.

The states resolutely refused to pay any kind of national tax to help finance the war. On 22 November 1777 Congress, mired in a huge paper debt of $20 million, "recommended" that the states raise a "federal" tax to underwrite it, but the appeal failed. Patriot America was struggling to create nothing less than the machinery of independence. When the Board of War was created on 12 June 1776, John Hancock proclaimed to George Washington that the act was "a new and great event in the history of America." It was, in fact, one of the foundation blocks of the nation itself. In making the army, the patriots were making

America. Although Congress was often stumbling and bumbling, and overwhelmed at times by the gigantic task, there is something not short of heroic about its determination to grapple with the administrative structure it would need not only to support its army but also to launch America as an independent republic.

The original Board of War was, in effect, a congressional standing committee charged with oversight of ordnance and military stores. As its members had obligations to other congressional committees, the vital work of supply control was often shortchanged. It was not until October 1777 that members were added who had experience of the Quartermaster and Commissary departments but who were not also members of Congress. Even so, there was little close supervision of these important departments by the board. In 1781 the board itself was superseded by the War Office under its first secretary of war, Major General Benjamin Lincoln.

On 16 June 1775 Congress began the business of bolting together the structure needed to supply the army of which the departments of the quartermaster general and the commissary general were key. The quartermaster general, working closely with a commanding general, helped plan marches, select campsites, survey roads and bridges, and, where necessary, repair them. He was responsible for all transportation for the army on the march and for providing it with shelter in camp. It was an intricate and demanding administrative role requiring an intimate knowledge of military and business affairs. In Washington's words, it required a man of "great resource and activity, and worthy of the *highest* confidence." Four men carried the burden during the war: Thomas Mifflin was appointed on 14 August 1775; Stephen Moylan (5 June 1776) lasted only briefly following his ineptitude during the New York campaign; Mifflin was recalled for another eighteen months and then handed the position over to a reluctant Nathanael Greene on 22 March 1778. Greene resigned to return to active service and passed on the responsibility to the equally reluctant Timothy Pickering on 22 September 1780.

The commissary general's primary role was to feed the army. Like the quartermaster general's job, it was something of a poisoned chalice.

The responsibility was massive, but the resources provided by Congress were inadequate, and the situation was compounded by frequent reorganizations and reforms, sometimes at the most inappropriate times, such as in the middle of a campaign season. Four men held the post during the course of the war: Joseph Trumbull (son of the governor of Connecticut, Jonathan Trumbull) from July 1775 until August 1777, William Buchanan until April 1778, Jeremiah Wadsworth until December 1779, and Ephraim Blaine thereafter.

Procurement of supplies was often financed by merchants, who charged a percentage for the service. During the French and Indian War they charged 5 percent for purchases, 2.5 percent on money, and 7.5 percent for storage and sale of goods sent by the colonies to defray their debt. The agent personally financed all procurements and personally carried any debt. During the War of Independence this system persisted until 1781, when Robert Morris, superintendent of finance, introduced contracts. It was a graft (in more senses than one) of the private and public that has something of a modern resonance: the eighteenth-century version of Halliburton meets the Pentagon.

Without an adequate bureaucracy or ready credit, Congress had little option but to rely on the relatively sophisticated commercial organization and financial clout that leading merchants, like Robert Morris of Willing and Morris, Philip Livingston, Thomas Mifflin, and Jeremiah Wadsworth, could offer. This was not an arm's-length relationship with Congress. Powerful merchants were also members of the Secret Committee that was responsible for overseas procurement. Thomas Mifflin became the first quartermaster general, and Jeremiah Wadsworth, the most successful merchant in Connecticut, was commissary general in 1778 and 1779. That the system opened itself up to conflicts of interest is hardly surprising. Robert Morris, Nathanael Greene, Wadsworth, and Mifflin, for example, all dabbled in private side deals in military supplies. Alexander Hamilton discovered Greene's and Wadsworth's shenanigans (buying up flour that was needed for the French fleet and raising the price "an hundred percent" for their own profit) and lambasted Greene in a letter of 26 October 1778.

I say when you were doing all this, and engaging in a traffic infamous in itself, repugnant to your station, and ruinous to your country, did you pause and allow yourself a moment's reflection on the consequences? Were you infatuated enough to imagine you would be able to conceal the part you were acting? Or had you conceived a thorough contempt of reputation, and a total indifference to the opinion of the world? Enveloped in the promised gratification of your avarice, you probably forgot to consult your understanding and lost sight of every consideration that ought to have regulated the man, the citizen or the statesman.[1]

Panicked but not deterred, Greene wrote to his coconspirator in profiteering, Jeremiah Wadsworth, who would become enormously wealthy on the back of his procurements for the French army.

You may remember I wrote to you sometime since that I was desirous that this copartnership between Mr Dean, [Silas Deane, principal American agent to France] you and myself should be kept a secret. I must beg leave to impress this matter upon you again; and to request you to enjoin it upon Mr Dean. The nearest friend I have in the world shall not know it from me; and it is my wish that no mortal should be acquainted with the persons forming the Company except us three.[2]

The temptations were too great, and corruption was endemic throughout the chain of procurement and supply. Quartermasters and commissaries sold government supplies and colluded with crooked suppliers to buy substandard food, clothing, and gunpowder. During the British invasion of Pennsylvania in August 1777 Ephraim Blaine, then deputy commissary general for the Middle Department, complained that 100 head of the cattle he had ordered removed from islands in the Delaware were sold by one of his own agents to the British, with a contract to provide more![3]

Transport was the linchpin of supply and throughout the war was a logistical headache for the quartermaster general's departments of both armies. The road system was sparse, the terrain rugged. In bad weather the roads turned to quagmire. (During February 1778 not one wagon could reach the encampment at Valley Forge.) It took over two days to travel the ninety miles between New York and Philadelphia. There were few bridges, and none across the main rivers. The lumbering wagons that carried goods throughout the interior were in short supply. Finding forage to feed the wagon horses was a constant challenge.

The Continental army had to requisition much of its land transport, and Congress never had enough money to compete with merchants who were also in the business of hiring wagons and drivers. On 14 May 1777 Congress tried to get itself on an even keel by creating a separate wagon department. It was not a success. First, Washington could not find a suitable candidate to become wagonmaster general. (One refused, citing his advanced age; one died before he could take up the post; one begged off for reasons of ill health.) Second, wagon construction could never keep pace with demand. By the winter of 1777–78 there was such a shortage that at Valley Forge men yoked themselves to little carriages in order to haul firewood and food.

Wagoners abused their horses or let them destroy private property—"wherever our Baggage marches the Soldiers and Waggoners plunder all houses & destroy every thing," complained Colonel Henry Lutterloh. The teamsters were a rough bunch, a mix of enlisted men and contracted civilians (in 1778, for example, there were 104 military wagon-drivers and 272 civilians)[4] much given to frequent tavern stops, which slowed their daily mileage from the expected twenty to three or four. However, after they had fortified themselves they would often drive "as if the Devil was driving them," which only ruined the horses.

It was grueling work and badly paid (£10 a month in the army compared with £16–£20 a month in the civilian world). It is no wonder, then, that wagon-drivers frequently dumped their military supplies in order to hire themselves out to merchants and sutlers, especially if the quartermaster's department was slow in paying their wages. John Chaloner, a deputy commissary, complained to James Young, the wagonmaster general for Pennsylvania, "I am just now informed that a number of County wagons coming from Lancaster [on their way to the troops at Valley Forge] have laid down their loads on the Horse Shoe road, and gone home, a practice so destructive to the publick weal."[5] When hauling casks of brined pork—an essential component of the soldiers' diet—wagon-drivers were prone to drain off the brining liquid in order to lighten the load. Inevitably, the pork arrived spoiled.

Feeding a great army is a massive logistical task. Washington estimated that he would need 100,000 barrels of flour and 20,000 pounds of meat to keep his 15,000 men surrounding Boston in 1775 fed for one year. The British army consumed 300 tons of food each week and 360,000 gallons of rum each year, most of it shipped from Britain. The grocery list was staggering. In seven months of 1778 the British army consumed 3.5 million pounds of bread and flour, and 2 million pounds of meat *over and above* the amounts the British Treasury had forecast.

The American army was well supplied and well fed during the first year of the war. This was thanks largely to the talent of Joseph Trumbull, who had been Connecticut's commissary general at the outbreak of war and became Washington's first commissary general on 19 July 1775. It helped also, during the first year, that most of the troops besieging Boston were New Englanders, whose nearby states were the best provided and organized and could succor their fighting men relatively easily. The turning point would come after the battle for New York had been lost. The British captured huge quantities of foodstuffs and, during their occupation of New Jersey, consumed and destroyed much of that state's ability to support the Continental army.

Once an American national army had been created, a standard ration for each soldier needed to be established. It was substantial, but this was a diet for men whose lives were, by modern sedentary standards, intensely physical. It was a diet that delivered calories in abundance.

1 lb beef or ¾ lb pork or 1 lb salt fish per day
1 lb bread or flour per day
1 pint of milk per day
1 quart spruce beer or cider per day
¾ pint of molasses per day

3 pints of peas or beans per week (or the equivalent weight in
   vegetables)
½ pint of rice or 1 pint "indian meal" (cornmeal) per week

The peas or beans were issued dried, so each soldier would have been allocated three pounds dry weight of legumes each week, which, when rehydrated, constituted approximately twelve cups per week. One half pint (one half pound) of dry rice would convert to two cups of cooked rice per week. Of course, the ration was not neatly portioned per day, and Congress's standard ration was theoretical. The milk, for example, was never supplied, and at Boston spruce beer was not available. When the army went to New York in 1776, molasses was unavailable but beer was. Occasionally there were a few extras. Joseph Trumbull, for example, managed to add six ounces of butter per man per week.[6] But as the war went on, deficiency rather than sufficiency became the rule. The army would never eat as well as it did in those early months.

The standard ration compared well with that available to British troops. Each day a British soldier was to receive:

1 lb beef or ½ lb pork
1 lb bread or flour
⅓ pint peas
1 ounce butter or cheese
1 ounce oatmeal
1½ gills rum

The standard ration (American or British) delivered more than enough calories and twice as much protein as was needed.[7] It contained adequate minerals but was low on vitamins A and C, and this deficiency had serious health implications of which Washington was keenly aware. In 1777 he lobbied Congress to broaden the standard ration, and it responded on 25 July with a directive to the Board of War to supply vinegar and sauerkraut (which were thought to be effective in combating scurvy; in fact they were only mildly useful compared with lemon or lime juice), vegetables, beer, and cider. When the board was tardy in complying, Washington took it upon himself to direct regimental quartermasters to gather sorrel and watercress for salads, which, he reported, had a "most salutary effect."[8] The British army also encouraged the growing of vegetables and salad greens. During the siege of Boston the British Treasury informed General Gage that "a good quantity of the small Salled [salad] Seed will be sent out, as it will grow, on being sown, almost anywhere on a little earth, and may be raised by the Soldiers on a little Space by each Mess, in sufficient quantities for their refreshment and use."[9] Washington also had interesting ideas about the place of alcohol in a healthy diet. He bemoaned "devouring large quantities of animal food, untemper'd by vegetables or vinegar, or by any other drink but water . . . Beer or cider seldom comes within the verge of the camp, and rum in too small quantities." One gill (one-quarter pint) of rum was meant to be issued on fatigue days.

Dr. Benjamin Rush's characteristically forceful views on the soldiers' diet would have been applauded today. For example, he wrote in his *Directions for Preserving the Health of Soldiers* of April 1777 that porridge was the answer.

> *The Diet of soldiers should consist chiefly of vegetables. The nature of their duty, as well as their former habits of life, require it. If every tree on the continent of America produced Jesuits bark, it would not be sufficient to reserve or restore the health of soldiers who eat two or three pounds of flesh in a day. Their vegetables should be well cooked. It is of the last consequence that damaged flour should not be used in camp. It is the seed of many disorders. It is of equal consequence that good flour should not be unwholesome by an error in making it into bread. Perhaps it was the danger to which flour was always exposed of being damaged in a camp, or being rendered unwholesome from the manner of baking it, that led the Roman generals to use wheat instead of flour for the daily food of their soldiers. Caesar fed his troops with wheat only . . . It was prepared by being well boiled, and was eaten with spoons in the room of bread. If a little sugar or molasses is added to wheat . . . it forms not only a most wholesome food but a most agreeable repast.[10]*

Despite the good Dr. Rush's admonitions concerning the danger of bread, it remained, with meat, the most important element of a soldier's diet. Rush, though, had a point. A regimental baker (usually simply a designated soldier who had some experience of baking in civilian life) could make a nice profit for himself. He took the soldier's one pound of flour and delivered back one pound of bread. One pound of flour makes approximately one and one-third pounds of bread—and so the extra 30 percent of flour became the baker's "perk," which he could sell to local civilians. Additives, some good, some decidedly less so, found their way into the soldiers' flour ration. Rye gave a distinctive and not unpleasant acidity; bran slowed digestion, bulked up the bread, and decreased hunger. Underbaked bread produced stomach acidity and promoted flatulence. However, white bread was highly prized, and unscrupulous bakers were known to add chalk and plaster of Paris to the flour. This practice was not confined to the military. The eighteenth-century demand for white bread led Tobias Smollett in *Humphrey Clinker,* written in 1771, to complain: "The bread I eat in London is a deleterious paste, mixed up with chalk, alum, and bone-ashes; insipid to the taste and destructive to the constitution. The good people are not ignorant of this adulteration; but they prefer it to wholesome bread, because it is whiter than the meal of corn." The bread supplied to the British army was often inedible. The victualler at New York in 1776 described it as "very old Bread, Weavile Eaten, full of Maggots, Mouldy, musty and rotten and entirely unfit for men to eat."[11]

Early in 1777 Congress had appointed Christopher Ludwick as superintendent of bakers for the Continental army. Ludwick and his bakers established themselves at the public ovens at Morristown and by 1780 were turning out 1,500 loaves of "hard" bread. Hard bread (as opposed to the soft bread the soldiers made themselves) was more like the hardtack biscuit, so familiar to sailors in the age of sail. It lasted longer and took up less space in a soldier's knapsack, which was a great advantage for troops on campaign. Private Joseph Martin called it "sea-bread" and, when given the chance, stocked up.

At the lower end of the street [he was in New York City prior to the battle of Brooklyn in August 1776] were placed several casks of sea-bread, made I believe, of canel and peas-meal, nearly hard enough for musket flints; the casks were unheaded and each man was allowed to take as many as he could, as he marched by. As my good luck would have it, there was a momentary halt made; I improved the opportunity thus offered me, as every good soldier should upon all important occasions, to get as many of the biscuits as I possibly could . . . I filled my bosom, and took as many as I could hold in my hand . . . they were hard enough to break the teeth of a rat.[12]

Baking for soldiers on the move was difficult, and in July of 1777 Washington ordered that small portable ovens constructed of sheet iron be issued, one per brigade. But by the time the ovens arrived at the end of the year when the army was going into winter quarters at Valley Forge, the supply of flour was extremely limited. During hard times soldiers were reduced to making crude "fire cakes," a simple paste of flour and water griddled on hot stones. "Fire-cake and water for breakfast! Fire-cake and water for dinner! Fire-cake and water for supper! The Lord send our Commissary for Purchases may have to live on fire-cake and water!" wailed Dr. Albigence Waldo from the Valley Forge camp. Without salt, it was dreary food indeed.

Salt was the refrigeration of the pre-refrigeration age. (Chemical refrigeration—dissolving saltpeter in water—was discovered in Italy in the 1540s, but the first commercial refrigeration plant was built by Alexander Catlin Twining at the Cuyahoga Locomotive Works, Cleveland, Ohio, in 1850.) "Salt," said Joseph Martin, "was as valuable as gold with the soldiers." And when he stumbled on a barreful during one his frequent foraging expeditions, he filled his pockets with it. Used for preserving as well as flavoring meat and fish, "it was as essential in the Revolutionary War as gunpowder and almost as scarce."[13] Before the Revolution it had been imported, mainly from Turk Island (in the Turks and Caicos group at the southernmost extent of the Bahamian archipelago) and Bermuda—both British possessions. Even though a barter trade continued throughout the war between the islands and the colonies, there were still grave shortages and prices had to be regulated to try and curb profiteering, but without much success. Congress, in an attempt to alleviate the situation, ordered that all incoming ships bringing in cargoes bought with congressional money should be ballasted with salt. The saltworks in Virginia and New Jersey became targets for British raids, and of the ships specifically chartered to import salt, half were intercepted.

The lack of salt to cure meat as well as the cost of transporting barrels of cured meat meant that cattle had to be driven considerable distances to be close enough to the army before they could be slaughtered, and the shortage of forage en route inevitably resulted in emaciated beasts ("quite transparent," wrote Joseph Martin. "I thought at the time what an excellent lantern it would make"). Eating meat unseasoned with salt was a common complaint, but even salted meat had limited appeal, especially if it was old. After a year in brine it had lost much of its nutritional value, and some of it was much older. A Hessian private en route to America described just such an ancient delectable: "The pork seemed to be four or five years old. It was streaked with black towards the outside and was yellow further in, with a little white in the middle. The salt beef was in much the same condition."[14] A British soldier described it "as hard as wood, as lean as carrion, and as rusty as the devil."[15] Tainted meat was sometimes rescued, if that is the right word, by boiling it with charcoal, ashes, lye, or potash.[16]

Most of the cattle and hogs eaten by the Continental army came from the plentiful livestock resources of New England, but organizational problems in the Commissary Department resulted in severe shortages not only of beef on the hoof but also of salted pork by the winter of 1777. The army at Valley Forge had practically no meat and only twenty-five barrels of flour. On Christmas Day 1777 there was not a single animal in the camp to be slaughtered. By 4 January 1778 the camp commissary reported that "on this day [the army] will consume the whole of the salt provisions Fish &c and God only knows what will be for the troops tomorrow."[17] Dr. Albigence Waldo gives some idea of what the troops were reduced to: "Poor food . . . nasty cookery—vomit half my time . . . There comes a bowl of beef soup, full of burnt leaves and dirt, sickish enough to make Hector spue."[18] Joseph Martin arrived at Valley Forge already famished, thirsty, and exhausted. Unable to find water, he was forced to buy some with his last money. He endured two nights and one day with nothing at all to eat except "half a small pumpkin," which, he reported, "I cooked by placing it upon a rock, the skin side uppermost, and making a fire upon it; by the time it was heat through I devoured it with as keen an appetite as I should a pie made of it some other time."[19] The Christmas of 1779 at Morristown was, if anything, more desperate. "Our magazines," wrote Washington, "are absolutely empty everywhere and our commissaries entirely destitute of money or credit to replenish them. We have never experienced a like extremity at any period of the war."[20] Even by the end of May 1780 Lafayette looked pityingly on "an army that is reduced to nothing, that wants provisions, and has not the necessary means to make war."

Hunger and hardship forced the soldier to make desperate culinary innovations. In the fall and early winter of 1775 Benedict Arnold's little army carried out one of the most heroic expeditions in American military history: a journey from Massachusetts up through the wilderness of Maine to Quebec, navigating the rivers where possible, manhandling their unwieldy bateaux in exhausting portages where not. Conditions were appalling, the men often starving. One of them, Dr. Isaac Senter, described some "sumptuous eating": "Our bill of fare for last night and this morning consisted of the jawbone of a swine destitute of any covering. This we boiled in a quantity of water, that with a little thickening constituted our sumptuous eating." By 1 November he was reporting:

> *Our greatest luxuries now consist in a little water, stiffened with flour, in imitation of shoemakers' paste . . . Instead of the diarrhea, which tried our men most shockingly in the former part of our march, the reverse was now the complaint, which continued for many days . . . The voracious disposition many of us had now arrived at rendered almost anything admissible. Clean and unclean*

were forms now little in use. In company was a poor dog [who had] hitherto lived through all the tribulations, became a prey for the sustenance of the assassinators. This poor animal was instantly devoured without leaving any vestige of the sacrifice. Nor did the shaving soap, pomatum, and even the lip salve, leather of their shoes, cartridge boxes etc., share any better fate.[21]

Captain Simeon Thayer was also on the march to Quebec and recorded his men "taking up some rawhides that lay for several days in the bottom of their boats, intending for to make them [into] shoes or moccasins [but instead] chopping them to pieces, singeing first the hair, afterwards boiling them and living on the juice or liquid that they soaked from it for a considerable time."[22]

On Nathanael Greene's march that led to his defeat at the battle of Eutaw Springs in South Carolina in 1781, the men were starving. Some of them had covered over 300 miles in twenty-two days: "Rice furnished our substitute for bread . . . Of meat we had literally none . . . Frogs abounded . . . and on them chiefly did the light troops subsist. Even alligators were used by a few."[23] A good soldier improvises and puts the best face he can on it, as James Collins did just prior to the battle of Kings Mountain, South Carolina, in October 1780.

Everyone ate what they could get . . . sometimes eating raw turnips and often resorting to a little parched corn, which, by the by, I have often thought, if a man would eat a mess of parched corn and swallow two or three spoonfuls of honey, then take a good draft of cold water, he would pass longer without suffering than with any other diet he could use.[24]

The main meal for both British and American soldiers was around midday (as it was for most civilians). The men ate in messes of about twelve, with one of their number designated as cook—an often thankless job. There were no mess kitchens and very few utensils. A nine-quart iron kettle was issued to every six to ten men. (Officers were issued the luxury of lids, while enlisted men had to make do with open kettles.)

The kettle weighed about three pounds, and lugging it on a march was a chore every man contrived to avoid, but with a little guile it might be surreptitiously infiltrated into the baggage wagon. The culinary wisdom of the day frowned on certain cooking methods. Food, pronounced Henry Knox, with a culinary confidence surprising in an artillerist, "ought always to be boiled or roasted, never fried, baked or broiled, which modes are very unhealthy."[25] It would take genetic reengineering to prevent an American man from grilling, and revolutionary soldiers were no exception. Ramrods made excellent spits, and old iron barrel hoops were transformed into gridirons.

The Continental soldier often had to provide his own eating utensils, but on occasion they came as standard issue. Maryland troops, for example, were provided a wooden trencher (plate), and bowl, as well as wooden and pewter spoons. Each man would have his knife, of course; and for quaffing his rum, cider, beer, or whiskey, a horn cup, which was extremely light compared with pewter or ceramic. Officers, as might be expected, had more refined utensils. George Washington's mess kit, for example, was a very elaborate affair housed in a handsome fourteen-compartment wood chest lined with green wool. It contained six tin plates, three tin platters, four tin pots with detachable handles, two knives, four forks, a gridiron, two tinderboxes, two glass bottles for salt and pepper, and eight cork-stoppered glass bottles for spirits.[26]

The problems of supply were equally problematic for the British, but they took a different shape. The original hope of the British strategists was that the army would be able to live off the land, sustaining itself from foodstuffs it would find in America. The opposite proved to be the case. Because it could never break into the hinterland and hold substantial areas, and thus provision itself, the British army was effectively contained geographically to peripheral "garrisons" (Boston, New York, Philadelphia, Charleston, Savannah, and finally, Yorktown) and so denied access to sources of local supply. It was always somehow stuck out on the margins. Cornwallis recognized this when he railed at the strategy of "posts" and urged his commander in chief, Clinton, to break out into rich, and relatively untouched, Virginia. The strategic flow was, however, against Cornwallis, and one of the ironies of the war was that he ended up (and ended the war) penned in the rat hole of Yorktown. So, unable to sustain itself entirely in-country, the British army was dependent on supplies from the British Isles.

Cork in western Ireland was the great depot into which beef, flour, pork, rum, munitions, and clothing flowed on their way to the army in America. The quantities were prodigious. In 1776 alone 1.8 million pounds of beef, 7 million pounds of pork, almost 10 million pounds of flour, and 1.7 million pounds of oatmeal and rice were shipped. Each British soldier in North America consumed one-third of a ton of food per year, much of it sent from Britain.[27] Fodder was too bulky to justify the cost of shipping it from Britain, and this in itself had strategic consequences. For example, General Howe's decision to maintain wide-flung cantonments like Trenton and Princeton in New Jersey in 1776 was largely dictated by the need to find provisions locally, particularly forage, but it exposed a vulnerability that Washington exploited brilliantly. Time and again British commanders complained to London that their strategic options were severely limited by shortages of supply. In September and October 1778 Sir Henry Clinton, for example, wrote to Lord Germain: "Your Lordship will be startled when I inform you that this Army has now but a fortnight's Flour left . . . Our Meat with the Assistance of Cattle purchased here will last about forty days beyond Xmas, and a Bread composed of Peas, Indian Corn and Oatmeal can be furnished for about the same time. After that I do not know how we shall subsist."[28] It was a sentiment that Washington would have heartily endorsed.

The quartermaster general for the British army in North America was Sir William Erskine (until June 1779, Lord Cathcart thereafter). The Commissary Department was split between Canada and the rest of Britain's North American possessions from Nova Scotia south to Florida. Nathaniel Day was commissary general for Canada, while three men held the post for the rest of North America: Daniel Chaumier (February 1774–February 1777), Daniel Weir (February 1777–September 1781), and Brook Watson (March 1782–December 1783). Many factors conspired to make life immensely difficult for the commissaries. As in the American army, corruption was widespread

(Chaumier was certainly involved in some very dubious business, and was dismissed because of it), and theft and fraud were rife. Nathaniel Day railed at the light-fingered Canadians he employed: Their "propensity for pilfering is such that it obliges me to send Conductors [inspectors] to protect the provisions. Notwithstanding all my efforts to protect it, I have had the Mortification to see the Butter taken out of the Firkins [casks] and Stones etc. put in lieu to compleat the Weight and so dexterously headed [resealed] that the best eye could not perceive the deception."[29] Back in Britain contractors took their piece of the action at the source: short-weighting, hiding defective provisions under good, and so on with impressive creativity: "The merchants begin to sniff the cadaverous *haut goût* of lucrative war," observed Edmund Burke, one of the leaders of the opposition to the war in the House of Commons.

Glut and shortage often threw the shipments out of kilter. Sometimes there was not enough butter but too many peas; or a mountain of flour but insufficient beef. Careless stowage on the transports and shoddy packaging accounted for grievous losses. And when the provisions finally did arrive, they could sit around on the wharves rotting for lack of transportation to the interior.

If the American supply machine was severely hampered by the failure of a central authority, so too was the British. Its bureaucracy was simply not sufficiently evolved to cope with the logistical complexities of the war in America as well as other parts of its empire simultaneously. In Britain three boards shared responsibility for transporting the myriad requirements of the army across the Atlantic. The Treasury Board undertook shipping all provisions through private contractors; the Ordnance Board carried artillery, engineers, and munitions; the Navy Board transported infantry, cavalry, hospital stores and, after 1779, grudgingly took on the responsibilities of the Treasury Board. As can be imagined, there was constant infighting and finger-pointing, especially between the Navy and Treasury boards.

The shortage of shipping was crippling, and the length of time transports were held in America aggravated the situation. John Robinson in England implored Howe in April 1776: "The immense Quantity of Tonnage employed and wanted, have drained this Country of Ships so that without . . . distressing the Commerce of the Kingdoms, Transports are not to be had . . . The Distress for Want of Transports is so great that it would be of the utmost Benefit . . . if it should be possible for you . . . to spare some of your largest Ships and return them home to be employed as Victuallers."[30] But it was not only the quantity but also the quality of the ships that exacerbated the issue. They were, more often than not, banged-up old tubs, and to travel in them for the six to eight weeks of an average transatlantic voyage was to experience a circle of hell that would have turned Dante green. Captain Jacobs, a British Guards' officer, left this animated description: "There was continued destruction in the foretops, the pox above-board, the plague between decks, hell in the forecastle, the devil at the helm."[31]

The Things They Carried

WEAPONS, EQUIPMENT, AND CLOTHING

The technological restrictions, and possibilities, of the eighteenth-century soldiers' weaponry lie at the heart of any understanding of the nature of the combat in the War of Independence. The range, accuracy, and, above all, lethality (or lack of it) of the musket and cannon to a large extent determined not only the tactics of the battlefield but also the soldiers' training.

> *The range and accuracy of firearms, for instance, set the distance at which opposing forces could engage each other. The ability of the individual soldier to carry his personal necessities on his body directly affected the need for a baggage train and consequently the speed with which a unit could maneuver . . . It is impossible to understand a war without understanding the armies which fought it, and it is impossible to understand an army without a knowledge of the physical objects with which it functioned.*[1]

It was a handmade war. In the pre-mass-manufactured world of the late eighteenth century, every musket, rifle, and cannon; every canteen, cartridge, and cartridge box; each piece of the soldier's uniform was handcrafted. Each musket or rifle, although sharing close similarities, was unique, and this individuality was, in a profound way, an expression of the society that produced it. Throughout the eighteenth century there was a tension between the idiosyncrasy of the unique and the drive for uniformity. The highly formalized nature of volley firing and the often intricate regimentation that had to be instilled in the soldier through his training was a response to the waywardness, the unpredictability, the resistance to standardized performance, of his musket. If the soldier could be trained to work on the battlefield as a coherent mass, the weaknesses inherent in individual man and musket could be overcome. At another level, the same tension between individuality and homogeneity can be seen at work. George Washington's insistence on a centralized, "national" army rather than a disparate conglomeration of state militias and provincial regiments was inspired by the drive toward uniformity and the predictability in performance that it was thought to guarantee. His passion for standardized drill, even for a standardized uniform, sprang from the same impulse.

So it was with weaponry. The American forces at the beginning of the conflict were armed with a hodgepodge collection of muskets of various calibers, some dating back to the French and Indian War; a scattering of rifles, mainly from the Pennsylvania, Maryland, and Virginia backcountry (the rifle was virtually unknown in the New England colonies; John Adams called it "a peculiar kind of musket"); fowling pieces, and even blunderbusses of antique origin. The value put on a gun by the provincial assemblies of Pennsylvania, Massachusetts, and Virginia ranged from £3 to £5, when £5 represented about two weeks' work for a laborer, so for many it represented a considerable investment.

At the beginning of the war, firearms were, more often than not, provided by the men themselves. It was an individualistic practice Washington wanted to replace with standard, "national" weapons: "King's Musquets, or Guns as near that quality as can be had."[2] Just as he sought to emulate the British army in organization and training, he wanted his army equipped with British-style firearms. The individuality of the American soldier in the early months of the war expressed itself in another way. When their term of service expired the soldiers took their guns home with them, against Washington's express orders. He had intended to reimburse them for their privately owned weapons, but in January 1776, after many men had returned home, he found himself with only 1,620 serviceable muskets. Apart from those that left with the men, many were rejected because they were worn out. In February 1776 2,000 soldiers lacked a musket because many recruits had turned up without a weapon.[3]

It may have given Washington succor if he had known that his predicament was shared by the British army. Many redcoat regiments had old and defective firelocks. (The term *firelock* is interchangeable with *musket*.) On the eve of the rebellion the 7th Foot was armed with thirty-six-year-old muskets; those of the 37th Foot had been issued in 1742, and of the 390 firelocks of the 34th Foot, 351 were defunct.[4] Another deficiency shared by both armies was the lack of artificers—arms repair technicians. (The Prussian army of the period had sufficient artificers, a major factor in its superior musketry on the battlefield.)

The American army relied on three sources for its muskets: domestic production, "inherited" British firelocks, and imports from France. Domestic manufacturers, understandably, conformed as closely as possible to the musket they had known down the years—the British "Brown Bess." On 4 November 1775 Congress issued to each colony a specification that was a close replica of the British musket and known as the "Committee of Safety" model.

> *Resolved. That it be recommended to the several Assemblies or conventions of the colonies respectively, to set up and keep their gunsmiths at work, to manufacture good fire locks, with bayonets; each fire lock to be made with a good bridle lock, ¾ of an inch bore [same as the British Brown Bess], and of good substance at the breach, the barrel to be 3 feet 8 inches in length [Brown Bess was 3 feet 6 inches], the bayonet to be 18 inches in the blade [British 17 inches], with steel ramrod, the upper loop thereof to be trumpet mouthed: that the price to be given be fixed by the Assembly or convention, or committee of safety of each colony.[5]*

Not many Committee of Safety muskets survive. They quickly wore out and were superseded after 1777 when the French shipped 23,000 Charleville muskets, which became the standard for domestic American manufacture. Charlevilles were inexpensive ($5) compared with the domestically manufactured musket ($12) because they were turned out in great numbers from the manufactures at St. Etienne, Maubeuge, and Charleville (all in the iron-and-coal region of northern France), and over the course of the war, more than 100,000 were imported. They were valued for their sturdy construction, though their smaller caliber (0.69 inch) and therefore lighter ball (twenty to the pound compared with fourteen to the pound for the Brown Bess) did not pack the same punch as the Brown Bess. Another shortcoming was that it fouled more quickly than its British counterpart, requiring cleaning every fifty discharges or so. Nevertheless, some historians have claimed that the greater range of the Charleville made it superior to the English model. Contemporary accounts suggest otherwise. The proof of the pudding was in the eating. William Lloyd, a New Jersey militiaman, was at a skirmish at Freehold, New Jersey, in 1778: "The enemy then retreated precipitately, throwing away many of their guns. I was, I believe, the foremost in following, got as many of their guns as I could conveniently manage on my horse, with their bayonets fixed upon them. Gave them to the [American] soldiers as they stood in rank. They threw away their French pieces, preferring the British."[6] John Henry was part of the American force that stormed Quebec in 1775 and recounted that at a certain point in the battle "many of the party . . . threw aside their own [muskets] and seized the British arms. These were not only elegant, but were such as befitted the hand of a real soldier."[7]

The truth was that the Brown Bess, with all its faults, was the most successful firearm of the eighteenth and early nineteenth centuries. During the 140 years of its production more than 7.8 million were made. The very name became synonymous with British power, but the origin of the name itself has become something of mystery. One American historian has suggested that the Bess refers to "Good Queen Bess," Queen Elizabeth I, in whose reign the predecessor to the firelock, the matchlock, was introduced. Another theory is that the word Bess is a corruption of the German *Büsche* gun; another that the *Brown* refers to the pickling of the barrel to retard corrosion, yet another that the stock, originally painted black, was, in the eighteenth century, made of walnut, hence *brown*. In any event, it probably would not have been called by its famous nickname during the American war. (The first use in print was not until 1785.)[8] Contemporaries knew it as the King's, or Tower, musket. The Tower of London was the Royal Armoury and issued patterns to the myriad small gun-making shops that clustered around it. (London and Birmingham were the largest gun-making centers in Britain.) Because production of the parts, rather than the whole gun, was carried out by hundreds of small workshops, uniformity and interchangeability of parts were almost impossible to guarantee, and the main bottleneck to production was the assembly of parts from various subcontractors.

The flintlock musket derived both from the snaphaunce, a sporting gun developed in Holland and France in the early seventeeth century (the name comes from the Dutch *snaphaan,* meaning "pecking hen," a description of the action of the beak-shaped hammer or cock striking the frizzen), and, more important (because snaphaunces were rare), from the more numerous English lock gun, also from the early 1600s. The introduction of the English lock changed radically how battles could be fought. In matchlock firearms, the forerunner to the flintlock, the pan holding the priming powder was open and therefore prone either to dampness or to accidental discharge from a carelessly handled "match" (a slow-burning wick). Placing a movable cover (frizzen) over the pan could lessen both these disadvantages. Now, when the cock struck the frizzen it automatically pushed it forward, exposing the powder in the pan and allowing a spark struck when the flint of the cock hit the frizzen to ignite the powder in the pan, which, in turn, ignited the powder charge in the barrel. Safety was further enhanced by providing a "half-cock" position for the hammer, so that if it was accidentally triggered it would not have the power of a hammer at "full cock" to discharge the gun (hence the term *to go off at half cock,* meaning prematurely and ineffectively).

The earliest version of the Brown Bess, the Long Land Service pattern (its barrel was three feet ten inches), is dated 1718,[9] but most historians date its general issue in the British army from the late 1720s to early 1730s. Around 1740 the barrel was shortened to three feet six inches to give the soldier increased mobility, and it was this model—the Short Land Service—that was the primary British army musket in the American war. The Long Land Service was relegated to use in Loyalist regiments.

How effective was the musket? It is a crucial question because from it flows the whole "architecture" of eighteenth-century battle: the way troops were trained, how they were disposed on the battlefield, the rate and range of fire, and, most important, the lethality of the weapon. It was certainly inaccurate, and that inaccuracy was, in a fascinating self-fulfilling cycle, reinforced by training and tactics. In order to counteract inaccuracy, much emphasis was put on maximizing the number of troops delivering fire and the rate at which they could deliver it. In order to increase the rate of fire, the loading sequence had to be made as swift as possible, and in order to speed it up, the ball had to be able to be inserted easily. The difference between the internal diameter of the barrel (the bore) and the external diameter of the ball is called *windage*; the greater the windage, the easier it is to insert and seat the ball in the barrel. (The windage on a Brown Bess was 0.05 inch.) The problem with windage is that it allows the gases from the fast-burning powder on which the ball sits in the barrel to escape around the edges of the ball, thus reducing range (the power of the propelling explosion is dissipated) and accuracy (the ball is buffeted by the gases in a ricocheting pattern down the length of the barrel). If the ball was a tight fit in the bore, as it was in the rifle, it cut down the gases escaping through windage and increased range and accuracy, but it took much longer to load. This, anyway, is the theory. A dissenting voice comes from the War of Independence historian and reenactor Lawrence E. Babits, whose personal experience has been radically different: "Even with undersized bullets it is possible without ramming to hit a man-sized target eighty yards away with five out of six shots in one minute . . . the extreme windage [a 0.63-caliber ball in a 0.75-caliber barrel] did not cause any loss of accuracy."[10]

The loading sequence for a musket was intricate. The soldier raised the gun vertically until the hammer was about level with his shoulder and pulled back the hammer to the half-cock position, which opened the pan by activating a spring that lifted the frizzen. He lowered the gun until the butt was grounded somewhat behind him (thereby pointing the barrel away from him) and steadied it with his left hand. Next, he reached into a cartridge box or pouch slung on his right hip or to his front and took out a paper roll about two inches long and three-quarters of an inch wide, sealed at both ends (with twine or paste), which contained powder and a ball. He bit off the soft end (American recruits were required to have at least two upper and two lower teeth which had to meet to enable biting)[11] and poured a little of the powder into the pan. He then poured the rest of the powder into the barrel and inserted the ball, shoving the remains of the paper roll in as wadding to prevent the ball from rolling out. He then slid out his rammer (an iron rod housed in a tube on the underside of the barrel), inverted it in order to place the trumpet-shaped end in the mouth of the barrel, and tamped down the contents of the barrel (the harder he did it, the better the result). He then withdrew the rammer, inverted it, and slid it back into its housing. He raised the weapon and nestled the butt into his shoulder. He closed the frizzen and pulled the hammer back to the full-cock. He was now ready to fire.

In parade ground conditions a well-trained soldier could fire four, maybe five, times a minute, but battle conditions were very different. There was the cacophony and chaos of officers shouting orders, muskets discharging as the historian Richard Holmes has evocatively described it "with the flat boom of a modern shotgun," and men screaming and huzzahing. At Cowpens, for example, the Americans almost lost the battle because the order for the right wing to "refuse" its flank (turn outward to repel a flanking attack) was misheard in the din as an order to turn about and retreat.

The stink of the smoke, the watering eyes, the adrenaline coursing through him, perhaps the unnervingly resonant thump of a one-ounce ball hitting the man next to him, the thirst created by fear, and a mouth fouled with gunpowder from the bitten cartridges were all powerful distractions. With shaking hand, the intricate coordination of placing the ramrod into the barrel mouth took much longer than it was supposed to. There was not enough time to replace the ramrod in its sleeve, so he simply stuck it into the earth beside him. He would regret it later, because earth would go into the barrel on the next loading, fouling it. Distracted, he might forget to retract the ramrod altogether and fire it. Sometimes he would not even bother with the whole paraphernalia of ramming. Instead, he emptied the contents of the cartridge into the barrel and banged the butt once or twice on the ground to settle powder and ball.

> In combat, distances seem foreshortened. When a person is in desperate straights, time seems to slow down; action seems to occur in slow motion. Other thoughts intrude as the musketeer, under the eyes of watchful sergeants, mechanically follows the manual of exercise that will guarantee his survival. During loading and firing, soldiers notice little increments of their task. The dry taste of black powder and waxed paper cartridges was one step. Then, a rattle of ramrods in the barrels as new charges of buck and ball were forced home against the breech plug with a distinctive ping. Platoon and division volleys crashed with bright yellow flashes from pan and barrel, highlighting the firing sequence. The blast of noise and light was so dramatic a soldier could not tell if his own musket fired. During priming, only a wisp of smoke coming out of the barrel's touch hole would show that the gun went off. The acrid smell of burnt powder, greasy black smears on hands and face from ramrods grown slick with sweat and powder residue, and cut thumbs from mishandling the musket's cock added to the individual's perceptions of the fight.[12]

The eighteenth-century and early nineteenth-century musket had a relatively low muzzle velocity (about 600–800 feet per second, compared with modern centerfire velocities of 2,400–4,000 feet per second) and an excessively parabolic trajectory at anything other than point-blank range. The ball dropped about five feet over 120 yards, and as a result "it was just possible for a good marksman to hit a man at 100 yards: a volley could be fired with some chance of obtaining hits on a mass of troops at 200 yards: but at 300 yards fire was completely ineffective and the bullet was no longer lethal."[13] The British major George Hanger, who fought in the War of Independence, wrote a classic account of the accuracy of a Brown Bess in 1814.

> *A soldier's musket, if not exceedingly ill-bored as many are, will strike the figure of a man at 80 yards . . . but a soldier must be very unfortunate indeed who shall be wounded by a common musket at 150 yards, provided his antagonist aims at him, and as for firing at 200 yards you might as well fire at the moon.*[14]

Accuracy trials of the black-powder period suffer from the distortion of parade-ground formalities rather than battleground realities and tend to err on the side of optimism. Trials in France in 1800 indicated 60 percent hits at 82 yards, 40 percent at 164 yards, and an unbelievable 20 percent hits at 328 yards.[15] William Müller, in his *Elements of the Science of War* (1811), stated that a well-trained soldier had a 53 percent chance of making a hit at 100 yards, 30 percent at 200 yards, and 23 percent at 300 yards. W. W. Greener, in *The Gun and Its Development* (1881), put the accuracy on a twenty-foot-by-six-foot target at 75 percent at 100 yards, 42 percent at 260 yards, and 16 percent at 300 yards. In 1782 the Prussian army tested musket fire against a target ten feet wide and six feet high. At 100 paces (about 80 yards) the soldiers achieved 60 percent hits, at 200 paces 40 percent, and at 300 paces 25 percent.[16]

If these hit rates are applied to a theoretical battle situation, they indicate casualty rates far higher than those actually experienced. For example, if two 500-man battalions were facing each other, one advancing, the other defending, by the time the advancing battalion had covered the final 100 yards (it would have taken about a minute to cover the ground in orderly fashion), it could reasonably have been expected to absorb two volleys from the defenders. (Highly experienced soldiers were expected to get off four discharges a minute, but in reality the fire rate was perhaps two or three.) In theory the attackers would have been totally destroyed.

In *Firepower: Weapons Effectiveness on the Battlefield, 1630–1850* Major General B. P. Hughes makes the reasonable assumption that about 25 percent of muskets (125 muskets in a 500-man battalion) would misfire. In addition, of the 375 successfully discharged muskets, Hughes calculates that "at least half of those that left the muzzle failed to reach their mark through a combination of inefficient distribution of fire, inaccurate aiming and, to the greatest extent of all, the eccentricity of the bullet." Hughes's calculation would result in 187 hits in one volley (37 percent of the total battalion). If the attackers had to endure two volleys, they would, theoretically, have lost 374 men, almost three-quarters of their strength, a massive casualty rate which finds no support in the actual rates of eighteenth-century warfare. Marshal Maurice de Saxe of France in his *Rêveries on the Art of War* (1757) was also skeptical about the lethality of the musket-volley: "Powder is not as terrible as believed. Few men in these affairs are killed from in front while fighting. I have seen whole salvoes fail to kill four men."[17] The eighteenth century was not alone in the inefficiency of its rounds-to-kill ratio. In Vietnam American troops fired 50,000 rounds for every enemy killed.[18]

The relative inefficiency of the musket imposed its own tactical boundaries when it came to battle. If individually aimed fire was unreliable, mass and density, numbers of muskets, and speed of firing and reloading would compensate. By the end of Marlborough's wars British infantry had "a sound appreciation of the supremacy of firepower over all other forms of combat . . . By the time of Blenheim a new fire-tactics—the famous platoon-fire system—had established itself as the forte of the British foot."[19] "Battles," declared Frederick the Great, "will be won through superior firepower . . . the infantry that can load the fastest will always defeat those who are slower to reload."[20] The emphasis on unified and coordinated bodies of musketeers delivering what the historian Robert L. O'Connell calls "corporate" fire[21] made the European model "crystalline": effective as long as its integrity and coherence could be maintained. The corollary, however, was fragility. Sometimes a blow against the tiniest flaw in crystal can cause it to shatter. An oblique blow could destroy an enemy's coherence—hence the emphasis on flank attack. The British center at Bemis Heights, for example, was almost destroyed by being knocked off kilter, and the Highlanders at Cowpens came close to collapsing the American right with enfilading fire from the right flank before they too lost their unity and were in their turn routed. In eighteenth-century warfare balance and timing, control and order, were critical.

The "corporate" tactics of European warfare in the eighteenth century were emulated by patriot America during the course of the war. The "individualized" tactics of the militia at the beginning of the conflict gave way to the corporate tactics of the Continentals (strongly influenced by the militarily most corporate of all European armies— Prussia—via Friedrich von Steuben's drill). This transition away from the individualistic and discrete toward the centralized and unified was true in both the military and the civil sphere. Washington, among others, saw it clearly and promoted it vigorously. If independence came from the muzzle of a musket, the weapon's own deficiencies in a profound way helped shape the nature of the independent state itself.

⊷ ⬥ ⬦ ⬥ ⊷

Much has been written about the long gun, the famed rifle of the American frontier sometimes known as the Pennsylvania-Kentucky rifle (though very few were made in Kentucky). Among the Scotch-Irish frontiersmen of North Carolina, it was known as the "Deckhard" rifle, after the Pennsylvania gunsmith Jacob Dickert of Lancaster, but it is not within the scope of this book to deal with the complex history of the American rifle. Very briefly, it was a refinement of the hunting rifle introduced to America by German and Swiss immigrants to Pennsylvania around 1710, and a weapon of considerable elegance, handcrafted to exacting specifications. Its spiral-grooved barrel bore was of a smaller caliber (0.55–0.60 inch) than the musket, though the octagonal barrel was somewhat longer than its smoothbore counterpart. Although about ten feet of muzzle velocity is gained for every extra inch of barrel length, the same can be achieved by adding more powder, so there was no particular ballistic benefit to the longer barrel; thirty inches would have delivered an equal perfomance. Fashion, balance, and convenience of loading when the butt was grounded seem to have been the governing factors.[22]

Its great advantage over the musket was range and accuracy. The rifle, in the hands of a skilled marksman, was accurate at over 300 yards, about three times the distance of the effective range of the musket. In 1775 six companies of riflemen drawn from Pennsylvania, Maryland, and Virginia were the first troops Congress sent up to George Washington at Boston. The process of selecting the riflemen was competitive and illustrates the impressive accuracy of the weapon and the skill of their owners. John Harrower, an indentured servant, watched just such a competition in Virginia. The captain of the company

> took a board of a foot square and with chalk drew the shape of a moderate nose in the center and nailed it up to a tree at one hundred and fifty yards distance, and those who came nighest the mark with a single ball was to go. But by the first forty or fifty that fired, the nose was all blown out of the board, and by the time his company was up, the board shared the same fate.[23]

For any British soldier, being the object of a rifleman's attention could be unnerving, to put it mildly, especially for officers, who were particularly sought out. At the battle of Brooklyn a British officer sought refuge in a house declaring "he'd be d——d if he was going to expose himself to that fire, that the d——d rascals picked out all the officers."[24] Major George Hanger, himself one of the most accomplished marksmen in the British army, was reconnoitering with Banastre Tarleton (he was Tarleton's second-in-command), when a rifleman fired at them.

> Colonel, now General Tarleton, and myself were standing a few yards out of a wood, observing the situation of a part of the enemy which we intended to attack . . . our orderly bugler stood behind us about three yards . . . A [American] rifleman passed over the milldam . . . and laid himself down on his belly; for in such positions, they always lie, to take a good shot at a long distance. He took a deliberate and cool shot at my friend, at me, and at the bugle-horn man. Now observe how well this fellow shot. It was in the month of August, and not a breath of wind was stirring. Colonel

Tarleton's horse and mine, I am certain, were not anything like two
feet apart; for we were in close consultation . . . a rifle-ball passed
between him and me . . . I directly said to my friend, "I think we
had better move, or we shall have two or three of these gentlemen
shortly amusing themselves at our expense." The words were hardly
out of my mouth when the bugle-horn man behind me, and directly
central, jumped off his horse and said, "Sir, my horse is shot." The
horse staggered, fell down, and died . . . I can positively assert that
the distance he fired from at us was full 400 yards.[25]

The disadvantages of the rifle as far as eighteenth-century warfare
was concerned were several. It took longer to load than the musket and
left the rifleman vulnerable to an attacking enemy during the extended
loading period. The problem with loading was, ironically, a by-product
of the rifle's accuracy. The ball had to be thumped down with a rammer
against the spiral grooves inside the bore, and it was the grooving that
imparted to the ball the spin that made it more accurate. The rifleman
did not have recourse to the preprepared cartridge. He took a pinch of
powder and primed the pan. He then took a powder horn and poured
some powder down the barrel. From a recessed compartment in the
butt of the rifle (the "patch box," which was closed with a movable
plate, usually of brass) he took a greased patch of linen or paper,
centered it over the mouth of the barrel and placed a ball on it, and then
rammed patch and ball down against the grooving. The greased patch
helped ease the passage of the one-half-ounce lead ball and reduced the
windage (because there was less escape of explosive gases, a rifle ball
exited at 1,800–2,000 feet per second, compared with about 900 feet per
second for the musket),[26] but it took time, and in that time the rifleman
was vulnerable to counterattack. Another disadvantage was the lack of
a bayonet, either for defense or for pressing the attack.

The main tactical problem faced by riflemen was, then, something
they shared with musketmen: timing. Reload time was the Achilles' heel.
The Hessian colonel Von Heeringen described the patriot rifleman's
dilemma at the battle of Brooklyn in 1776: "These terrible men deserve
more pity than fear—they want nearly fifteen minutes for loading

their pieces, and during that time they feel our balls and bayonets."[27] Major George Hanger described the tactic his friend Colonel Sir Robert Abercrombie had used against Morgan's riflemen in Pennsylvania: "They [the riflemen] marched to attack our light infantry ... The moment they appeared before him he ordered his light troops to charge them with the bayonet; not a man out of four had time to fire, and those that did had no time given them to load again."[28] Morgan himself recognized their vulnerability. At the battles of Saratoga, he said, "My Riflemen would have been of little service if we had not always had a line of Musquet and Bayonette men to support us, it is this that gives them confidence. They know if the enemy charges them they have a place to retreat to."[29] Another way of dealing with the problem was to double up so that the man reloading was covered by another who was ready to fire. At Cowpens, American riflemen were ordered to have every third man fire while two were kept in reserve.[30] At the small but bloody battle of Oriskany (6 August 1777), the militia commander Nicholas Herkimer, ambushed by Indians and Loyalists while on his way to relieve Fort Stanwix in northwestern New York State, ordered his men to double up—one firing, the other ready to dispatch the Indians who rushed in expecting to make a kill during reloading.

The frontier rifleman held, and still does hold, a special place in the history of the war. Even solid and respectable historians can become misty-eyed: "Out of the west came tall men," says one, as though John Wayne clones were striding out of the setting sun. George Washington was not so enamored. He detested their individuality, their resistance to discipline, their irrelevance to the formalized warfare he wanted to adopt. They may have been useful as skirmishers and snipers, but Washington and Congress wanted dense formations of muskets. They believed fervently that it would be mass that would win the war, not individual marksmen. When the Maryland assembly proposed sending a corps of riflemen to Washington's army outside of Philadelphia in 1776, Congress gave it the cold shoulder.

If muskets were given them instead of rifles the service would be more benefitted, as there is a superabundance of riflemen in the

Army. Were it in the power of Congress to supply muskets they would speedily reduce the number of rifles and replace them with the former, as they are more easily kept in order, can be fired oftener and have the advantage of Bayonetts.[31]

It was a view that found sympathetic echo among many field commanders. General Peter Muhlenberg wrote to Washington on 23 February 1777: "The whole regiment consists at present of riflemen; and the campaign we have made to the southward last summer fully convinced me that on the march, where soldiers are without tents, and their arms continually exposed to the weather, rifles are of little use. I would therefore request your Excellency to convert my regiment into musketry."[32] Washington agreed. In 1778 General Anthony Wayne left no doubt whatsoever about his choice of firearm.

I don't like rifles—I would almost as soon face an Enemy with a good Musket and Bayonet without ammunition—as with ammunition without a Bayonet; for altho' there are not many instances of bloody bayonets yet I am Confident that one bayonet keeps off an Other . . . The enemy knowing the Defenseless state of our Riflemen rush on—they fly—mix with or pass thro' the Other Troops and communicate fears that is ever Incident to a retiring Corps—this Would not be the Case if the Riflemen had bayonets—but it would be still better if good muskets and bayonets were put in the hands of good Marksmen and rifles entirely laid aside. For my own part, I never wish to see one—at least without a bayonet.[33]

Riflemen did not fit the standardized mold of a national army. General John Thomas wrote of them from Boston in 1775: "as Indifferent men as I ever served with, their Privates mutinous and often Deserting to the Enemy, unwilling for Duty of any kind, exceedingly vitious and I think the army here would be as well without them as with them."[34] They seemed to invite trouble. Israel Trask, a cook and messenger, recorded a fracas at the Cambridge camp in the fall of 1775.

A rifle corps came into camp from Virginia, made up from recruits from the backwoods and mountains of that state, in a uniform dress totally different from that of the regiments raised on the seaboard and interior of New England. Their white linen frocks, ruffled and fringed, excited the curiosity of the whole army, particularly to the Marblehead regiment [Rhode Island] . . . Their first manifestations were ridicule and derision, which the riflemen bore with more patience than was their wont, but resort being made to snow, which then covered the ground, these soft missives were interchanged but a few minutes before both parties closed, and a fierce struggle commenced with biting and gouging on the one part, and knockdown on the other part with as much deadly fury as the most deadly enmity could create. Reinforced by their friends in less than five minutes more than a thousand combatants were on the field.[35]

Apart from the German jäger rifle corps, the British army put its faith in massed muskets. But one British innovation is interesting: Major Patrick Ferguson's breech-loading rifle. Although breech-loading and repeating firearms had been known, and manufactured, in America and Britain since the late seventeenth century,[36] they were intricate and therefore very expensive to make. Ferguson's invention was simpler and more efficient and seems to have been the only breech-loader to have been made in any quantity, or any rifle to have a bayonet. Ferguson himself demonstrated it at the Royal Arsenal at Woolwich on 1 June 1776 in a performance that "astonished all beholders."

Notwithstanding a heavy rain and the high wind, he fired during the space of four or five minutes at the rate of four shots per minute, at a target two hundred yards distance. He next fired six shots in one minute, and also fired (while advancing at the rate of four miles an hour) four times in a minute. He then poured a bottle of water into the pan and barrel . . . and in less than half a minute he fired with it as well as ever . . . Lastly, he hit the bull's eye lying on his back . . . He only missed the target three times during the whole course of the experiments.[37]

Only 200 or so of the rifles were issued, and although they had little impact on British tactics at the time, they were "the first standard military weapon capable of delivering accurate, aimed fire.[38] Sturdier and more powerful than the Pennsylvania-Kentucky rifle, they had the considerable additional advantages of being made of standardized parts and being able to carry a bayonet. The reasons that Ferguson's rifle did not become standard issue were several: at £4 it was twice as expensive as a Brown Bess to manufacture; there was weakness in the stock where it had to be cut away to accommodate part of the specialized mechanism; it could not take a regulation cartridge; and it required operating skills beyond those of standardized training. Nevertheless, it would point the way to the future of infantry warfare. Ironically, Ferguson, no longer in command of his rifle corps and not armed with his own rifle, was cut down at Kings Mountain on 7 October 1780, riddled with balls from patriot rifles.

Whether it was musket or rifle, each depended on gunpowder. It was as precious as gold and, for the patriot cause, almost as rare. Gunpowder is the mixture of saltpeter (potassium nitrate), charcoal, and sulfur (in the ratio 75:15:10). Potassium nitrate is a by-product of the bacterial decay of organic matter, particularly dung and urine. A sixteenth-century treatise particularly approved of urine. Especially "of those persons whiche drink either wyne or strong beers. Then Dong specially of those of horses, which be fed with oates, and be always kept in the stables."[39] It was fitting that even puissant Mars had his feet firmly planted in the ripe ordure of the eighteenth century.

Saltpeter forms naturally in warm climates that have a regular dry season that evaporates the solution (usually rainwater) in which the nitrate has been suspended and turns it into salts deposited on the soil surface.

America, although having some of the climatic preconditions, especially in the South, did not have a developed gunpowder industry and relied almost entirely on imports, either from Europe or from the French and Dutch West Indies, particularly Martinique, Santo Domingo, and St. Eustatius. The stocks of powder available to Washington at the start of the war were so neglible (about one half pound per man) that, according to Brigadier General John Sullivan, when the commander in chief discovered the state of affairs "he did not utter a word for half an hour." By the end of 1775 he was still troubled by their "want of powder [which] is inconceivable. A daily waste and no supply administers a gloomy prospect." By the end of January 1776 the army was practically without powder. Had Howe known, "he could have marched out to Cambridge and crushed the newly recruited colonial army."[40]

British troops were supplied with sixty cartridges per man, and to match that, Washington would have needed four hundred barrels of powder. He had thirty-eight barrels, or fifteen to twenty rounds per man.[41] John Adams was acutely aware of the crisis, writing to James Warren in October 1775: "We must bend our attention to salt petre. We must make it. While Britain is Mistress of the Sea and has so much influence with foreign courts we cannot depend upon a supply from abroad. It is certain it can be made here . . . A gentleman in Maryland made some last June from tobacco house earth . . . that the process is so simple a child can make it. It consists of nothing but making a lixivium from the earth which is impregnated with it, and then evaporating the lixivium . . . I am determined never to have salt petre out of mind, but to insert some stroke or other about it in every letter for the future. It must be had."[42]

The aptly named Joseph Barrell, who had made saltpeter, explained its manufacture in November 1775.

> Take the earth from under old houses, Barns, &c., & put it lightly into a hogshead or Barrel; and then fill it with water, wch immediately forms a lie. This lie he then puts into an ashes leach that has all the goodness extracted before, this being only as a strainer. After it is run thro' w^{ch}, he boils the Lie so clarified to a certain Consistance, & then puts it to cool, when the saltpetre forms, & is immediately fit for use; & from every Bushel of earth he produces ¾ lb. of saltpeter.[43]

Congress jump-started domestic production, particularly in the southern states because the earth on the floors of tobacco warehouses was rich in nitrate, and by the fall of 1777 115,000 pounds had been harvested, mostly between January and November 1776 when "hard money" was still available and revolutionary passions still burned bright. Despite these attempts to stimulate home-produced gunpowder, imports still accounted for 90 percent of America's needs (most of them from France via the West Indies), and, as Adams implied, if the British navy had managed to impose a truly effective blockade, the rebellion may well have fizzled out.

Black powder was susceptible to damp, despite attempts to make it damp-resistant. (The mixture of saltpeter, charcoal, and sulfur was moistened and formed into round cakes, dried, and stone-ground into granules. The granules, in turn, were polished in tumblers together with graphite, which gave them a slight water-resistant glaze.) And damp powder played a major role in determining tactics. Torrential rain, for example, forced Washington to give up on his plan of 16 September 1777 to attack the British converging on Philadelphia. Henry Knox wrote to his wife: "We came in sight of the enemy, and drew up in order of battle . . . but a most violent rain coming on obliged us to change our position, in the course of which nearly all the musket cartridges of the army that had been delivered to the men were damaged, consisting of about 400,000. This was a most terrible stroke to us . . . This unfortunate event forced us to retire."[44]

On the march, soldiers tried to protect the locks of their weapons under the "lappets" (lapels) of their coats. Cartridges were protected to some extent in leather-covered "cartouche" (cartridge) boxes that contained wooden blocks with holes drilled out to receive the paper cartridges (usually fifteen to twenty-four per box). A leather flap, sometimes with an inner one as extra protection, closed the top of the box and, in theory, kept out the rain. Cartridge boxes were crucial to the soldier's fighting capability, and Washington took them very seriously. On 3 October 1777 he wrote to Congress:

> With respect to Cartouch boxes, without which it is impossible
> to act . . . I would advise that much care should be used in
> choosing the Leather. None but the thickest and best is proper for

the purpose, and each Box should have a small inner flap for the greatest security of the Cartridges against rain and moist weather. The Flaps in general, are too small and do not project sufficiently over the ends or sides of the Boxes. I am convinced of the utility nay necessity of these improvements.[45]

This complaint would have found its echo in the British army. Cartridge boxes were the responsibility of a regiment's colonel, who, with an eye to economy, would often buy inferior goods. Only half of the 33rd Foot, for instance, had any cartridge boxes at all in 1775. The 42nd Foot fought the whole of the War of Independence with one issue, which, by 1783, was "wore out on Actual Service."[46] Sometimes the holes in the blocks were too small to accommodate the cartridges, or the leather was cheap and quickly rotted away. The box could not have held all cartridges issued (the British, for example, regularly issued sixty per man), and men would have to improvise. The American general John Ashe described how his men carried their ammunition on their way to the battle of Briar Creek, Georgia, in March 1779: "We immediately beat to arms, forming the troops into two lines and served them with cartridges, which they could not have prudently been served with sooner, as they had several times received cartridges which had been destroyed and lost for want of cartouche boxes. We marched out of lines to meet the enemy, some carrying their cartridges under their arms, others in the bosom of their shirts."[47] Riflemen, ever individualists, were known to carry lead balls in their mouths, as did Private James Collins on his way to the battle of Kings Mountain, South Carolina, in October 1780, "to prevent thirst, [and] also to be in readiness to reload quick."

The balls themselves were cast in individual molds made of brass, iron, soapstone, or, occasionally, wood—one for approximately forty soldiers. "Gang molds" that could produce multiple balls were preferred by musketeers, although riflemen tended to have an individual mold that best suited the exact gauge of their weapon. The process was relatively crude. Molten lead was poured into the top of a hinged mold and ran through small channels into a series of spherical cavities formed when the two halves of the mold were brought together. When the mold cooled, it was opened and the balls fell out. "Double-shotted" muskets were often used; they were loaded not only with the large ball but with 0.30-caliber buckshot (usually three or four in addition to the main ball), which could be lethal at ranges under fifty yards. Occasionally, muskets were loaded with "modified" or "dum-dum" balls, which then, as now, caused outrage because they were altered to cause grotesque wounds. They were considered beyond the pale of the rules of war. General William Howe had cause to complain to Washington in 1776: "My Aid de Camp charged with the delivering of this Letter will present to you a Ball cut and fixed to the Ends of a Nail, taken from a number of the same Kind, found in the Encampments quitted by your Troops . . . I do not make any comments upon such unwarrantable and malicious Practices, being well assured the Contrivance has not come to your Knowledge."[48] Howe, likewise, was ignorant of similar practices in the British infantry.

Nothing represents the bloody intimacy of the eighteenth-century battlefield more than the bayonet, and no weapon was so powerfully invested with mystique, fascination, and horror. Its impact on tactics was profound, though often more in the anticipation than the execution. From the introduction of the matchlock on the sixteenth-century battlefield, shooters were most vulnerable during the elaborate procedure of reloading. In the early years of firearms, pikemen took care of their colleagues by fending off enemy cavalry or counterattacking infantry. But, as is the way with most organizations, if separate functions can be combined, efficiencies result. During the 1640s–60s the musketman began to be given the wherewithal for his own defense. Bayonets were developed that "plugged" into the barrel—good for fending off unfriendlies but bad for shooting the same. In the British army the Tangiers Regiment was the first to be issued plug bayonets, in 1663,[49] and by the late 1690s the plug had been replaced throughout the army by the "socket" version, which, because it was attached to the barrel's exterior with a simple ring mechanism, enabled the soldier to fire and stab at the same time.

Bayonet lengths varied slightly between the combatants of the American war, but they were usually in the range of fourteen to seventeen inches of blade, usually triangular in section, and slung in either a waist belt or a shoulder belt scabbard when not in use. Some historians have suggested that a shorter blade made loading less hazardous, but it is difficult to see the logic. In fact, the opposite would seem to be true; the longer the blade, the less likely the soldier was to kebab his hand, as the point would be farther from the muzzle.

The bayonet was important to eighteenth-century infantry tactics; but how important? There is some confusion. Frederick the Great, for example, could declare that firepower was the deciding factor and then advise commanders to "fire as little as possible with the infantry in battle: charge with the bayonet."[50] A French ordinance of 1755 emphasized that infantrymen were expected to fight with the bayonet without shooting.[51] There are two schools of thought. One contends that bayonet fighting was so intense, so deeply and instinctively revolting, that it was extremely effective as threat but not important in lethality. The evidence is drawn from wars across centuries. Napoleon's surgeon in chief, Dominique-Jean Larrey, for example, encountered only five cases of bayonet wounds throughout his career. Sidney George Fisher interviewed a Civil War veteran: "I said it seemed to me that the most terrible thing in battle must be a charge of bayonets, that a confused melee of furious men armed with such weapons, stabbing each other & fighting hand to hand in a mass of hundreds, was something shocking even to think of. He said it was so shocking that it very rarely happened that bayonets are crossed, one side always giving way before meeting."[52]

Close-quarters combat was so rare that when it did happen, it was worthy of note. A Hessian officer at the battle of Bennington (16 August 1777) remarked that "men fell, as they rarely fall in modern warfare, under the direct blows of their enemies."[53] Throughout military history, asserts Lieutenant Colonel Dave Grossman, "bayonet combat is extremely rare . . . wound statistics from nearly two centuries of battles indicate that what is revealed here is a basic, profound, and universal insight into human nature. First, the closer the soldier draws to his enemy the harder it is to kill him, until at bayonet range it becomes extremely difficult."[54]

The truly dangerous time for the infantry was mostly when it broke and ran. Fleeing seemed to excite the homicidal instinct of the pursuer, and the bayonet was used mainly against the already defeated. Jabez Fitch, a Continental lieutenant, described the aftermath of the battle of Brooklyn: "Capt: Jewett had Recd: two Wounds with a Bayonet after he was taken, & Strip'd of his Arms & part of his Clothes, one in the Brest & the other in the Belly, of which he Languished with great pain until the Thursday following when he Died; Sargt: Graves was also Stab'd in the Thigh with a Bayonet, after he was taken."[55] As the British captain Frederick Mackenzie chillingly put it after the fall of Fort Washington in November 1776: "Very few prisoners have been taken. The soldiers generally make use of their bayonets."

On the other side of the coin, commanders on both sides often exhorted their troops to use the bayonet rather than shoot. Before the British amphibious assault on Manhattan at Kips Bay (15 September 1776), General Howe "recommends to the troops an entire dependence upon their bayonets, with which they will ever command that success which their bravery so well deserves." Before the attack on Trenton on 25–26 December 1776, one of Washington's aides, Colonel John Fitzgerald, recorded Washington as saying, "Tell General Sullivan to use the bayonet. I am resolved to take Trenton." The American advance guard at Germantown on 4 October 1777 was ordered to take out the British pickets with "charged bayonets without firing." Burgoyne, after his defeat at Freeman's Farm on 19 September 1777, noted that "the impetuosity and uncertain aim of the British Troops in giving their fire, and the mistake they are still under, in preferring it to the Baynotte [sic] is much to be lamented."

When rain disabled muskets, as it did to both armies as they maneuvered after the battle of Brandywine in September 1777, the bayonet came into its own. Washington, however, although spoiling for a fight, was in a bind. As the Continental artillery lieutenant Samuel Shaw noted, "Our army was formed in order of battle; but rain, coming on very fast, the General filed off, choosing to avoid an action which the discipline of the enemy in the use of their bayonets (the only weapon that could then be of any service, and which we were by no means generally supplied with) would give them too great a superiority."[56] This, it seems, confirms the general view that the British soldier was skilled with the bayonet.[57] Others disagree. The historian J. A. Houlding's benchmark survey of training in the British army of the eighteenth century argues that the redcoat was decidedly ill-prepared for bayonet warfare.

> *Bayonet drill was, curiously, rather neglected in the eighteenth century. Buried away in the manual exercise, it was performed in the most unrealistic fashion . . . Not until 1805, when Capt. Anthony Gordon published his systematic* Science of Defence, for the Sword, Bayonet, and Pike, in Close Action, *was the subject treated in depth by anyone. Many writers lamented this lapse . . . Bennett Cuthbertson, writing in 1768 . . . deplored the fact that British troops in training were seldom even "permitted to fix them." . . . From Marlborough's campaigns onwards it was the touchstone of British tactical thinking that heavy fire was all important.*[58]

However, three significant engagements of the American war were carried out almost entirely at bayonet point. The first was a nighttime British attack on General Anthony Wayne's sleeping encampment at Paoli, Pennsylvania, on 21 September 1777. The British commander of the attack, General Charles Grey, had an aide, the ill-fated Captain John André (who would later die on the gallows, involved in Benedict Arnold's treason and condemned as a spy), who kept a diary.

> *No soldier . . . was suffered to load; those who could not draw their pieces [extract the charge] took out the flints . . . On approaching the right of the Camp we perceived the lines of fires, and the Light Infantry being ordered to form to the front, rushed along the line putting to the bayonet all they came up with, and, overtaking the main herd of the fugitives, stabbed great numbers and pressed on their rear till it was thought prudent to order them to desist. Near 200 must have been killed.*[59]

Of Wayne's 1,500 men, 13 percent (200) were killed and another 7 percent (100) wounded. (This was a shocking inversion of the usual ratio of killed to wounded; in most battles it was about three wounded to every man killed.) At Stony Point, on the west bank of the Hudson in New York, on 16 July 1779 Wayne took some measure of revenge during a night attack against the British garrison of about 600 men. "The whole business," reported Nathanael Greene, "was done with fixed bayonets." And done quickly—as the patriot light infantry "rushed on." The British lost 63 killed, 71 wounded, and 442 captured. Almost exactly one month later the Americans did it again, this time to the British garrison at Paulus Hook, New Jersey: "We gained their works," reported Captain Levin Handy, "and put about fifty of them to the bayonet."

The key factors that tactically linked all three actions were speed and surprise. A bayonet attack against well-prepared troops was an invitation to disaster, as Major General James Abercrombie discovered when he tried to take the French fort at Ticonderoga in July 1758 by bayonet assault: the "result was carnage." British troops would also learn the same lesson at, for example, Bunker's Hill and Guilford Courthouse. The "bayonet could be the final arbiter in an action . . . but it was generally impossible to assault unshaken infantry with the bayonet alone. Firearms had to reduce the cohesion of the defending ranks first and shake the resolution of the men forming them."[60]

If bayonets were an evolution of the pike, and the pike of the tribal spear, they illustrate how eighteenth-century warfare was a transitional halfway house between the feudal and the industrial. "Pole arms," spearlike weapons, were still used by both armies. Sergeants carried halberds, seven- or eight-foot-long pikes with elaborate spear-cum-ax points that, by the middle of the eighteenth century, were more ornamental designations of rank than fighting weapons. By the end of the French and Indian War the British army had begun to abandon them, favoring swords and short bayonet-equipped muskets (fusils) for its sergeants, but they were still in use at the battle of Waterloo.[61] It is often claimed, however, that halberds did have a practical purpose in controlling volley fire by "tapping-up" the barrels of muskets that might be sagging and therefore firing too low. Again, it is difficult to imagine how this could have been done effectively. Sergeants stood on the ends of firing lines of perhaps twenty men and could have reached only the three or four men nearest them.

The battalion officers' spontoon, a six-foot "half pike," still retained a practical martial purpose. Although the British army tended to replace spontoons with fusils during the war, Washington was particularly keen on them as arms à l'outrance, and it is interesting that he shared this enthusiasm with Anthony Wayne, one of the most aggressive officers in the Continental army and one of the foremost proponents of the bayonet. In December 1777 Washington issued a general order to his officers: "As . . . the officers derive great confidence from being armed in the time of action, the General orders every one of them to provide himself with a half pike or spear." Fusils, he felt, were too distracting. The spontoon, on the other hand, was a weapon of "involvement," occasionally in ways that could be quite spectacular if unconventional. At the battle of Cowpens the commander of General Morgan's main line, Lieutenant Colonel John Eager Howard, saw a chance to capture a piece of British artillery: "[I] called to Captain Ewing who was near me to take it. Captain Anderson, hearing the order, also pushed for the same object; and both being emulous for the prize kept pace until near the first piece, when Anderson, by putting the end of his spontoon forward into the ground, made a long leap which brought him upon the gun and gave him the honor of the prize."[62] Pole-vaulting to glory!

In a fascinating echo of the time when history's first musketeers had been defended by pikemen, Washington ordered Morgan's corps of riflemen to arm themselves with spears. On 13 June 1777 he wrote Morgan, "I have sent for Spears, which I expect shortly to receive and deliver you, as a defense against Horse."[63] He had very specific design criteria, which included "a spike in the butt end to fix them in the ground" as a sort of chevaux de frises against cavalry. It was an ancient, but still useful, notion.

Hatchets and tomahawks, standard issue for boarding parties in the Royal Navy, were also used in the Continental army, particularly by riflemen, but as the war progressed most infantrymen preferred the bayonet to the tomahawk. Swords were not carried by infantry privates on either side. They were officially discontinued for British privates in 1768, but the trend was already in motion during the French and Indian War. The only exception was the Scottish Highland regiments, but even they gave up their claymores for bayonets after the battle of Brooklyn in 1776. In the infantry swords were primarily weapons of the officer and NCO class. Rankers in the artillery retained them, and, of course, they were essential to the cavalry. Infantry officers tended to use short swords for close-quarter fighting. Some were saberlike "hangars," others rapierlike and as much a status symbol as an effective battlefield weapon.

The saber was *the* primary weapon of the cavalry. It is true that cavalrymen also had a variety of firearms—pistols, musketoons, and carbines—but the blade was paramount, as an experienced cavalryman, Captain Epaphras Hoyt of Massachusetts, stressed.

> *It is generally agreed by experienced officers, that fire arms are seldom of any great utility to cavalry in an engagement, while they are drawn up in regiments, squadrons, or other considerable bodies: Indeed there is little hope of success from any who begin their attack with the fire of carbines or pistols . . . It is by the right use of the sword they are to expect victory: This is indisputably the most formidable and essentially useful weapon of cavalry: Nothing decides an engagement sooner than charging briskly with the weapon in hand.*[64]

Heft was important, and cavalry swords, whether of the curved-blade saber type used primarily for slashing, or the straight-bladed broadsword for piercing, were heavy and long—the blades about three feet.

The history of uniforms in this period is complex and tends to attract those with the obsessive passion of bird-watchers or train-spotters. The minutiae of uniforms—the different regimental facings and other myriad details that are so fascinating (to some)—is not an area that can be adequately dealt with here. But there is a broader context to the uniform that makes it highly relevant to the central purpose of this book—how the war was fought. Uniform is the public proclamation of identity and allegiance, to both regiment and nation.

The British red coat (which gave the British soldiers the soubriquet of "lobsterback") had been instituted in 1660 and was not to leave the battlefield until 1882. It was the national "corporate logo," and arrayed beneath it were subordinate brands—the regiments with their facing colors (the contrast color of the lapel and cuff), connected to the mother brand but differentiated. It was both practical and symbolic, and it was well established. Patriot America, on the other hand, was betwixt and between. Colonial militias and provincial regiments had a tradition of uniforms, but at a national level, of course, there was none at the beginning of the war. And so what evolved was a meshing together of the individual and national. The clothing of the soldiers symbolized a society in transition: aspiring to a national uniform but having to make do with homespun or a uniform sent to it by a foreign state, France. American soldiers dressed accordingly. Some wore civilian clothes, while some, not as many as Washington would have liked, sported "corporate" colors. In the early stages of the war men and officers were often indistinguishable in dress. Captain Alexander Graydon wrote of the patriot army in New York in 1776: "I speak particularly of the officers, who are in no single respect distinguishable from their men, other than the coloured cockades, which, for this very purpose, had been prescribed in general orders; a different colour being assigned to the officers of each grade."[65]

Washington, preoccupied as he always was with the need to stamp a national identity on his army, at first had to work with what was · available. Woolens were in short supply because they had always been imported from Britain, and the trade had collapsed. In 1774 £650,000 worth of woolen goods came into the colonies. By 1776 the amount had fallen to a mere £2,500.[66] The linen hunting shirt of the riflemen offered an ad hoc solution, and Washington, although not a great admirer of the freebooting frontiersmen, thought it might make for an interim national uniform. The shortage of woolens, though, had drained the market of linen "tow cloth," and Washington's plans were thwarted. Nevertheless, his instinctive grasp of the psychology of uniform was right on target. The rifleman's hunting garb carried with it a message the British in Boston well understood. Riflemen, expert snipers, were the "widow makers," and Washington knew hunting shirts would put the fear of God into the redcoats: "It is a dress which is supposed to carry no small terror to the enemy, who think every such person a complete marksman."

This use of uniform for its demoralizing impact on an enemy is, of course, ancient. Elaborate face and body painting, extravagant headgear to emphasize height, gaudy dress to overawe—all were echoed in eighteenth-century warfare. Patriot and Loyalist irregulars often adopted Indian dress and body painting. The grenadier's tall cap exaggerated his height. Silas Deane recorded of American militia: "Their coat is made short, falling but little below the waistband of the breeches, which shows the size of a man to great advantage." The flashiness of the uniform (about as far removed from the drab camouflage of today's soldiers as could be imagined) was designed not only to inspire fear in the enemy but also to instill confidence in the soldier wearing it. To look out on rank after rank of men dressed in the same color is to recognize, symbolically, a unified enemy, and one with a disconcertingly common goal: your destruction. The language is subtle: all in the nuance. For example, the British officer's gorget (the small metal plate suspended from a chain worn around the neck) was vestigial and purely symbolic (a hangover from medieval armor). It had no function other than to send the message that here was a knight, superior in arms and honor.

Unlike the British army, the American did not have a unified uniform at any point in the war; the blue-and-buff of popular imagination was never truly universal. Individual states dressed their men in state colors. George Washington himself wore the blue-and-buff of his prewar Virginia colonelcy. At the beginning of the war there was a complete mishmash of "uniform." Many men wore civilian clothes; others came in their militia uniforms, and yet others in old uniforms dating back to the French and Indian War. It pained Washington to see his men in motley, for one very practical reason: "It creates much irregularity; for when a soldier is convinced that he will be known by his dress to what Corps he belongs, he is hindered from committing many faults for fear of detection," wrote Washington in December

1776.[67] In the fall of 1775 Congress chose brown as the original color for the national uniform. In October 1778, however, France sent uniforms with a predominance of blue coats (although there were also brown), and so on 2 October 1779 Congress switched to blue as the official color. Even so, there was much variation among the states, and there were demoralizing shortages throughout the war. For example, the return of August 1781 for a brigade stationed in the High Hills of the Santee in South Carolina showed 367 men fit for duty. They had between them thirty-seven coats, of which thirty-four needed repair. Just over a third of the men had shoes. There were 198 blankets.[68]

Uniform has, broadly speaking, three functions. The first and most basic is to clothe; the second is to identify; the third is to impress. Clothing as a purely practical issue was an important incentive for enlistees to the Continental army, and it was built into the bounty process. In 1776 a clothing bounty of $20 (rising to $48 in 1777) was offered. This was to provide two linen hunting shirts, two pairs of overalls, a waistcoat with sleeves, a pair of breeches, a hat, two shirts, two pairs of stockings, and two pairs of shoes.[69] The shortages of materials, however, soon resulted in serious clothing arrears, and throughout much of the war America staggered from one crisis to the next. Transportation was a problem (particularly during those periods when the British held the major ports of Philadelphia and New York), and the misappropriation of Continental clothing by individual states for use by their troops was a constant irritant. Simple theft also took its considerable toll.

As Washington's defeated army retreated across New Jersey in the fall of 1776, it was ragged. The march took a particular toll on shoes: "Our soldiers had *no* shoes to wair; was obliged to lace on their feet the hide of the cattle we had kill'd the day before,"[70] said Sergeant John Smith. At Trenton at the end of the year many soldiers were literally shoeless and in rags. A year later, at Valley Forge, von Steuben recorded, "The men were literally naked . . . The officers who had coats had them in every color and make. I saw officers . . . mounting guard in a sort of dressing gown made of an old blanket or woollen bed-cover."[71]

Footwear played a crucial role in the soldier's ability to fight, as a British manual of 1758 emphasized.

*Two pair of good shoes are indispensably necessary for a Soldier,
as he must otherwise be obliged (if depending on one pair) after
a wet day's march, to give them a hasty drying by the fire which
not only cracks the leather, but is the certain method of shrinking
them in such a manner, as to give the greatest pain and trouble to
the wearer: the best shoes will always be found the cheapest [in
the long run], and it will be necessary to strengthen the heels, with
some small nails: the toes should be round and flat: the straps full
large enough to fill the buckle.*[72]

The Hide Department was created by Congress to be responsible for
procuring leather for the soldiers' shoes. It was both corrupt and inept.
Washington knew how important shoes were to the fighting potential
of his army, and bitterly complained to the Board of Trade against
the department's cheating "by putting in small scraps and parings of
Leather and giving the Shoes the appearance of strength and substance,
while the Soals were worth nothing and would not last more than a day
or two's march."

As America struggled to clothe its soldiers, the diversity of uniforms at
times made identification difficult. In the early phase of the war the choice
of uniform colors was left to the discretion of regimental commanders,
not always with satisfactory results. The colonel of one cavalry regiment
chose an almost exact replica of his British counterpart, the Queen's
Dragoons, and Washington literally had to cover up the potential
confusion by having the Americans hide their scarlet uniforms under
"frocks" (hunting smocks).[73] The distinctiveness of uniform could bring
with it its own liabilities. British officers' marks of distinction, particularly
epaulettes, invited unwelcome attention from patriot marksmen. The
American captain John Chester, on his way to the battle of Bunker's Hill,
also knew the perils of uniform as identification. On 22 July he wrote to
a friend, "We soon marched, with our frocks and trowsers on over our
clothes (for our company is in uniform wholly blue, turned up with red),
for we were loath to expose ourselves by our dress."[74]

At the beginning of the war, when regular uniforms for patriot
troops were in particularly short supply, occasionally the only means

of identification was a simple piece of paper or cockade attached to the hat. Just before Montgomery and Arnold's attack on Quebec in 1775 the surgeon Isaac Senter recorded, "To discriminate our troops from the enemy in action, they were ordered each officer and soldier to make fast a piece of white paper across their caps from the front to the acmé of them."[75] Misidentification could be disastrous. The Loyalist lieutenant colonel John Simcoe described how he bagged a patriot militia unit who "were deceived by the dress of the Rangers [Queen's Rangers] and came to Lt. Col. Simcoe, who immediately reprimanded them for not coming sooner, held conversation with them and sent them prisoners to General Arnold."[76] The British lieutenant William Digby recorded a close call at Freeman's Farm in 1777: "On its turning dusk we were near firing on a body of our Germans, mistaking their dark clothing for that of the enemy."[77] Confusion over uniform could be lethal, as the Loyalist cavalry commander Colonel John Pyle discovered on 18 February 1781 when his detachment came upon a cavalry unit on the same road. It was, in fact, a patriot force led by Henry "Light Horse" Lee. Pyle mistook Lee for the British cavalry leader Banastre Tarleton, since both Lee and Tarleton's men wore short green coats. Playing the part, Lee invited Pyle to draw up his men by the side of the road so that he could pass. Pyle complied, and when the patriot force was level with the misguided Loyalists, it wheeled on them, slashing and shooting them from their saddles: "There was no fight; it was simply a massacre."[78] The anecdote is a striking illustration of mistaken identity, but it may be simply Loyalist propaganda.

Uniforms brought other much more common but equally fatal perils. They were unhealthy in a number of ways. Given the general absence of cleaning facilities, many communicable diseases were spread by infected clothing. Scabies—"the itch" to soldiers of the time—headed the list. The scabies mites caused vesicles on the skin which, when scratched, became infected with staphylococcus. Infection reached epidemic proportions in the patriot army at Valley Forge. Rejecting the "official" treatment of sulfur in hog's lard, some soldiers, half demented by the itch, tried their own remedies. One covered his body in mercury ointment and died of poisoning. Another drank one and a half pints of rum and also died.[79] Stripping the dead was common, and a dead officer's kit was usually auctioned off, all of which only served to spread disease.

The estimable Dr. Benjamin Rush had forthright views on uniforms and health. He did not like linen, and he did not approve of Washington's enthusiasm for the linen hunting shirt.

> *It is a well-known fact that the perspiration of the body, by attaching itself to linen and afterwards by mixing with rain, is disposed to forming miasmata which produces fevers. Upon this account I could wish the rifle shirt was banished from our army. Besides accumulating putrid miasmata, it conceals filth and prevents a due regard being paid to cleanliness. The Roman soldiers wore flannel shirts next to their skins.[80]*

The British uniform looked the part but was hardly functional. Heavy cloth, coarse canvas, cumbersome leather shoulder and waist belts, tight "rolers" at the throat, restrictive spatterdashes, and leggings were hell to fight in during hot weather. For example, fifty-nine British and German soldiers died of heatstroke at the battle of Monmouth on 28 June 1778. On a march from Bedford to Flushing in August of the same year, nine soldiers died and sixty-three more were felled by sunstroke.[81] Medical opinion at the time only aggravated the soldiers' suffering: "They were told never to expose their chest to the air, because this would result in pulmonic and bowel disease. A soldier was prohibited from drinking water when he perspired. If he did and felt bad, he was given whiskey."[82]

The whole issue of the weight of the British soldier's kit is, for some, another illustration of the crassness of British military management. How could British officers have loaded their men like mules, while light-footed Americans cut them to shreds? Christopher Ward's influential *The War of the Revolution* (1952) described redcoats going into battle at Bunker's Hill carrying "an estimated weight of a hundred pounds for each man." Twenty years after Ward's book, another popular history has the same troops going up Bunker's Hill with "a load reckoned at

120 pounds."[83] Not to be outdone, a more recent history makes the absurd claim that the redcoats had to hump "about the same weight as if they carried a good sized deer [that would be 200–300 pounds!] on their backs."[84] British commanders, like their American counterparts, were certainly capable of "absurdities" (as military hierarchies before and since have been), but a moment's consideration should give pause. First, contemporary accounts give a fairly clear indication of the weight of a soldier's gear. Captain Alexander Baillie of the 60th Foot, who was stationed in America in 1762, carefully weighed the equipment of one of the grenadiers of his regiment.

	LBS	QTRS
Regimental coat	5	2
Waist coat	2	1
Pair of breeches	1	2
Hat with cockade	1	0
Shirt	1	0
Knee buckles	0	3
Firelock with sling	11	0
Shoe buckles, stockings, garters	1	2
Waist belt and buckle	0	2
Hangar, sword knot, scabbard	2	3
Bayonet and scabbard	2	1
Cartridge pouch and belt	3	0
24 cartridges	2	0
Oil bottle	0	1
2 flints and steel	0	1
Haversack and strap	0	3
6 days' provisions	10	1
Full canteen	3	1
Total	47	26
	[53 ½ lbs][85]	

Second, British soldiers of the day (or American, for that matter) could hardly have physically managed 100–125 pounds even in noncombat conditions, and certainly not in battle. A modern, superbly fit, special-forces soldier can routinely hump 125 pounds carried in a scientifically designed pack. The eighteenth-century soldier, by contrast, was usually in relatively poor physical condition and had only a knapsack (of goatskin with the hair left on, or of painted canvas) in which to carry his gear. It simply did not have the capacity to carry the enormous loads some historians have suggested.[86]

Even if he dropped his knapsack prior to battle (which was the usual drill), there was still a good deal of equipment about his person: "Cross-belts compressed the chest and, if the knapsack-straps were connected by a breast-strap, could seriously impair breathing. Unless the cross-belts were connected at the back the cartridge-box and bayonet could become entangled and the soldier thrown off-balance."[87]

7

The Big Guns

ARTILLERY

Artillery had an almost parasitic relationship with the musket. Because muskets were individually so inefficient, they had to be massed in compensation, and it was this density of men that gave eighteenth-century artillery its raison d'être. Lacking accuracy itself, it was a beast of omnivorous and indiscriminate appetite, guzzling, like some Cyclops, the herds of men conveniently marching in dense formation toward its greedy muzzle.

The guns of the War of Independence were, like muskets, all smoothbore and, like muskets, lacked efficient aiming mechanisms. Although gunners thought of themselves as a cut above other branches of the army because their business carried with it the aura of "science," it was as much a craft, even an art. The gunner's "eye" and "feel" for his weapon mattered more than the complicated mathematical tables of barrel elevation, shot weights, and powder charges. The field gunner had recourse to three ammunition types. The first and most popular was simple round or solid shot: iron balls of different weights. Artillery firing round shot was graded by the weight of the ball rather than the caliber or diameter of the muzzle. The second was canister or case shot: tin cans filled with musket balls or any old pieces of iron junk. (The Prussian captain Johann Heinrichs described American case shot loaded with bits of "old burst shells, broken shovels, pickaxes, hatchets, flat-irons, pistol barrels, broken locks etc etc.")[1] The case disintegrated as it emerged from the muzzle to spread its projectiles in a spray pattern, shotgun fashion. The third was spherical shell: hollow cast-iron globes filled with gunpowder and fitted with a fuse which, if cleverly adjusted, could be configured to explode just over the heads of advancing infantry. (In 1804, it would evolve into the famous shrapnel shell.)

Solid shot accounted for about 70 percent of the ammunition carried by the artillery and was used for medium-range fire. Ranges and lethality varied according to the size of ball and powder charge. William Müller, an officer in the King's German Legion, made extensive tests of artillery accuracy which he published in *The Elements of the Science of War* (1811). Müller's tests, although from a slightly later date than the War of Independence, would have been applicable to the earlier time frame. As with all such tests, there is some dislocation from the realities of the battlefield, but they are nevertheless useful as a range finder of lethality. For example, the 6-pounder (the workhorse of field artillery) firing at a cloth target six feet high by thirty feet wide (roughly representing a company of infantry) scored 100 percent hits at 520 yards and 31 percent hits at 950 yards. At 1,200 yards, however, it scored only 17 percent hits. Müller's test does not quite tell the whole story because gunners firing a round shot tried to get two or even more bites at the cherry through ricochet: the effect of a cannonball bouncing before coming to rest. A 6-pounder might have a range of about 1,200 yards, but the ball had several phases of lethality over that distance. At zero degrees elevation (parallel with the ground) the first bounce ("graze") would have been at about 400 yards; it would then travel for another 400 yards at shoulder height before making its second graze, and then on for a further 100 yards at about three feet from the ground before rolling to a halt. The knack was to pitch the ball just in front of the enemy's first rank and have it skip and rise through the subsequent ranks. If the ground was hard and stony, each graze would kick up splinters, which, in their turn, became secondary projectiles. Even in its last phase, a rolling ball could be deceptively harmful. There are many instances of men being badly wounded, even killed, by trying to "catch" what seemed to be spent cannonballs, innocently rolling toward them. John Trumbull was with the American army besieging Boston when rewards were given for salvaged British cannonballs, but

> *it produced also a very unfortunate result; for when the soldiers saw a ball, after having struck and rebounded from the ground several times (en ricochet), roll sluggishly along, they would run and place a foot before it, to stop it, not aware that a heavy ball long retains sufficient impetus to overcome such an obstacle. The consequence was that several brave lads lost their feet, which were crushed by the weight of the rolling shot.*[2]

Private Edward Elley of Virginia described another incident, at the siege of Yorktown: "The works of the battery were thrown up by the militia soldiers, and whilst they were cutting brush a cannonball came bounding along on the ground, and a youngster put his heel against it and was thrown into lockjaw and expired in a short time."[3]

The most destructive potential for solid shot was when it was fired "in enfilade": into the side of a rank of men. Müller estimated that one ball, fired in enfilade at effective range, would kill three men and wound four or five, but greater numbers were often recorded. The weight of the ball was an important factor in determining lethality because although the muzzle velocity of most field guns was about equal at approximately 900 feet per second, the heavier the ball, the more velocity it retained over longer distances. For example, at a 1,000-yard range, an eighteen-pound ball traveled at 840 feet per second, compared with 450 feet per second for a six-pound ball.[4]

A six- or nine-pound ball of iron traveling at anything up to 900 feet per second could do terrible damage to human flesh. Peter Brown, a patriot soldier crossing the Neck onto Charlestown Peninsula during the battle of Bunker's Hill, described the effect of British gunships firing in enfilade: "One cannon [ball] cut off 3 men in two [cut them in half] on the neck of land." James Duncan of the Pennsylvania Line at Yorktown recorded that on 3 October 1781 "four men of [his] regiment . . . were unfortunately killed . . . by one ball."[5] Even small-caliber guns like the three-pound ball from a "grasshopper" (a mobile gun sometimes referred to as a "galloper") could pack a punch. For example, at Monmouth in June 1778 Joseph Plumb Martin described how the British "occupied a much higher piece of ground than we did and had a small piece of artillery, which the soldiers called a 'grasshopper.' We had no artillery with us. The first shot they gave us from this piece cut off the thigh bone of a captain, just above the knee, and the whole heel of a private in the rear of him."[6]

Even near misses could be fatal. A large ball created potentially devastating shock waves, with sometimes macabre results, as Joseph Martin witnessed at Yorktown.

> *I was sitting on the side of the trench, when some of the New York troops coming in, one of the sergeants stepped up to the breastwork to look about him . . . at that instant a shot from the enemy (which doubtless was aimed at him in particular, as none others were in sight of them) passed just by his face without touching him at all; he fell dead into the trench; I put my hand on his forehead and found his skull was shattered all to pieces, and the blood flowing from his nose and mouth, but not a particle of skin was broken.[7]*

A gun crew loading with ball could get off two or three rounds a minute; the heavier the ball, the slower the process (firing canister speeded up the rate). A six- or nine-pound gun would normally have a specialist crew of a minimum of five men (often supplemented by infantrymen to help move the gun and fetch ammunition). One man stood to the right of the muzzle with a combination rammer-sponger; the man to the left of the muzzle was the ammunition loader; a man to the left of the vent hole at the rear of the barrel carried a slow-burning match on a forked rod ("linstock"); opposite him stood the "ventsman." At the rear stood the gun chief, who aimed the piece and gave the order to fire.

The loading sequence for the first discharge started with the crew chief directing aim ("laying" the gun) and moving it on the horizontal plane by having it manhandled with poles ("handspikes"). The elevation of the barrel was controlled by a screw mechanism at the rear of the barrel (or perhaps on older pieces by inserting a wooden wedge, the "quoin"). The loader now slid a cartridge consisting of a flannel or paper bag of powder and ball into the muzzle, and the rammer pushed it down the length of the barrel. (On larger pieces, the powder charge and ball were more often separate.) If the barrel was depressed below the horizontal, a wad (of straw, hay, a coil of rope, even turf) was rammed down to prevent the ball from rolling out. The ventsman now inserted a "pricker" into the vent to puncture the powder bag. He then inserted a quill or paper tube filled with "quick match" (cotton strands soaked in saltpeter and alcohol). When the order to fire came and the men had stood clear, the firer extended his linstock across to the vent (being careful to keep clear of the wheel when it recoiled) and lit the quick match.

Before the next shot could be loaded, the gun had to be relaid because the recoil would have thrown it back several feet. (Not until 1897 would the recoil problem be solved by the hydraulic "antirecoil" mechanism of the famous French 75.) The rammer reversed his pole and used the end covered in sheepskin and soaked in water to swab out the barrel. When the loader inserted the next powder bag, the ventsman covered the vent with his thumb (protected by a leather "thumb stall") to prevent any accidental discharge. If a smoldering piece of the powder bag or wadding remained, the rammer would use a pole with a corkscrewlike end (the "wormer") to extricate so it too wouldn't create an accidental discharge. But, of course, in the heat of battle accidents did happen. At the siege of Charleston in May 1780 Lieutenant John Peebles of the Royal Highland Regiment (42nd Foot) recorded, "An artillery man lost an arm and an assistant killed by one of our own guns hanging fire and going off when they put in the spunge."[8]

Canister or case shot (sometimes referred to as grapeshot, which was made up of larger three-ounce balls and was primarily a naval warfare munition) was reserved for relatively close-up work. It made, said the American artillery sergeant White at the battle of Princeton, "a terrible squeaking noise" as it flew. Characteristically, each canister contained 85 balls. Tests carried out in 1810 indicated that 55 of those balls (65 percent) would make hits at 200 yards; 36 (42 percent) at 400 yards, but only 6 (7 percent) at 600 yards.[9] These are hit patterns that one would expect from a shotgun spread. (The balls spread to thirty-two feet over the first 100 yards.)[10] At 80–100 yards (the effective range of a musket) it would be reasonable to assume that the hit rate of canister might have risen to at least 80 percent (44 balls). Theoretically, a standard battery of six guns firing at this range would have delivered 264 balls on target, compared with the 188 of a 500-musket battalion (the guns being 71 percent more effective). Even if we take into account men hit by multiple balls, no such level of casualties was inflicted by canister in any American battle (perhaps because there were few such concentrations of guns on the battlefield). The effect of concentrated case shot, however, could be withering. In 1793 near Tournai (in what is now Belgium) the Coldstream Guards were caught unawares by a French battery firing case at close range: "The fire was so sudden that almost every man by one impulse fell to the ground—but immediately got up again and began a confused fire without orders—The second discharge of the French knocked down whole ranks."[11] At Waterloo men and horses fell "like that of grass before the mower's scythe." An artillery officer at Waterloo described "four or five men and horses piled up on each other like cards, the men not even having been displaced from the saddle, the effect of canister."[12]

Spherical shell was the lob shot of eighteenth-century warfare. It was usually fired from short-barreled howitzers and mortars. (Regular cannons could not elevate their barrels sufficiently.) The skill of the gunner lay both in setting the trajectory and in trimming the fuse (which was lit automatically by the cannon's flash) to achieve either an airburst just above enemy troops or an explosion as the ball bounced into the enemy's lines.[13] If the fuse or trajectory was miscalculated, the shell would fizz on the ground and could be extinguished by an intrepid soldier, but it was a hair-raising business. Tolstoy in *War and Peace* describes the (highly) disconcerting effect of a fizzing shell, pregnant with destruction, "whirring like a bird in flight . . . spinning like a top." When it went off it made "a splintering sound like a window-frame being smashed" followed by "a suffocating smell of powder." Some gunners firing shell were true master-craftsmen and could place their projectile with unnerving accuracy. The Continental army regimental surgeon Dr. James Thacher reported from Yorktown in October 1781:

> *It is astonishing with what accuracy an experienced gunner will make his calculations, that a shell shall fall within a few feet of a given point, and burst at the precise time, though at a great distance. When a shell falls, it whirls round, burrows, and excavates the earth to a considerable extent, and, bursting, makes dreadful havoc around. I have more than once witnessed fragments of the mangled bodies and limbs of the British soldiers thrown into the air by our bursting shells.[14]*

Attacking infantry typically had to pass through three killing zones, characterized by the three types of artillery ammunition usually employed. When the enemy was about 1,000 yards out, the medium guns (mainly 6- and 9-pounders) would engage, as would the howitzers with their shells. It would take the infantry about four to five minutes to make it to the 400-yard mark, during which time each gun could have fired about twelve to fifteen times. At 400 yards the attackers would also start to receive canister in addition to solid shot. It would take about three minutes for the attackers to get to the 100-yard mark, the start of the final and most intense killing zone, which would bring them into musket range as well as more canister and shot.[15]

The patriot cause had few artillery pieces at the beginning of hostilities: forty-one cannons of various calibers, fourteen mortars, and three howitzers. A substantial number of guns, particularly of the larger sizes, had fallen into American hands with the capture of Ticonderoga and Crown Point in May 1775. They remained there until Henry Knox, that extraordinary autodidact bookseller who became the head of Continental artillery, managed to transport thirty-nine brass and iron cannons, fourteen mortars (three of which weighed over a ton), and two howitzers by sled through the wild winter terrain in one of the most extraordinary efforts of the war. (It ranks alongside Arnold's march to Canada as a feat of staggering endurance.) More pieces were picked up when the British evacuated Boston, and another forty-nine guns when Burgoyne surrendered in October 1777. On the debit side, however, many guns were lost with the defeat at Long Island and the subsequent surrender of Forts Washington and Lee, where 146 pieces were taken by the British.[16]

Even though colonial America manufactured 30,000 tons of pig iron a year, making it the seventh largest producer in the world, there had been little call before the war to create domestic cannon-foundries, and those guns that were made tended to be of heavier caliber, more suited to defense installations rather than smaller and mobile infantry support weapons. In early 1776 congresses began to encourage and finance foundries in Pennsylvania, New Jersey, New York, and Connecticut, but it proved slow going. Many of the cannons were substandard (due mainly to inferior metallurgy) and had to be rejected, and then there was the old monkey-on-the-back of a depreciating Continental dollar, which quenched the mettle of even the most patriotic ironmasters.

8

The Sanguinary Business

WOUNDS, DISEASE, AND MEDICAL CARE

W arfare is not chess or a TV reality show. It is about killing
and disabling as effectively as possible, and its history is
a steeply rising graph of lethality. We are now very good
at killing and, in an obverse and perverse way, very good at rescuing
and repairing the soldiers who have fallen into the ever-more-efficient
meat grinder. It is as though a black angel and a white angel contested
the battlefield. By way of illustration: In World War II 4.5 percent of
hospitalized American soldiers died. In the Korean War that figure
dropped to 2.5 percent, and in Vietnam it fell to 1.8 percent.[1] In the
War of Independence about 25 percent of hospitalized casualties died.
This is known as progress. Eighteenth-century warfare was not a
particularly efficient killing machine, but neither was it equipped to
deal with the victims of its crude weaponry. In our age the black angel
and the white hover close together over the battlefield. Back then, they
were far apart.

Musketry accounted for approximately 60 percent of *battle* casualties
during the War of Independence, although the soldier stood a much
higher chance of dying from disease, which accounted for perhaps 80–
90 percent of all casualties. During the Seven Years' War, soldiers dying
from disease outnumbered those who succumbed to weaponry by 800

percent. Although the musket was a strikingly inefficient killing tool, when its heavy ball did find a target at anything close to its effective range, the damage was extensive. A wound profile test of a 0.69-caliber lead ball (about the same as a Pennsylvannia-Kentucky rifle) shot into 10 percent ordnance gelatin (simulating flesh) at 540 feet per second shows a surprisingly deep penetration of seventy-three centimeters (twenty-nine inches), which is comparable to the penetration of a conical bullet fired from a modern 0.45 automatic pistol.[2]

The large contact surface of a musket ball (almost three-quarters of an inch in diameter) would have inflicted an extensive area of damage through what in wound ballistics is known as *crushing* (the initial impact), but, unlike the modern bullet, it would not have "yawed" on penetration. Although the modern military rifle bullet has great aerodynamic integrity in flight, its delicate balance is literally knocked for a loop when it hits flesh. It enters the body point forward, but after traveling for approximately five inches (the distance will vary depending on the bullet model), it will tumble 180 degrees as its center of mass, the base, shifts to the front. It may also fragment.

Both of these characteristics, "tumbling" and fragmentation, create extended areas of damage or, in the euphemism of the ballistics expert, are extremely "disruptive of tissue." The entry wound is characteristically small, but the exit wound large (*stellate* is the ironically poetical technical term). By comparison, a large musket ball, because it cannot tumble, will do most of its damage "up front" through crushing. And assuming it does not hit bone that will cause the ball to fragment, it creates a relatively straightforward wound profile. It depends, of course, where the ball hits. "Elastic" tissue, such as muscle, skin, bowel wall, or lung, is good at absorbing the kinetic energy of the projectile; tissue in bone, or the walls of the arteries, or the liver, is another matter altogether. (A misconception is that high velocity is necessarily more lethal than low velocity because more kinetic energy is delivered to the target. The critical factor is where in the body that energy is delivered and how the projectile behaves on entry.)

A man hit in the stomach or head by a musket ball at full velocity almost invariably died. (Belly wounds, if not immediately fatal, usually led to peritonitis when food in the punctured intestine was released.) Corporal Roger Lamb of the 23rd Foot was with Burgoyne's advance to Saratoga when "a man, a short distance on [Lamb's] left, received a ball in his forehead which took off the top of his head."[3] Others were luckier. Also at Saratoga, Dr. James Thacher recounted: "A brave soldier received a musket-ball in his forehead, observing that it did not penetrate deep, it was imagined that the ball rebounded and fell out; but after several days, on examination, I detected the ball lying flat on the bone, and spread under the skin which I removed. No one can doubt but he received his wound while facing the enemy, and it is fortunate for the poor fellow that his skull proved too thick for the ball to penetrate."[4] A more likely explanation for this soldier's reprieve was that the ball was spent when it struck.

A soldier of the Napoleonic era described the pitiful reaction of men hit by a musket ball.

> *I have observed a Soldier, mortally wounded, by a shot through the head or heart, instead of falling down, elevate his Firelock with both hands above his head, & run round & round, describing circles before he fell, as one frequently sees a bird shot in the air . . . Men, when badly wounded, seek the shelter of a stone or a bush, to which they betake themselves, before they lie down, for support & security, just as birds, or hares do, when in a similar state of suffering.[5]*

The heavy cavalry saber was capable of inflicting frightful wounds. Dr. Robert Brownsfield described the aftermath of Banastre Tarleton's infamous attack at the Waxhaws, South Carolina, on 29 May 1780.

> *A furious attack was made on the [American] rearguard, commanded by Lieut. Pearson. Not a man escaped. Poor Pearson was inhumanly mangled on the face as he lay on his back. His nose and lip were bisected obliquely; several of his teeth were broken out in the upper jaw, and the under completely divided on each side . . . Capt. John Strokes . . . was attacked by a dragoon, who*

aimed many deadly blows at his head, all of which by the dextrous use of the small sword he easily parried; when another on the right, by one stroke, cut off his right hand through the metacarpal bones. He was then assailed by both, and instinctively attempted to defend his head with his left arm until the forefinger was cut off, and the arm hacked in eight or ten places from the wrist to the shoulder. His head was then laid open almost the whole length of the crown to the eye brows. After he fell he received several cuts on the face and shoulders.[6]

For good measure, poor Stokes was then bayoneted several times. Eventually, he had his wounds roughly dressed by a British surgeon, who filled them "with rough tow, the particles of which could not be separated from the brain for several days." Against all the odds, he survived and went on to live a full and successful life.

Because medical services were crude for both armies, the extrication of the wounded from the field could take many hours. The British lieutenant Thomas Anburey described the scene after the battle of Freeman's Farm on 19 September 1777.

[The] friendly office to the dead, though it greatly affects the feelings, was nothing to the scene of bringing in the wounded; the one was past all pain, the other in the most excruciating torments, sending forth dreadful groans. They had remained out all night, and from the loss of blood and want of nourishment, were upon the point of expiring with faintness; some of them begged they might lay and die . . . some upon the least movement were put in the most horrid tortures and all had near a mile to be conveyed to the hospitals . . . These poor creatures, perishing with cold and weltering in their own blood, displayed such a scene, it must be a heart of adamant that could not be affected by it.[7]

If Anburey could have looked into the American lines, he would have seen a mirror image. Samuel Woodruff fought at Bemis Heights on 7 October 1777 (the follow-up battle to Freeman's Farm) and recorded that once the fighting had petered out (about 8:00 A.M.), the American wounded were brought in: "About two hundred of our wounded men, during the afternoon, and by that time in the evening, were brought from the field of battle in wagons, and for want of tents, sheds, or any kind of buildings to receive and cover them, were placed in a circular row on the naked ground. It was a clear, but cold and frosty, night. The sufferings of the wounded were extreme, having neither beds under them nor any kind of bed clothing to cover them."[8] By the next morning, over a third of them were dead.

At the finale of Bemis Heights, the British wounded were in a pitiful state. Without adequate shelter and under constant fire they had to endure the sort of horrors Goya would later depict in his *Disasters of War* etchings. Baroness von Riedesel, the wife of the Hessian commander, cowered with her young children in the cellar of a house while cannonballs smashed through it.

Eleven cannon balls went through the house, and we could plainly hear them rolling over our heads. One poor soldier, whose leg they were about to amputate, having been laid upon a table for this purpose, had the other leg taken off by another cannon ball, in the very middle of the operation. His comrades all ran off, and when they again came back found him in one corner of the room, where he had rolled in his anguish, scarcely breathing.[9]

After Cornwallis's self-destructive victory at Guilford Courthouse on 15 March 1781, a British participant reported that "the cries of the wounded and dying who remained on the field of action during the night, exceeded all description. Such a complicated scene of horror and distress, it is hoped, for the sake of humanity, rarely occurs, even in a military life."[10]

The burial of the dead after battle followed a melancholy and time-honored pattern. The common soldiery was simply tipped into pits, as often were the officers (separate pits, but anonymous holes nevertheless). After the battle of Bemis Heights Thomas Anburey remembered the sloppiness with which the British dead were jumbled together with "heads, legs, and arms above ground." Sometimes the dead would not stay put. Anburey's brother-in-arms at Bemis Heights, Lieutenant William Digby, recorded that "during the night it rained heavy, and on the 26th, many bodies not buried deep enough in the ground appeared, (from the great rain) . . . and caused a most dreadful smell." After the battle of Kings Mountain the slain Loyalists also got short shrift. A patriot, James Collins, described how the burial of the dead "was badly done; they were thrown into convenient piles, and covered with old logs, the bark of old trees, and rocks; yet not so as to procure them from becoming a prey to the beasts of the forest, or the vultures of the air; and the wolves became so plenty, that it was dangerous for anyone to be out at night . . . also, the hogs in the neighborhood gathered in to the place to devour the flesh of men."[11]

Although British officers killed at Bunker's Hill were brought back to Boston for individual burial, sometimes circumstances precluded the perquisites of rank. Anburey recorded that after Bemis Heights "the [British] officers were put in a hole by themselves. Our army abounded with young officers, in the subaltern line . . . three of the Twentieth Regiment [the East Devonshires] were interred together, the age of the eldest not exceeding seventeen."[12] Senior officers also suffered their ignominies. At Kings Mountain Patrick Ferguson, the British commander, was stripped naked after death by the less-than-deferential soldiery, as was the American general Johann de Kalb at Camden. High and low invariably shared the same fate.

But it was not as though this was a particularly callous age (far from it, compared with the staggering scale of savagery of the twentieth century). Even that crusty martinet Frederick the Great was responsive to the appeals of humanity: "One army surgeon per battalion must remain with the battle lines; the other, however, must stay with the regimental surgeons in the *wagenburg* [wagon-park] so they can tend the wounded properly and therefore better. Humanity and gratitude toward those who have staked their lives . . . bid you care for them as a father."[13]

With some exceptions, the medical personnel of both armies extended their care to friend and foe alike. Billy Lunsford, for example, one of Colonel William Washington's dragoons, was determined to shoot a British sentry so he could boast of having killed "one damned British son of a bitch," but he found the tables turned. The sentry shot Lunsford clean through the body with "as pretty a shot as could have been made in daylight." The sentry then carried his would-be assassin to the regimental doctor, who cared for him all through the night. Lunsford eventually recovered.[14]

The medical infrastructure of both patriot and royal armies was woefully inadequate. In both there were attempts to establish centralized organizations, but they often simply tightened the knot of bureaucratic muddle. It was a classic and characteristic challenge for the Americans: how to create a medical structure that had the coherence of centralized control yet was flexible enough to meet demands across a widening geographic involvement from Canada down to Georgia. It also needed to be designed to meet the needs of soldiers requiring immediate care close to the action at the regimental level as well as create cost-effective general hospitals. It was the old problem of centralization versus localism. And the patriot cause had to create its medical bureaucracy on the wing, in the eye of the storm.

In July 1775 Congress drew up a sketchy plan for the Hospital Department, to be controlled by a director general. Three of the four men who held this post were not a resounding success. The first, Benjamin Church (July–September 1775), was a traitor and was banished. (He went down with his ship on his way to the West Indies.) The second, John Morgan (October 1775–January 1777), not only was run ragged by the unraveling of the war for the patriots in the first phase of the conflict but was also unhinged by his paranoia and obsessive political feuding. The third, William Shippen (April 1777–January 1781), was urbane and politically well connected but attracted the enmity of two powerful enemies: Dr. Benjamin Rush and John Morgan. Accused of corruption, Shippen was court-martialed in June 1780 and although acquitted was reprimanded for speculating in hospital supplies. The last director general, John Cochran (January 1781 until the end of the war), was probably the most efficient (and, ironically, the only one of the four who had not earned an M.D.). At regular intervals throughout the war, Congress reorganized the Hospital Department, as though reorganization could on its own address the deep-seated problems of lack of supplies and the collapse of the Continental currency.

Understandably, both armies relied primarily on regimental surgeons and surgeon's mates—the front-line caregivers. Levels of skill varied widely. It has been estimated that there were 3,500 physicians in America in 1775, of which 400 held M.D. degrees. Of the 1,200 who served in the American military, only 100 or so were qualified doctors with degrees either from a European university or from fledgling American schools. (The first medical school had been founded by Dr. Thomas Bond in Philadelphia in 1751; the second, in New York, in 1771.)[15] Some had seen service as surgeons or surgeon's mates with the British during the French and Indian War. Most were apprentice-trained. As the loquacious diarist Dr. James Thacher records, each surgeon's mate underwent an examination supervised by a board of physicians (and by Thacher's estimation quite rigorous). In any event, George Washington did not trust them. In his view they were "great rascals countenancing the men in sham complaints to exempt them from duty and often receiving bribes to certify indisposition with a view to procure discharges or furloughs."[16]

In Britain an act of George II required the College of Surgeons to examine all army and navy surgeons and surgeon's mates and to offer them free lectures on practical surgery, but not until 1858 was an army medical school created; and British medical personnel of the eighteenth century were poorly trained.[17] It is surprising, therefore, that Dr. Benjamin Rush praised his enemy's medical facilities and practices, having visited the British camp after the battle of Brandywine: "There is the utmost order and contentment in their hospitals. The wounded whom we brought off from the field were not half so well treated as those whom we left in General Howe's hands. Our officers and soldiers spoke with gratitude and affection of their surgeons. An orderly man was allotted to every ten of our wounded, and British officers called every morning upon our officers to know whether their surgeons did their duty." But, Rush added, "you must not attribute this to their humanity. They hate us in every shape we appear to them. Their care of our wounded was entirely the effect of the perfection of their military establishment."[18]

In its anxiety to have a centrally controlled medical service Congress created an organizational nightmare. Regimental surgeons were obliged to send men back to a regional general military hospital, even though it would have been much more effective to have dealt with many of the sick and wounded at the regimental level. It was a circular argument that cost many men their lives. The regimental surgeons were starved of supplies by the general hospitals, and thus were forced to send men to hospitals where filth, overcrowding, and infection made them merely "way stations to the grave." There was "no reason," complained Nathanael Greene in October 1776, "either from policy or humanity, that the stores from the General Hospital should be preserved for contingencies which may never happen and the present regimental sick left to perish for want of proper necessities."[19]

Continental surgeons were particularly short of drugs, which before the war had been largely imported from Britain. Most medical chests were required to have Peruvian or Jesuit's bark, potassium nitrate, and camphor for the treatment of fevers; pain-controlling opium (laudanum) in the form of gum and tinctures; purges (which have been described as the "chicken soup" of eighteenth-century medicine) like jalap, senna, castor oil, and Epsom salts; emetics like ipecac and potassium tartrate; red mercuric oxide and mercurial ointment for the treatment of wounds and venereal disease; and sulfur in hog's lard for the "itch." In addition, the medical chest would have had probes and ball extractors; a range of lancets, splints, saws, and tourniquets for use during amputations; trepanning instruments; suturing needles; dry and wet sutures; and bandages.[20] In the field a medic (or someone co-opted to the role) might be forced to fall back on folk medicine. In September 1780 a British force in the Carolina backcountry came across a "Rebel militia-man that got wounded in the right arm . . . The bone was very much shattered. It was taken off by one Frost, a blacksmith with a shoemaker's knife and carpenter's saw. He stopped the blood with the fungus of the oak."[21]

Dr. John Jones's *Plain, Concise, and Practical Remarks on the Treatment of Wounds and Fractures* . . . (published in Philadelphia in 1776) was the standard work in America on wounds. Jones was professor of surgery at King's College, New York, and his work was heavily influenced by one of the great military physicians of the eighteenth century, Sir John Pringle, whom Jones had met in London in the 1760s. Balls were extracted with the index finger or an extractor, but if deeply embedded, they were usually left in place. A musket ball made a narrow opening with considerable surrounding bruising. Hemorrhaging was stemmed with styptic or by tie, and then the wound opening was widened for more effective drainage. Often it was not the projectile itself that proved fatal, but what came with it. A leading cause of death from penetrating wounds in a world before antibiotics was clostridium myositis— gangrene. It killed 5 percent of men wounded in pre-antibiotic World War I but only 0.7 percent in World War II.

If a projectile caused compound fractures, particularly if the heads of bones were broken or ligament was destroyed, the limb was usually amputated. In 1718 the French surgeon Jean-Louis Petit had transformed the business of amputation with the invention of his compression tourniquet (a screwlike device that controlled arterial blood flow). Surgeons who carried out amputations were "taught to show no emotion in the face of the screams of the patient" and had to work fast. Robert Morton went to see an amputation in Philadelphia in October 1777: "I went to see Doc. Foulke amputate an American soldier's leg, which he completed in 20 minutes, while the physician at the military hospital was 40 ms. Performing an operation of the same nature."[22] Joseph Townsend was an American Quaker who viewed the aftermath of the battle of Brandywine from behind the British lines. It is a remarkable illustration of sangfroid.

> *The wounded officers were first attended to . . . After assisting in carrying two of them into the house I was disposed to see an operation by one of the surgeons, who was preparing to amputate a limb by having a brass clamp or screw [Petit's tourniquet] fitted thereon a little above the knee joint. He had his knife in his hand, the blade of which was . . . circular . . . and was about to make the incision, when he recollected that it might be necessary for the wounded man to take something to support him during the operation. He mentioned to some of his attendants to give him a*

little wine or brandy . . . to which he relied, "No, doctor, it is not
necessary, my spirits are up enough without it."[23]

Apart from alcohol, opium, and Peruvian bark, there were no really effective analgesics. Sometimes the soldier had to resort to folk remedies to relieve pain. David Freemayer, who was wounded during Sullivan's campaign against the Six Nations in the fall of 1779, wrote in his pension application how he dealt with the severe pain of a leg wound.

> *Here affiant [Freemayer] and Murphy had their wounds dressed*
> *for the first time after their infliction, except the wound on the side*
> *of the affiant's leg, which gave affiant so much pain in traveling*
> *the day before that he was compelled to do something if possible to*
> *relieve it, which he done by killing a striped squirrel and putting*
> *the brains of the squirrel on the wound and fastening them on the*
> *skin thereof.[24]*

And, of course, there were no anesthetics. (Ether was introduced in 1846 and chloroform the following year.) The unfortunate soldier awaiting amputation or trepanning may have been given some opium or rum and his ears may have been filled with lamb's wool to deaden the hideous sounds of bone being sawed.[25] Shock would take the lives of many amputees, and in a world ignorant of aseptic procedures, wound infections like tetanus ("lockjaw") took a huge toll. Amputation at the mid-thigh resulted in death for 45 to 65 percent of patients.[26] (For example, 70 percent of amputees at Waterloo died of infection.)

Conditions in the military hospitals of the period promoted the lethal incubation of infection: "the sinks of human life," as Corporal William Wheeler's description of one in the Peninsular War gruesomely illustrates.

> *During the five weeks I was in it, what numbers I had seen die*
> *under the most writhing torture, and their places filled again by*
> *others, who only come to pass a few days in misery, and then to*

be taken to their last home. The bed next mine were occupied by
six soldiers, five died, the sixth I left in a hopeless state. One of
those men I knew, he was a sergeant in the 82nd, his wife was
nurse to the ward, she pricked her finger with a pin left in one of
the bandages, caught the infection, her finger was first amputated,
then her hand, the sluff appearing again in the stump, she refused to
undergo another operation, the consequence was she soon died.[27]

Anthony Wayne wrote to Horatio Gates in December 1776: "Our hospital, or rather house of carnage, beggars all description, and shocks humanity to visit. The cause is obvious; no medicine or regimen on the ground suitable for the sick; no beds or straw to lie on; no covering to keep them warm, other than their own wretched clothing."[28] Dr. Lewis Beebe described the hospital—"this dirty stinking place"—of the American expeditionary army to Canada in 1776: "Language cannot describe nor imagination paint the scenes of misery and distress the soldiery endure . . . One—nay two—had large maggots, an inch long, crawl out of their ears, [and] were almost on every part of their body."[29] In November 1779 J. Mervin Nooth, the superintendent for hospitals for the British army in North America, reported to Sir Henry Clinton that men in some hospitals were "frequently so shamefully nasty and lousy and so much in want of necessaries that it is out of the power of the persons attending them to clean them."[30] The physician visiting a military hospital was advised to avoid infection by wearing a "waxed linen coat," take tincture of bark, stuff his nostrils with lint soaked in camphorated spirits, and hold his breath when examining a patient.[31]

Of the approximately 25,000 American military deaths during the war, 10,000 died of disease. Contributing factors were poor nutrition, inadequate shelter, unsuitable clothing, disregard of hygiene in camp and hospital, medical ignorance of the causes and treatment of disease, and concentrations of young men who had not previously been exposed to diseases. It started with the squalor of camp life. At the beginning of the war the Massachusetts Provincial Congress tried to forewarn Washington of the filthy conditions of the camps sprawled around

Boston that he was about to inherit: "Although naturally brave," it declared, " . . . the youth of America are not possessed of the absolute necessity of cleanliness in their dress and lodging . . . to preserve them from diseases frequently prevailing in camps." Washington, with his experience of managing a large plantation, was a stickler for order and hygiene, and one of his first general orders (4 July 1775) addressed the problem:

> *All officers are required and expected to pay diligent Attention to keep their men neat and clean; to visit them often at their quarters, and inculcate upon them the necessity of cleanliness, as essential to their health and service. They are particularly to see, that they have Straw to lay on . . . They are also to take care that Necessarys [latrines] be provided in the camps and frequently filled up to prevent their being offensive and unhealthy.*[32]

When the reforming von Steuben arrived at Valley Forge, he found latrines and rotting animal carcasses close by cooking sites, and infected men living cheek by jowl with the healthy. With characteristic energy and good sense he had the infected quarantined and instructed quartermasters to dig latrines at least 300 feet from the tents. Latrines were to be filled after four days' use, and new ones dug. Dysentery (the "bloody flux" or "putrid diarrhea") always accompanied the unsanitary conditions of the camp, and men would "melt away from running off at the bowels." The usual treatment was often merely a mirror of the disease itself: bleeding, purges, and emetics until the condition had run its course. One doctor advised inserting lard and a hard-boiled egg in its shell into the rectum, but some more enlightened physicians recognized the need for cleanliness, the fresh air of the countryside, and a light diet of meat broths, fruit, and milk.

Typhus ("hospital fever," "jail fever," "camp fever") also visited those soldiers forced to live in crowded and filthy camps and hospitals. Like scabies, it was carried by the human louse, which laid its eggs (about sixty every week) in the seams of clothing or the hair of its human host. In the fall of 1777, following the battle of Brandywine, 500 American casualties were lodged with the religious sect of Dunkards in Ephrata, Pennsylvania. Of 700 soldiers crammed into buildings designed for 400, over 300 died of typhus by the end of 1777. Of the 1,500 soldiers who were at Ephrata until June 1778, one-third died.

Smallpox was a great scourge in the seventeenth and eighteenth centuries, not as dramatic as the plague but consistent. In Europe it killed 400,000 a year.[33] When it was introduced to the New World by European colonists, it had a massive impact. For instance, Mexico lost half its population—3.5 million—in the six months after Cortés landed. By the early eighteenth century inoculation (the introduction of a small amount of the pus from a recovering victim through a cut in the skin) was known to give lifelong protection, although the practice was still viewed with misgiving. In 1776 the American army of the Canadian expeditionary force had been devastated: "ten times more terrible than Britons, Canadians, and Indians together," pronounced John Adams. Some men inoculated themselves by introducing pus under their fingernails, but because it was not done under controlled conditions of isolation, it posed a grave threat to men who had not previously been infected. Their commander, General John Thomas, forbade inoculation. Ironically, Thomas, a doctor, died of the disease shortly after. His successor, John Sullivan, had 2,000 out of his 8,000 command incapacitated by smallpox. The British had been badly hit by it during their incarceration in Boston, and Washington was concerned to prevent the disease from spreading to his own army. In January 1777 he decided to attack the problem systematically and ordered William Shippen to undertake inoculation of every soldier who had not already had the disease: "Necessity not only authorizes but seems to require the measure, for should the disorder infect the Army, in the natural way [which resulted in a 16 percent death rate compared to 0.33 percent for inoculation] and rage with its usual Virulence, we should have more to dread from it, than the Sword of the Enemy."[34] It would save his army.

Previously, patients underwent two weeks of preparation for inoculation, during which time they were put on a light diet, dosed with mercury, bled, and purged. The skin (usually of the leg) was punctured and the virus introduced. Washington's army could not afford the luxury of prolonged preparation, and soldiers were simply given a dose of jalap to purge them prior to inoculation. They were then kept in isolation until the scabs fell off.

Scurvy, usually associated with the vitamin-C-deprived diet of eighteenth-century sailors, was also prevalent among the soldiery of the period. The symptoms were alarming: "His gums became itchy, swelled, bled, and became red and spongy with associated bad breath. The gums became putrid and developed fungal growths . . . the victim's skin was covered in red, blue and black spots . . . the patient's ankles swelled . . . His skin broke down, and ulcers developed . . . Any bruise developed into an ulcer. When the patient moved he could suddenly expire. Old wounds and fractures broke down. The skin could burst from swelling. The patients started to bleed from all orifices and they developed jaundice, dropsy, melancholia, colic, pain in the chest . . . and ultimately died."[35] The effectiveness of lemon juice had been known since the late seventeenth century and had been proved scientifically by the British naval surgeon Dr. James Lind in 1754 (although it was not until 1795 that lemon juice was made part of the regular issue in the Royal Navy, with miraculous results), but the deficiencies of a diet so heavily dependent on salted meat and bread were hard to rectify in wartime conditions. One remedy was of great benefit to the 12,000 British troops cooped up in Boston during the winter of 1776–77. Almost a half million gallons of porter, a strong ale (about 7 percent alcohol by volume) were shipped in between October 1775 and March 1776 (about a quart per man per day). Although it was ineffective against scurvy, it must have given morale a wonderful uplift. But, as all soldiers know, the "brass" is averse to a good thing, and so the delicious and inspiring porter was replaced with three to four quarts per day per man of spruce beer, an obnoxious concoction made from the extract of boiled spruce needles mixed with molasses and water—like an unholy marriage of turpentine and root beer.[36]

9

"Trulls and Doxies"

WOMEN IN THE ARMIES

George Washington had a problem with women. They got in the way and caused a considerable nuisance. But the men of his army, to His Excellency's chagrin, seemed to find them necessary, even indispensable. He issued a stream of orders intended to get the women in line: "The multitude of women in particular, especially those who are pregnant, or have children, are a clog upon every movement . . . get rid of all such as are not absolutely necessary."[1] On 4 July 1777 he proclaimed "that no women shall be permitted to ride in any wagon, without leave in writing from the Brigadier to whose brigade she belongs . . . Any woman found in a wagon contrary to his regulation is to be immediately turned out." He might as well have saved his breath. Two years later he was again huffing that the "pernicious practice of suffering the women to encumber the Waggons still continues notwithstanding every former prohibition." In the end he could not deny the women their place in his organization and "was obliged to give Provisions to the extra Women in these Regiments, or loose by Desertion, perhaps to the Enemy, some of the oldest and best Soldiers In the Service."

At the beginning of the war women did not trouble the commander in chief. Few had attached themselves to the army besieging Boston, and although this may have relieved the American high command, it had a bad effect on the men. Without women to keep them on the straight and narrow, the lads went to the dogs: "They have no women in the camp to do washing for the men, and they in general not being used to doing things of this sort, and thinking it rather a disparagement to them, choose rather to let their linen, etc., rot upon their backs than to be at the trouble of cleaning 'em themselves."[2]

As the war progressed the number of women attached to the American army fluctuated. The December 1777 return for the main army at Valley Forge noted 400 women, which would have given a ratio of women to enlisted men of 1:44. The January 1783 return at New Windsor indicated the ratio had risen in favor of women to 1:30, and in the intervening years it was probably in the 1:35 range, or about 3 percent of the total manpower.[3] In the British army the ratio was much more generously weighted with women. The May 1777 return for the British army in New York was 1:8 (the German, 1:30), and by August 1781 it had risen to 1:4.5 (German, 1:15). The same pattern is true for both armies: women were accumulated as the war progressed.[4]

Just as many men were driven into enlistment by economic necessity, so too were the women camp followers. For those not in the (relatively small) middle and upper classes, necessity was the unforgiving mistress of the eighteenth century. If a family man went off to the army, the outlook for the women and children left behind could be grim. This was a society almost without safety nets, and the prospect of starvation if the fragile network of family support was broken was a stark reality. ("I am without bread, & cannot get any, the Committee will not supply me, my Children will Starve . . . *Please Come Home,*" wrote a desperate wife to her husband at Valley Forge.)[5] Many women of the lower classes needed a connection to the army simply to survive. In return for a ration a day ("without Whiskey"), they washed and cooked and, occasionally, endured the dangers of the battlefield. There were no free lunches, as the general orders of the 2nd Pennsylvania Regiment for autumn 1778 made clear: "Should any woman refuse to wash for a soldier at the [army] rate he must make complaint to the officers commanding the company . . . who [if he] finds it proceeds from laziness or any other improper excuse he is immediately to dismiss her from the regiment . . . no woman shall draw rations . . . unless they make use of their endeavours to keep the men clean."[6] No doubt other services of a more intimate nature were sometimes provided, but it is clear that a woman's best chance of survival was through an attachment to one man, very often a husband whom she followed into the service. Prostitutes, much more tolerated in the British army than the American (with its preoccupation with republican "virtue"), were, of course, in attendance, but a clear distinction needs to be made between them and camp followers.

During battle camp followers were kept in the rear with the baggage, but there were occasions when women accidentally found themselves, or chose to be, in the action. For instance, at the battle of Brandywine women attached to the 6th Pennsylvania took "the empty canteens of their husbands and friends and returned with them filled with water . . . during the hottest part of the engagement, although frequently cautioned as to the danger of coming into the line of fire."[7] British and German women also did similar water duty at Saratoga. Cooking was not always safe, as a soldier's wife at Brandywine discovered. She was preparing a "neat's tongue" (ox tongue) when the British artillery opened up: "Unfortunately one of the enemies shot dismounted the poor camp kettle with the fier [sic] and all its contents away with it."[8] She, unlike the much anticipated meal, survived.

Some women took an even more active part in combat. The legendary "Molly Pitcher" (who brought water to the American artilleries at the battle of Monmouth Courthouse, 28 June 1778) may have been a composite of several women rather than a specific individual.[9] Private Joseph Martin described a "Molly Pitcher" of his own at Monmouth.

> One little incident happened, during the heat of the cannonade, which I was eyewitness to . . . A woman whose husband belonged to the Artillery, and who was then attached to the piece in the engagement attended with her husband at the piece the whole time; while in the act of reaching a cartridge and having one of her feet as far before the other as she could step, a cannon shot

from the enemy passed directly between her legs without doing any other damage than carrying away all the lower part of her petticoat,—looking at it with apparent unconcern, she observed, that it was lucky it did not pass a little higher, for in that case it might have carried away something else, and ended her and her occupation.[10]

Others were not so lucky. Margaret Corbin fought alongside her husband during the defense of Fort Washington on 15 November 1776 and was badly wounded. Some paid the ultimate price. After the battle of Freeman's Farm (19 September 1777) Lieutenant Thomas Anburey came across some American dead: "I met with several dead bodies belonging to the enemy, and amongst them were lying close to each other, two men and a woman, the latter of whom had her arms extended, and her hands grasping cartridges."[11] At the British retreat from Lexington and Concord "even women had firelocks. One was seen to fire a blunderbuss between her father and husband from their windows. There they three, with an infant child, soon suffered the fury of the day [that is, they were all killed by British soldiers]."[12]

A few women joined the army disguised as men, the most famous being Deborah Sampson, who served as "Robert Shurtliff,"[13] enlisted on 20 May 1782 at Worcester, Massachusetts, and served in the light infantry company of the 4th Massachusetts. She was wounded (twice) at Tarrytown. She then became a waiter to General John Paterson at New Windsor until her discharge on 23 October 1783 at West Point.[14] After the war she married and had children and as Mrs. Gannett petitioned and was eventually awarded a veteran's disability pension of $4 per month. Most other women who were widows of common soldiers were refused federal pensions (officers' widows were excepted) until 1832, by which time, of course, few would have been alive to receive them.

Although subsequent histories have dutifully wrapped the women of the camp in the flag of patriotic heroism, contemporaries were often appalled by them. "Refined" observers like Hannah Winthrop reviewed with horror the British camp followers after Burgoyne's surrender at Saratoga.

I have never had the least Idea that the Creation produced such a sordid set of creatures in human Figure . . . great numbers of women who seemed to be the beasts of burthen, having a bushel basket on their backs, by which they were bent double . . . children peeping through gridirons and other utensils . . . the women bare feet, cloathed in dirty rags, such effluvia filled the air while they were passing, had they not been smoaking at the time, I should have been apprehensive of being contaminated by them.[15]

Washington was desperately concerned that no women camp followers should be seen during his army's parade through Philadelphia on 23 August 1777: "Not a woman belonging to the army is to be seen with the troops on their march thro' the city." And even to an ordinary country boy like Joseph Martin the women of the camp seemed, literally, alien beings. At Tappan on the west bank of the Hudson in 1780 he watched them go by: "A caravan of wild beasts could bear no comparison with it. There was 'Tag, Rag, and Bobtail;' 'some in rags and some in jags,' but none 'in velvet gowns.' . . . They 'beggared all description;' their dialect, too, was as confused as their bodily appearance was odd and disgusting; there was the Irish and Scotch brogue, murdered English, that insipid Dutch and some lingos that would puzzle a philosopher to tell whether they belonged to this world or some 'undiscovered country.'"[16] It is a fine snapshot of the mob—the slobbering, foul-smelling, and potentially murderous beast that a terrified "polite" eighteenth-century society tried to control with religion, the noose, and the whip.

In the British army a certain proportion of enlisted men's wives (about 10 percent, chosen by ballot) were allowed to travel to America with their husbands, and there may have been as many as 5,000 wives with the British army over the span of the war.[17] Some shared fully in the dangers their menfolk faced. Two were particularly notable. Baroness von Riedesel left a fascinating memoir of the horrors she and her three children endured as the wife of the commander of German troops of John Burgoyne's ill-fated expeditionary force that went down to defeat at Saratoga in 1777. Lady Harriet Acland (sometimes spelled Ackland), the young and pregnant wife of bluff and hard-drinking grenadier Major John Dyke Acland, was also at Saratoga and went to the American lines to nurse him when he was badly wounded and captured at Bemis Heights on 7 October 1777. Lieutenant William Digby paid tribute to Harriet's fortitude on an earlier occasion when the Aclands' tent burned down: "Lady Harriot Ackland, who was asleep in the tent when it took fire, had providentially escaped under the back of it; but the major was much burned in trying to save her. What must a woman of her rank, family and fortune feel in her then disagreeable situation; liable to constant alarms and not knowing the moment of an attack; but from her attachment to the major, her ladyship bore everything, with a degree of steadiness, and resolution, that could alone be expected from the experienced veteran."[18] The baroness had her own terrifying ordeals at Saratoga and in the subsequent march into captivity, and they too were borne with "steadiness and resolution."

Ordinary women also sacrificed. At the battle of Princeton (3 January 1777) several British women were taken prisoner; the wife of a grenadier was killed in the action prior to the British capture of Philadelphia, as was a woman "who kept close by her husband's side" at Fort Anne in July 1777.[19] But of their anonymous lives we know next to nothing. A tribute to the women of the British army in America by Sergeant Roger Lamb of the 23rd Foot serves as a memorial: "If war sometimes in bad men, calls forth all the viper passions of our nature, in women it is the obverse; it rouses into action an heroism otherwise unknown, an intrepidity almost incompatible with their sex, and arouses all the dormant susceptibilities of their mind."[20]

10

Cuff and Salem, Dick and Jehu

BLACKS IN THE WAR

The conventionally accepted figure for black men serving in the Continental army is 5,000, and one eminent historian has claimed that by 1780 one in six soldiers was a man of color.[1] These figures so baldly stated seem to imply that the patriot cause was, in an uplifting way that appeals so much to us, integrated: a brotherhood of blacks and whites fighting shoulder to shoulder for a shared freedom, a shared country: "The black soldier or sailor was in fact eager to fight on two fronts—for his own freedom and for the freedom of his country."[2] But on closer inspection the picture is not quite so heartwarming.

As we shall see with native Americans, patriot and British leaderships were in a dilemma when it came to the question of inducting blacks into their armies. On the one hand, the prospect of a plentiful supply of black men had its attractions; on the other, the possibility that such recruitment might give blacks ideas above their station ran the risk of unraveling the whole fabric of colonial society, particularly in the South. For their part, blacks, like Indians and indeed like the poor whites on whom the major burden of fighting rested, sought to use the war in whatever way they could to gain advantages within a system that was unapologetically exploitative.

At the very outset of the war the Massachusetts Committee of Safety explicitly denied the right of anyone except "freemen" to serve in its forces because it would "be inconsistent with the principles that are to be supported, and reflect dishonour on this colony; and that no slaves be admitted into this army upon any consideration whatever."[3] Four days after becoming commander in chief Washington, as might be expected from a Virginia slave owner, prohibited the enlistment of blacks. His attitude was representative of the white oligarchy. For example, Horatio Gates instructed recruiters in July 1775 to reject "any deserter from the Ministerial army, nor any stroller, negro, or vagabond,"[4] and General Philip Schuyler in 1777 lamented that the presence of "Negroes" did nothing but "disgrace our arms."[5] However, by the end of December 1775 Washington was forced to change his position in response to the British threat to arm Virginia slaves. In a panicky letter to the president of Congress on 31 December 1775, Washington warned that if blacks could not be brought into the Continental army, "they may seek employ in the ministerial army." Gingerly Congress conceded that *free* blacks could serve.[6]

Blacks had served in the New England army besieging Boston, as they had during the Concord and Bunker's Hill battles. Colonel William Prescott paid a particular compliment to Salem Poor, who at Bunker's Hill "behaved like an experienced officer as well as an excellent soldier."[7] It has been estimated that there were 4,500 free blacks in Massachusetts, of whom about 500 served in the Continental army.[8] For example, Peter Salem (who had been freed by his masters, the Belknaps of Framingham) served in Captain Simon Edgel's Framingham company of minutemen in the Concord and Bunker's Hill battles.[9] During the British retreat from Concord on 19 April 1775, Lieutenant Frederick MacKenzie of the 23rd Foot (Royal Welch Fusiliers) makes the particular observation that "as soon as the troops had passed Charlestown Neck the Rebels ceased firing. A Negro (the only one who was observed to fire at the royal troops) was wounded."[10] Did MacKenzie mean the only black man he had observed during whole retreat, or the only one to fire at that late stage of the battle?

The presence of blacks in the army besieging Boston offended the southern congressmen who visited the camps. And not only the southerners: Alexander Graydon of Pennsylvania observed "a number of Negroes, which to persons unaccustomed to such associations had a disagreeable, degrading effect."[11] It was all too much. Edward Rutledge of South Carolina introduced a resolution into Congress on 26 September 1776 to purge all Negroes from the army. This was opposed by New England and Pennsylvania on the grounds that such a measure was impracticable, and a compromise was reached in which black soldiers already in service could stay, but no new ones would be admitted.

As the war progressed, the severe manpower shortages began to put pressure on white leaders to admit blacks more officially into the army. On 2 May 1777 Connecticut passed legislation declaring that any *two* men who bought the services of a substitute were exempted from service themselves—and that substitutes could be black. To help promote the transaction, the legislators rewarded a slave master who freed a slave to become a substitute not only by exempting the master from the draft but also by absolving him of all maintenance costs for the slave who served.[12] The role of blacks as substitutes was widespread during the war. For example, in Concord, Massachusetts, where there were relatively few blacks, they made up 8 percent of the men Concord sent to the Continentals.[13] It is not hard to see the attraction, both for white men who could afford to buy black substitutes and for recruiting officers, who, in Massachusetts, got $10 per recruit and "were not overly squeamish about enlisting Negroes [so that] slaves and free Negroes unobtrusively filtered into states' levies, generally signing up for three years and receiving the same bounty as whites."[14]

By 1778 the Continental army has been described as being "well sprinkled with blacks." In the official return of 24 August 1778 there were 755 black men spread over fourteen brigades.[15] What ratio of black to white might this represent? We know, for example, that Washington had 15 brigades at Valley Forge with a paper total of 17,491 rank and file (of whom only 7,600 were fit for duty—but that is another story). With an average brigade strength of 1,166, it can be assumed that fourteen brigades would total somewhere in the region of 16,300; so the ratio of black to white was 1:21, or 4.6 percent black.[16]

In February 1778 Rhode Island authorized the formation of a battalion of freed slaves, the state footing the bill of buying their freedom. The first slaves to enlist were Cuff Greene, and Dick and Jack Champlin, for whom the state paid £120. White reaction to the law was so strong that it was repealed on 10 June 1778, but during the five-month period somewhere between 130 and 300 black men from Rhode Island had joined the Continentals. For some their service left them with a sense of betrayal, powerfully expressed with slashing irony by Jehu Grant in his pension application of 1836.

> *Their [British] ships lay within a few miles of my master's house . . . and I was confident my master traded with them, and I suffered much from fear that I should be sent aboard a ship of war. This I disliked. But when I saw liberty poles and the people all engaged for the support of freedom. I could not but like and be pleased with such thing (God forgive me if I sinned in so feeling) . . . These considerations induced me to enlist into the American army, where I served faithful for ten months, when my master found and took me home. Had I been taught to read or understand the precepts of the Gospel, "Servants obey your masters," I might have done otherwise, notwithstanding the songs of liberty that saluted my ear, thrilled through my heart.[17]*

In April of 1778 Massachusetts legalized the enlistment of blacks. In March 1781 New York promised a land grant of 500 acres to any master who would give up a black servant who would serve for three years. However, it would be misleading to imagine that these attempts to gain blacks for the army sprang from some kind of Yankee altruism. Blacks were as disparaged in the North as in the South. The French traveler Jacques Pierre Brissot de Warville, writing of the northern colonies, observed: "There still exists too great an interval between them [free blacks] and the Whites, especially in the public opinion. This humiliating difference prevents those efforts which they might make to raise themselves."[18]

Even in the South by the summer of 1780 the pressure of the war began to bend the will of those who had resisted black involvement. As early as November 1775 the royal governor of Virginia, Lord Dunmore, had lobbed something of a bombshell into the patriot camp when he issued an incendiary proclamation urging slaves of patriot masters to join his "Ethiopian" regiment. About 800 answered the call (including 100 from Loyalist masters).[19] General Sir Henry Clinton further set the cat among the pigeons by declaring in his "Phillipsburg Proclamation" of 30 June 1779 that any runaway slaves coming over to the British would be emancipated. Thomas Jefferson estimated that 30,000 slaves defected from Virginia alone (and perhaps 20,000 from South Carolina).[20] In fact so great was the response that it blew back in Clinton's face. Dismayed by the reaction he had triggered, on 30 August 1781 he instructed Cornwallis, commander of British forces in the South, to "make such arrangements as will discourage their joining us."[21]

Like the Americans, the British were reluctant to put armed ex-slaves in the ranks, preferring to use them as laborers, scouts, inland waterway pilots, and, occasionally, spies.[22] In fact, Loyalist regiments began to purge blacks from their rolls because they incited disrespect from regular British regiments.[23] On the limited occasions when ex-slaves were used by the British as soldiers, for example, at the siege of Savannah in 1779, it horrified the Americans and was, said Nathanael Greene on 30 March 1781, "sufficient to rouse and fix the resentment and detestation of every American who possesses common feelings."[24] Many Britons shared Greene's abhorrence. Edmund Burke, one of the leading opponents of the war in the House of Commons, warned that using armed slaves to suppress the patriot rebellion in the South would unleash "barbarian" mayhem as they would make "themselves masters of the houses, goods, wives, and daughters of their murdered lords."[25]

The debate in the South concerning black enlistment in the patriot cause was, as can be imagined, as painful as the horns of a dilemma could make it. One can trace something of this in the sequence of exchanges that took place between John Laurens, a Continental officer and son of Henry Laurens (a prominent South Carolinian plantation owner and from November 1777 until December 1778 president of the Continental Congress), Alexander Hamilton, and George Washington. The young Laurens (then twenty-four years old) wrote to his father on 14 January 1778 proposing that Continental forces should be augmented by southern slaves. Not only, he argued, would there be military benefit, but, perhaps more important in his view, his proposal "would bring about a twofold good": "first I would advance those who are unjustly deprived of the Rights of Mankind to a state which would be a proper Gradation between abject Slavery and perfect Liberty—and besides I would reinforce the Defenders of Liberty." There would be a substantial military benefit: "Men who have the habit of Subordination almost indelibly impress'd on them, would have one very essential qualification of Soldiers." But, as important, a great moral evil would be expunged: "I have long deplored the wretched State of these men . . . the Groans of despairing multitudes toiling for the Luxuries of Merciless Tyrants [and] have had the pleasure of conversing with you sometimes upon the means of restoring them to their rights."[26]

John Laurens put his argument to Alexander Hamilton, then aide to Washington, and Hamilton responded much more to the appeal of military expediency than any antislavery notions. In a letter to John Jay of 14 March 1779[27] Hamilton applauded Laurens's argument for its military rationality because it "promises very important advantages," and goes on to say: "It is a maxim with some great military judges, that with sensible officers soldiers can hardly be too stupid . . . Let officers be men of sense and sentiment, and the nearer the soldiers approach to machines perhaps the better."[28] Hamilton's response casts an interesting light not only on the question of using slaves in the army but also on the American military hierarchy's thoughts about soldiers in general.

Washington's response to Laurens's proposition is also illuminating. He wrote to Henry Laurens on 20 March 1779 that he had not ever given much thought to the idea of enlisting slaves (surely disingenuously), but his main concern was the preservation of the institution of slavery itself. Would not giving some the freedom to fight disaffect all the rest? "I am not clear that a discrimination will not render Slavery more irksome to those who remain in it."[29] John Laurens's plan, predictably, went nowhere, squashed by southern delegates in Congress. Washington wrote to him on 10 July 1782: "I must confess that I am not at all astonished at the failure of your Plan. The spirit of Freedom which at the commencement of this contest would have gladly sacrificed every thing to the attainment of its object has long since subsided, and every selfish Passion has taken its place; it is not the public but the private Interest which influences the generality of Mankind nor can the Americans any longer boast an exception; under these circumstances it would rather have been surprising if you had succeeded."[30] It was a breathtaking piece of humbuggery when one remembers that Washington did not free any of his many slaves to fight in "the spirit of Freedom." The young Laurens is one of the most attractive figures of the war; brave both physically (he was wounded at Germantown and Monmouth Courthouse) and morally. Not many scions of wealthy slave-owning families had the courage to confront the evil from which their fortunes flowed. Tragically, he would be killed in a minor action at the last gasp of the war.

Benjamin Quarles, author of *The Negro in the American Revolution* (1961), asserts that "most slave soldiers received their freedom with their flintlocks."[31] The idea of *manumission* (a word resonant of ancient Roman gladiators who could win their freedom through combat) has an appealing nobility. But the truth is somewhat less uplifting. Black men who fought in the Continental or British army were much more likely to be used either as personal servants (British field officers, for example, were allowed two) or as laborers (working on fortifications, felling trees, toiling in ironworks or lead mines) than toting a musket. Captured slaves of Loyalist masters were even offered as enlistment incentives to patriot whites. In South Carolina a white man enlisting as a private got one slave, a colonel three large and one small slaves. However, there were simply not enough to go around. At the end of the war Wade Hampton's regiment, for example, was owed ninety-three and three-quarters grown slaves and, intriguingly, "Three Quarters of a Small Negro."[32]

The "Proper Subjects of Our Resentment"

INDIANS

There were two declarations of independence on 4 July 1776. One was issued by a nascent imperial power that would go on to enjoy a world hegemony unparalleled in human history. The other was from a nation that also had once entertained, admittedly within the confines of its smaller sphere of influence, imperial ambitions. It too had enjoyed conquests and assimilated the defeated into its empire. Its fate, however, would be very different. The first declaration (which characterized the indigenous peoples as "savages") would write "finis" to the second.

A powerful consortium of Indian nations (Iroquois, Delaware, Ottawas, Cherokee, Wyandot, Mingo, and Shawnee) gathered in grand council at Muscle Shoals on the Tennessee River on or about 4 July 1776 to make their declaration of independence—independence from patriot America. Cornstalk of the Maquachake Shawnee (who would later be murdered by patriot militia while in captivity) spoke for them all: "It is better for the red men to die like warriors than to diminish away by inches. Now is the time to begin. If we fight like men we may hope to enlarge our bounds."[1]

The Iroquois Confederacy was, at the outbreak of war, a 200-year-old alliance composed of Mohawk, Seneca, Oneida, Onondaga, Cayuga, and, from 1722, the Tuscarora.[2] It had been a highly successful colonial power whose domain stretched from the Hudson Valley in the east to the shores of Lake Ontario in the west. Built into its "articles of confederation"—"The Great Law of Peace of the Longhouse People"—was an aggressive declaration of the right of conquest: "When the council of the League has for its object the establishment of the Great Peace among the people of an outside nation and that nation refuses to accept the Great Peace, then by such refusal they bring a declaration of war upon themselves from the Five Nations.[3] Then shall the Five Nations seek to establish the Great Peace by a conquest of the rebellious nation." But as with so many similar declarations, be they the Pax Romana, Pax Britannica, or Pax Americana, economic self-interest lay beneath the thin veneer of altruism. The Iroquois Confederacy was fighting to maximize its trading position with first the Dutch and then the English and, like any good corporation, saw acquisition as the way to add muscle to its negotiating arm. Between 1648 and 1656 the Iroquois had destroyed their competitors, the Huron, the Tobacco, the Neutral Nation, and the Erie, to the west. They would have certainly gone on to conquer the Mahican (Stockbridge) of the upper Hudson Valley, the Susquehannock of the Susquehanna Valley, and the Sokoki of the upper Connecticut Valley had not those tribes, like the Iroquois, been armed with European flintlocks.[4]

The patriot and British hierarchies had a problem with the Indians. And to some extent it was the same problem. On the one hand, the Indians had long been valued as light-infantry auxiliaries, as George Washington had realized in his early military career: "The cunning and vigilance of Indians in the woods are no more to be conceived, than they are to be equaled by our people. Indians are only match for Indians; and without these, we shall ever fight upon unequal terms."[5] On the other hand, they were held in considerable contempt and fear. To the majority of whites, whether American or British, they were savages, and to unleash them on white men, even enemies, was a profound moral betrayal. It was un-Christian. John Adams (whose grandfather had been killed by Indians) recoiled at "the incivility and Inhumanity of employing such savages."[6] As it turned out, a flimsy justification was constructed along the lines of "they did it first, so we can follow suit" to enable both sides to use Indians. For example, Lord Dartmouth, secretary for the colonies, wrote in July 1775: "the Rebels having excited the Indians to take part, and of their having actually engaged a body of them in arms to support their rebellion [even before the outbreak of hostilities, the Massachusetts Provincial Congress had called on the Stockbridge Indians to act as minutemen] justifies the resolution His Majesty has taken requiring the assistance of his faithful servants, the Six Nations."[7] At the end of the war both sides would betray their Indian allies and the patriot side would ruthlessly strip its Indian enemies of their lands.

The British had a considerable advantage over the "Bostonians," as Indians called the Americans. Their prewar policy had been designed to prevent, or at least retard, American incursions into tribal lands. The Exclusion Proclamation of 1763, cemented in 1768 by the Fort Stanwix Treaty, attempted to create no-go zones for land speculators seeking further westward expansion. (George Washington, for one, was disgruntled by the legislation.) They were in part good-faith measures and in part a self-interested attempt to limit Indian-white friction and thereby control the cost to Britain of frontier policing.

Given the bloody history between colonists and native Americans, the patriot hierarchy knew there was little chance of winning over the mass of Indians to the cause. If their neutrality could be secured, that was probably going to be the best result the Americans could expect, and indeed, a treaty of neutrality was agreed on between the Iroquois at German Flats in 1775. But continued encroachments on their ancestral lands and the inability of patriot America to supply them with the goods, particularly firearms and powder, on which they now relied pushed most of the tribes into the British camp.

The British colonial administration had been very successful in creating an infrastructure to support and supply many of the tribes east of the Alleghenies. The Indian Department, led by such inspired administrators as William Johnson, who was thoroughly integrated into

Indian society through marriage, had worked assiduously and wisely to forge and nurture alliances. The payoff for the British would be to profit from the Indians' passionate resentment of American settlers, an anger that had only intensified as the white population of America exploded during the eighteenth century. For example, in New York colony alone whites numbered little more than 20,000 in 1700; by 1740 they were 65,000, and in 1770 they had soared to 160,000—an 800 percent increase in seventy years.[8] The population of the whole of the eastern woodland Indians was about 150,000 in 1775.[9] In any event winning Indians to the patriot cause was flawed at its heart because American interests would be best served by fighting rather than affiliating with those Indians whose lands had long been targeted for expropriation.

The war patriot America fought against the Indians was like a Chinese box within the larger box of its war against Britain—a smaller replica, with Americans cast in the role of Britain. And to a large extent the tactics employed by the patriots were Europeanized, formalized, dependent on mass and weight. Indian tactics, on the other hand, were most successful when they followed the traditional forms of guerrilla warfare: the ambush, hit-and-run attack, firing from cover, retreat and counterattack, sniping, picking off weaker forces, and using the psychological impact of terror. It was an alien form of warfare that the early colonists cannily adopted for their own use: "In our first war [Pequot War] with the Indians, God pleased to show us the vanity of our military skills in managing our arms, after the European mode," wrote John Eliot to Robert Boyle on 23 October 1677. "Now we are glad to learn the skulking way of war."[10]

On those occasions when Indians were either headstrong enough or pressed by their British overlords to fight in static and compacted formation, they usually came to grief. For example, Burgoyne learned that at best they were a fifty-fifty proposition on his invasion from Canada. He had about 400 with him at the start, but whatever advantage they conferred as scouts and skirmishers was canceled by their refusal to conform to the hierarchical command structure of a European army. Their war-fighting tradition valued individualistic and opportunistic combat rather than "corporate" uniformity—a failure of "discipline" as far as Europeans were concerned. Forced into a role within a formalized army, Burgoyne's Indian auxiliaries often performed poorly. They ran away at Bennington: "The Indians to a Man, and most of the Canadians Ran away at first and got safe," wrote a disgusted British artilleryman, Lieutenant James Hadden. Lieutenant Thomas Anburey summed up Burgoyne's native auxiliaries.

> *They were of vast service in foraging and scouting parties, it being suited to their manner; they will not stand a regular engagement, either through the motives I formerly assigned, or from fear, but I am led to imagine the latter is the case, from the observation I have made of them in our late encounter with the enemy. The Indians were running from wood to wood, and just as our regiment had formed in the skirts of one, several of them came up, and by their signs were conversing about the severe fire on their right. Soon after, the enemy attacked us, and at the very first fire the Indians ran off.*[11]

Other aspects of Indian behavior also deeply offended Europeans, even those who were their allies. Their treatment of captives, for example, had more in common with the pre-Columbian rituals of the Aztecs and Maya. Excruciating tortures were inflicted to test the captive's soul, and cannibalism captured his strength.

> *Burning was a common element in torture. The victim was frequently made to walk barefoot over fires, as well as being slowly roasted in other ways. Hot knives and hatchets would be applied to his body till his skin was in shreds. His muscles would be pulled out and pierced. Hot irons or splinters would be thrust through his limbs. His fingernails would be wrenched out, his fingers crushed, his flesh cut, his scalp removed. The whole village—men, women, and children—would usually participate."*[12]

Although these gruesome rituals had a religious or spiritual basis—a gift to the god of war—they appalled eighteenth-century Europeans, who, nevertheless, were at peace with their own predilection for hanging-

and-quartering, the ritual execution of children, and, in their still fresh history, the rack, the stake (as recently as December 1774, black slaves who went on a murderous rampage in St. Andrew Parish, Georgia, were burned alive after capture),[13] and all those other ingenious instruments fertile Christian minds had devised to bring the benighted to a better understanding of God's mercy.

That other trademark of Indian warfare—scalping—had also been enthusiastically adopted by whites. There was a long tradition of paying bounties for Indian scalps: "This Scalping Business hath been encouraged, in the Colonies, for more than a century past. Premiums have been given, frequently, by the *Massachusetts* Assemblies, for the Scalps of Indians, even when they boasted loudest of their Sanctity," declared the Loyalist Peter Oliver.[14] The British paid bounties for live captives and scalps. As it was the same price for both, those captives who could not keep up during the forced marches were often dispatched and scalped.

Native Americans, contrary to the woolly-minded notion that they were simply put-upon protohippies, fought for their land ferociously. Women and children were killed as often as they were taken into captivity and adopted. The war was brutal, especially during the 1778 Cherry Valley and Wyoming Valley campaigns. For example, Colonel John Butler's Loyalist Rangers and his pro-British Indian allies, mainly Seneca and Cayuga, lured the 450-strong garrison of Forty Fort out into the open and either killed or wounded 300 of them—a number larger than the casualties of many of the "big" battles of the war. The fate of captive soldiers could be grim, to put it mildly. And it worked both ways; in March 1782 the patriot colonel David Williamson gathered ninety "friendlies"—Christian (Moravian) Delaware, men, women, and children—at Gnadenhutten, Pennsylvania, tied their hands behind their backs, and executed each one with a mallet blow to the head.[15] Despite an outcry, Williamson was never punished. To induce the British garrison at Vincennes to surrender, the leader of the besieging patriot force, George Rogers Clark, pulled out five captured and bound Indians and tomahawked them to death for the edification of the besieged.[16] And so it went on, in the immemorial way of war, the systole and diastole of savagery.

By the end of 1778 the northern and western frontiers had been ravaged by a series of Loyalist-Indian raids, and George Washington decided to launch a massive punitive expedition—in fact three expeditions acting in concert—to suppress the Iroquois of the Six Nations. The overall commander was John Sullivan, a man of indifferent military ability and cautious almost to the point of paralysis. Washington's orders were categorical: the land of the Iroquois was not to be "merely overrun but destroyed."[17] Sullivan, who relished the chance to "march against an enemy whose savage barbarity . . . has rendered them proper subjects of our resentment," painstakingly (excruciatingly so as far as Washington was concerned) gathered a huge column consisting of 2,500 men, 1,200 packhorses (twenty for Sullivan's personal baggage alone), and 120 boats for artillery and stores. He was to set off from Easton, Pennsylvania, move north through the Susquehanna Valley, and meet up at Tioga with a second column of 1,500 men (including a contingent of Oneida who had broken with the Iroquois Confederacy) under Brigadier General James Clinton which would be coming south from the Mohawk Valley. A third column of 600 men under Colonel Daniel Brodhead set off from Fort Pitt (modern Pittsburgh) up the Allegheny Valley with the goal of meeting up with Sullivan at Genesee. On 18 June 1779 Sullivan's leviathan left Easton, making only six miles a day, and not until 22 August did it meet up with Clinton's brigade at Tioga in Delaware country.

On 19 August the Delaware made a fateful decision. They would stand and meet the invading patriot army at Newtown. Captain Walter Butler of the Tory Rangers urged them to remain true to their tactical heritage and advised sending out strong parties "along the Heights to harass the Enemy upon their March and keep them in perpetual alarms." The Delaware were adamant, and Butler "was obliged to comply." Dug in behind log breastworks, they were no match for Sullivan's artillery and were soon in full retreat, "thoroughly intimidated." The Sullivan expedition certainly destroyed many villages and thousands of acres of crops, and notched up an impressive body count. But the more thoughtful patriot officers realized the expedition's strategic futility: "The question will naturally arise," wrote Major Jeremiah Fogg on the return of the expedition in September 1779, "what have you to show for your exploits? Where are your prisoners? ... When the querist will point out a mode to tame partridges, or the experience of hunting wild turkey with light horse, I will show them our prisoners. The nests are destroyed but the birds are still on the wing."[18]

The following year, 1780, the birds were truly on the wing. Led by the great Mohawk war leader Joseph Brant with 500 Indians and Tories, and Sir John Johnson with 400 Tories and 200 Indians, the Mohawk Valley was torched, Canajoharie overrun, a Pennsylvania militia column cut to pieces, and the Scoharie Valley in Tryon County, New York, ravaged. As late as June and August 1782 Indians were inflicting substantial defeats on patriot forces. In June Colonel William Crawford's force, sent to pacify the Delaware and Wyandot of the upper Sandusky, was destroyed and Crawford burned alive. In August the half-breed Simon Girty killed or captured 50 percent of Hugh McGary's Kentucky militia on the lower Blue Licks River.

Although the patriot cause was anathema to most Indians, one tribe—the Stockbridge—was as loyal as its fate was tragic. Lieutenant Johann von Ewald of the Schleswig Jäger Corps described the Stockbridge Indian contingent of the Continental army after the battle of White Plains on 31 August 1778.

> Their costume was a shirt of coarse linen down to the knees, long trousers also of linen down to the feet, on which they wore shoes of deerskin, and the head was covered with a hat made of bast [bark]. Their weapons were a rifle or musket, a quiver with some twenty arrows, and a short battle axe which they know how to throw very skillfully. Through the nose and in the ears they wore rings, and on their heads only the crown remained standing in a circle the size of a dollar-piece, the remainder being shaved off bare.[19]

What is chilling about Ewald's meticulous description is that he was reviewing the dead. The Stockbridge had just been killed almost to a man in a vicious skirmish with German and Loyalist forces at King's Bridge (now Van Cortlandt Park in the Bronx). A Loyalist officer recorded that "no Indians, especially, received quarter, including their chief called Nimham and his son, save for a few."[20]

The Stockbridge of western Massachusetts were part of the Mahican tribe and, since the arrival of the Europeans, had experienced an unrelenting dispossession of their lands. In the 1730s they had been converted to Christianity and during the French and Indian War had been loyal to the British and American cause, fighting with Rogers' Rangers. In the 1760s Chief Nimham protested the escalating rent demands being made on the Stockbridge by great Hudson Valley landowners like the Philipse family. In 1765 he went to New York to seek redress but could not persuade any lawyer to represent his people. In 1766 he even went to the authorities in London to make petition, but they merely referred him back to the authorities in New York, who threw out the case. By 1774 the 75 percent of land they had owned in the town of Stockbridge had dwindled to 6 percent. Even so, a number of Stockbridge volunteered as minutemen, thirty-five of them joining Captain William Goodrich's company, and in the spring of 1775 seventeen of them joined the Continental army at Cambridge; it would be their presence, noted by the British general Gage holed up in Boston, that gave him a convenient justification to bring on the vastly more numerous Indian allies of the British: "We need not be tender of calling upon the Savages, as the Rebels have shown us the example."

The Stockbridge's attempts to integrate themselves with, and perhaps appease, Americans were doomed. By war's end, their ancestral lands were taken by patriot speculators. The Stockbridge lamented, "The late unhappy wars have Stript us almost Naked of every thing . . . our Numbers vastly diminished [they were reduced to 450 souls], by being warmly engaged in favour of the United States Tho' we had no immediate Business with it . . . we are truly like the man that fell among Thieves, that was Stript, wounded and left for dead in the high way."[21] The same fate awaited the Oneida and Tuscarora, who also took up the tomahawk in the patriot cause.

One of the great ironies of the independence struggle is that while patriots had gone to war to free themselves from empire, they were bloodily engaged in creating their own at the expense of native Americans.

As General Nathanael Greene wrote on 4 January 1776, "Heaven hath decreed that Tottering Empire Britain to irretrievable ruin—thanks to God since Providence hath so determined America must raise an Empire of permanent duration."[22] Joseph Brant ironically borrowed classic Whig rhetoric when he exhorted an Indian gathering at Oquaga, near Niagara, in December 1776 to "defend their Lands & Liberty against the Rebels, who in a great measure began this Rebellion to be sole masters of this Continent."[23] Chief Cornstalk of the Maquachake Shawnee perhaps lacked Brant's Europeanized sophistication (Brant had been educated by missionaries, spent time in England, helped translate the Gospels into Mohawk, was an Anglican and Freemason, and the subject of one of the greatest portrait paintings of the period, by Gilbert Stuart), but he put the case with brutal clarity when on 7 November 1776 he sent this address to Congress via the patriots' Indian agent, George Morgan.

> *Now we & they [the Shawnee] see your people seated on our Lands which all Nations esteem as their & our heart—all our Lands are covered by the white people, & we are jealous that you still intend to make larger strides—We never sold you our Lands which you now possess on the Ohio between the Great Kenhawa & the Cherokee & which you are settling without ever asking our leave, or obtaining our consent . . . We live by Hunting & cannot subsist in any other way—That was our hunting Country & you have taken it from us . . . Now I stretch my Arm to you my wise Brethren of the United States . . . I open my hand & pour into your heart the cause of our discontent.*[24]

Thomas Jefferson did not mince his words when it came to Indians and American imperial ambition: "Nothing will reduce these wretches so soon as pushing the war into the heart of their country. I would never cease pursuing them while one of them remained on this side of the Mississippi."[25] This sentiment would be heartily endorsed by the newly installed president, George Washington: "The gradual extension of our settlements will as certainly cause the Savage as the Wolf to retire; both being beasts of prey tho' they differ in shape."[26] In the American mind

Indians had been transformed into animals, almost a kind of vermin—a convenient justification for the ethnic cleansing that was to follow. And when the inflatus of patriotic rhetoric began to drift away, like smoke from the battlefield, the bare bones of self-interest were revealed. James Duane, chairman of Indian affairs, declared that if Indian lands were not seized as the spoils of war, "this Revolution in my Eyes will have lost more than half its' [sic] Value."[27]

With the Peace of Paris that concluded the war in 1783, Britain completely abandoned its native allies, whose lands "as far west as the Mississippi" were ceded to the United States. Shelburne, the new prime minister, wrapped the cynicism in rhetorical packaging: "The Indian nations were not abandoned to their enemies; they were remitted to the care of neighbours."[28] It would be the tender care the wolf lavishes on the lamb.

PART TWO

THE GREAT BATTLES

The War in the North

Ambush

LEXINGTON AND CONCORD,
19 APRIL 1775

At 2:00 AM on Wednesday, 19 April 1775, 700 men under the command of Francis Smith, lieutenant colonel of His Britannic Majesty's 10th Regiment of Foot, a portly fifty-two-year-old with thirty-four years of solid if uninspired service under his ample belt, set off from Boston on the twenty-mile hike to Concord, Massachusetts. Smith's strike force was a mixture of twenty-one light infantry and grenadier companies of several regiments, unsupported by artillery in order to move swiftly.[1] It had been ferried over from Boston around 10:00 PM on the eighteenth in what its commander in chief, Sir Thomas Gage, had hoped would be an operation wrapped in deepest secrecy. Smith's mission was to locate and destroy the military stores that Gage's network of spies had told him lay in and around Concord. It was, in fact, a ridiculously optimistic notion that in a city teeming with patriots 700 men could be gathered, boarded on long boats, rowed across the Back Bay, and landed at Lechmere without raising suspicion. Notwithstanding Paul Revere's legendary "one-if-by-land, two-if-by-sea" warning lantern, it is hard to imagine that Smith's sizable force could have ever gotten off undetected. In addition, Gage's own security was as leaky as a colander.

The light companies of the 4th and 10th Foot led the way under the command of Royal Marine major John Pitcairn, a Scot with twenty years of service who was a year older than Smith. Ahead of them two young officers, Lieutenants William Sutherland and Jesse Adair, scouted on foot. Behind came the column of grenadiers under Smith, and a small contingent of Royal Artillery that had been sent out the previous afternoon with light chaises loaded with equipment to spike the cannon they anticipated finding at Concord. Various locals unlucky enough to blunder into the column were detained, but the word had already been quite effectively spread by Revere and other messengers. The redcoats were headed into a world of trouble.

When Pitcairn and the advance guard arrived at Lexington Green at about 4:30 AM, they were confronted by 130 or so armed patriots under the local militia commander, Captain John Parker. What happened next can only be surmised. There are plenty of possible reasons that one side or the other would have opened fire, and it matters little which did so first. This was an exchange waiting to happen, and had it not been here, it would have been elsewhere. John Pitcairn is usually portrayed as a levelheaded, humane, and diplomatic soldier, much revered by his regiment and certainly not one to ignite the tinder with a careless shot. But there was another side to him that may have some bearing on the showdown at Lexington. Pitcairn was a hard-liner when it came to suppressing the rebellion. A month before the confrontation at Lexington he had written to the earl of Sandwich, "I am satisfied that one active campaign, a smart action, and burning two or three of their towns, will set everything to rights."[2] Some historians have argued that Pitcairn would have had nothing to gain from killing patriots on Lexington Green, but it could also be argued that he had nothing to lose. A whiff of musketry, "a smart action," a crack of the whip, as he might have expressed it, may have had something to recommend it. Certainly he must have known even before the exchange of gunfire on the green that the game was up and all hope of surprise long gone. There was zero chance in those early hours that he and Smith would be able to hit Concord unannounced, destroy the patriot stores, and get off back to Boston unmolested. (In fact, Dr. Samuel Prescott had raised the alarm at Concord at about 1:30 AM.) Even before the opening shots, Smith, reacting to patriot alarm guns, had sent back to Boston for reinforcements.

Leaving eight patriot dead and nine wounded, the column, screened by elements of the light companies deployed on the flanks, pushed on to Concord six miles away. At around 7:00 AM the British approached Concord from the east, and the flankers chased off militiamen stationed on a ridge on the column's right. As Smith and Pitcairn entered the outskirts of the town, they were met on the road by a militia force under Colonel James Barrett, which, prudently, about-faced and marched back into the town, crossed the North Bridge, and took up position on Punkatasset Hill, which overlooked the bridge. Smith now secured his position by sending three light companies under Captain Pole to the South Bridge and seven companies under Captain Lawrence Parsons to the North Bridge. The grenadiers were kept in the town to search out military stores.

When Parsons arrived at the North Bridge, he left three companies (about 100 men) under Captain Walter Laurie and took the rest over the bridge to head off for Colonel Barrett's farm, which Gage's spies had identified as the main cache of patriot weapons and ammunition. Laurie sent two of his companies under the command of Lieutenant Waldron Kelly across the river and positioned them on the lower slopes of Punkatasset, keeping the remaining company on the town side of the bridge. Tactically it all seemed perfectly reasonable: a cordon defense that could retreat back over the bridge and create a "plug" that would defy any attacking militia. But tactics are only as good as the training and experience of the men who have to carry them out. Things were about to unravel.

Barrett, on Punkatasset Hill, was alarmed by what seemed to be the British busily setting fire to the town (in fact this was not the case), and started to descend from his perch en route to the North Bridge. By now his numbers had swelled to about 400 by militia reinforcements from surrounding towns like Acton, Carlisle, Chelmsford, Westfield, and Littleton. The two British companies across the river tried to funnel back across the bridge and, as they awkwardly backpedaled, were fired at and two redcoats were killed on the far side of the bridge.[3]

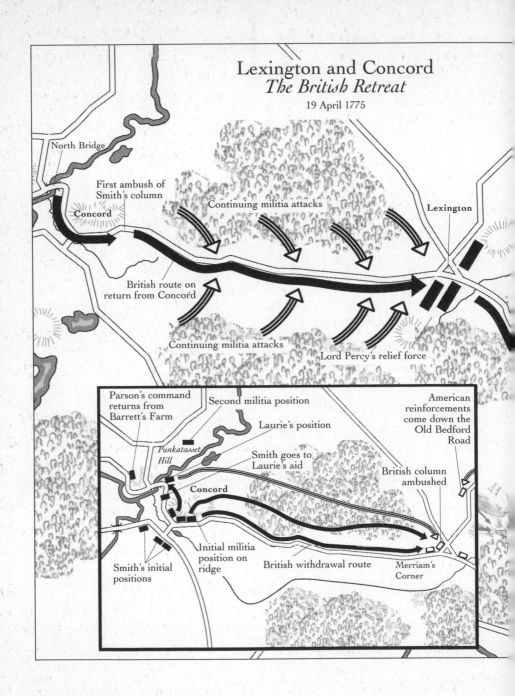

Lexington and Concord
The British Retreat
19 April 1775

North Bridge

First ambush of
Smith's column

Continuing militia attacks

Lexington

Concord

British route on
return from Concord

Continuing militia attacks

Lord Percy's relief force

Parson's command
returns from
Barrett's Farm

Second militia position

American
reinforcements
come down the
Old Bedford
Road

Laurie's position

*Punkatasset
Hill*

Smith goes to
Laurie's aid

British column
ambushed

Concord

Smith's initial
positions

Initial militia
position on
ridge

British withdrawal route

Merriam's
Corner

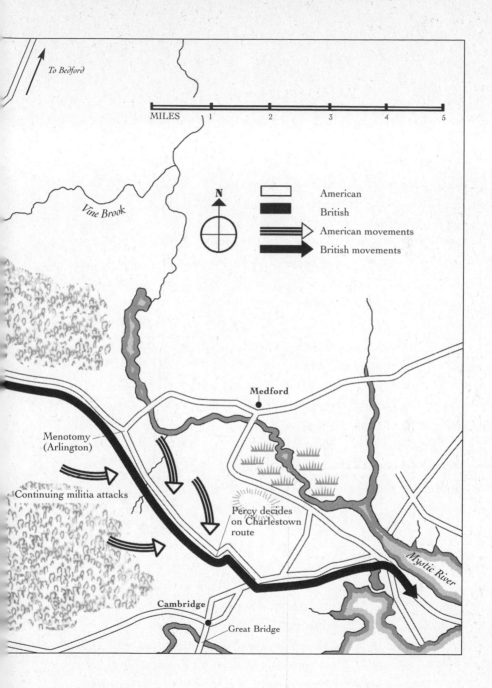

To Bedford

MILES 1 2 3 4 5

Vine Brook

N

American
British
American movements
British movements

Medford

Menotomy
(Arlington)

Continuing militia attacks

Percy decides
on Charlestown
route

Mystic River

Cambridge

Great Bridge

Laurie now tried to execute a complicated maneuver called "street fighting," whereby he would draw his companies into columns of twelve lines of eight men each. The theory was that the first line would fire and then peel off to each side and move to the rear, leaving the second rank to fire, peel off, and so on down the length of the column, exposing the patriots to continuous volley fire. It was an evolution that had been developed in Britain when the army acted as "police" putting down civil unrest. It may have worked effectively against unarmed strikers and rioters, but the realities of battle are not always sympathetic to complicated maneuvers carried out by frightened and inexperienced men. Under fire, the discipline of the redcoat front ranks collapsed, blocking the line of fire of the men stacked up behind them. Laurie's force was pushed back 400 yards until it collided with units of the grenadiers Smith had sent up in support. A bayonet charge by the grenadiers drove off the patriots, who withdrew back across the bridge and partway up Punkatasset Hill. Surprisingly, Parsons returned from his foray to Barrett's farm and recrossed the North Bridge without being molested by the militia on Punkatasset and the whole British force reassembled in preparation for their long march back to Boston. It was now close to noon.

Smith, sensibly, had flankers out on both sides of his column, where, recorded Lieutenant Sutherland, "a vast number of armed men drawn out in battalia order, I dare say near a thousand were approaching through the trees" on the column's right and "a much larger body drawn up to my left,"[4] but as the column approached a small bridge over Tanner's Brook at Merriam's Corner, the flankers were forced back down to the road in order to cross with the main force, and this afforded the militia, now massed on both sides of the road, the opportunity to close and fire on the redcoat units as they were funneling over the bridge. British casualties started to mount. They had gone only one mile, but it was the prelude to a bloody running of the gauntlet that would continue all the way back to Lexington, where a relief force of about 1,400 under Lord Percy was waiting for them. Panicked, dehydrated, burdened with wounded, and almost out of ammunition, the Smith force ran and stumbled into Percy's protective cordon at 2:30 PM.

After allowing the exhausted men of Smith's column thirty minutes' rest, Percy moved the whole British force back to Boston in a skillfully executed retreat, his flanking troops ruthlessly clearing houses that harbored patriot snipers as they went. (In Jason Russell's house, for example, they killed all the eleven "minute men" they found there, including Russell himself.) At a fork in the road near Cambridge Percy had to make a critical decision: to take a right turn in order to cross the Great Bridge over the Charles River that he had taken earlier that day to bring him to Lexington, or push on and cross Charlestown Neck and have the troops ferried back across the Charles to Boston. Percy, a distinguished veteran of the battle of Minden in 1759, knew what was probably waiting for him at the Charles bridge and made the correct call. As dusk fell he took his battered force on to the Charlestown Peninsula, closing the door on the pursuing patriots. Smith's column had marched fifty miles and had been in action for almost twenty-four hours without rest. The foray had cost the British 73 killed and 174 wounded, "many mortally." Including the casualties at Lexington, the patriot losses were 49 killed and 39 wounded, a number that, although accepted by most authorities, seems completely out of proportion to the usual ratio of wounded to killed (roughly 3:1).

The ill-fated expedition to Concord is often used to illustrate the inability of British tactical doctrine to adjust to American conditions. Louis Birnbaum, the author of *Red Dawn at Lexington* (1986), puts the conventional case, "From a tactical point of view, the British withdrawal from Concord was a contest between the tactics of the British, based on the formal practices of Frederick the Great, and the tactics of a light skirmish line founded on the pragmatic strategy of the American riflemen."[5] It is an analysis that has taken on a kind of genuflective inevitability. In fact, Smith dealt with the tactical problem intelligently. His primary challenge was *strategic*. He had been sent on an operation that would put him in the greatest peril through a combination of the inequalities of the contending forces and the nature of the terrain, which greatly favored the patriot gunmen. As far as he could, he fought against his strategic disadvantages with tactical intelligence. The idea that he employed "German" formal tactics against quick-footed

American irregulars is not supported by the facts. Smith had few tactical options given the situation in which he found himself, but he made the best of it. Far from deploying in "European" line, he put out his light company flankers to search out and destroy the patriot gunmen, which they did to some effect. (Most American casualties were inflicted by British light forces outambushing the ambushers.) As fatigue took its toll, however, the flankers were forced to work closer to the road, and so the opportunities for the militia increased and Smith's options shrank to one overriding imperative: drive the column forward and try to suppress the growing wave of panic.

The Americans, given their numerical and geographic advantages, had many more tactical options, which they found difficult to exploit, and British vulnerability was never turned into overwhelming victory. No force swept in front of the struggling column to cut it off and annihilate it.[6] Inexperience and a lack of leadership were to blame, rather than any failure of courage. Lord Percy would write privately to the administration in London: "Whoever looks upon them as an irregular mob will be much mistaken. They have amongst them those who know very well what they are about, having been employed as rangers among the Indians . . . Nor are several of the men void of spirit or enthusiasm . . . for many of them . . . advanced within ten yards to fire at me and other officers, though they were mortally certain of being put to death themselves in an instant."[7]

In this opening battle, sometimes dismissed as a mere skirmish, a fundamental equation of the war was being enacted. An occupying force was being swarmed, and that, essentially, was the stark arithmetic some strategists in England feared the most: numbers would determine the eventual outcome. Gage, in Boston, saw it clearly. To his political masters in London he wrote, "If force is to be used at length, it must be a considerable one, for to begin with small numbers will encourage resistance and not terrify; and will in the end cost more blood and terror."[8]

13

"A Complication of Horror . . ."

BUNKER'S HILL, 17 JUNE 1775

General Thomas Gage was, in the idiom of the day, "pent up." Boston was useless strategically, a suffocating little room in which the British army in North America found itself locked. "To keep quiet in the Town of Boston only," Gage wrote on 10 February, "will not terminate Affairs; the Troops must March into the Country." The first objective was to secure the two heights that overlooked the city: the Dorchester Peninsula to the southeast, and the Charlestown Peninsula to the northwest. In a council of war with the recently arrived triumvirate of generals—Howe, Burgoyne, and Clinton—Gage settled on a date to attack the Dorchester Heights. It would be Sunday 18 June, perhaps to take advantage of the Sabbath preoccupations of the patriots. Once captured, the Heights would afford a jumping-off point to attack the main patriot army at Cambridge.[1] Ironically, the British had already taken possession of Bunker's Hill and fortified it in the days after Percy and Smith's retreat from Concord, but Gage, ever fearful of stretching his defenses, had withdrawn the garrison back to the city. It was but one of many fateful mistakes.

Of course, nothing was truly secret in Boston, and the British plans were quickly known to the patriot command at Cambridge. Artemus Ward, the commander in chief of the New England troops around

Boston (11,500 from Massachusetts under Ward and John Thomas; 2,300 from Connecticut under Israel Putnam; 1,200 from New Hampshire under Colonel John Stark; 1,000 from Rhode Island under Nathanael Greene), conferred with the Massachusetts Committee of Safety a few days before Gage's target date and resolved to preempt the British attack with one of their own, aimed at diverting attention from Dorchester Heights. The committee issued its recommendation.

> *Whereas, it appears of Importance to the Safety of this Colony, that possession of the Hill, called Bunker's Hill, in Charlestown, be securely kept and defended; and also one hill or hills on Dorchester Neck be likewise Secured. Therefore, Resolved, Unanimously, that it be recommended to the Council of War, that the abovementioned Bunker's Hill be maintained, by sufficient force being posted there.*[2]

It seemed clear enough. But not to the man tasked with leading the expedition. William Prescott of Massachusetts would later state emphatically in a letter to John Adams on 25 August 1775, "On the 16 June in the Evening I recd. Orders to march to Breed's Hill in Charlestown, with a party of about one thousand men."[3] Prescott, accompanied by Colonel Richard Gridley, an artillerist and engineer who had surveyed Charlestown Peninsula early in May and had recommended constructing a redoubt on Bunker's Hill, led their men—about 850 from three regiments—out of Cambridge around 9:00 PM on the 16th, picking up about 200 Connecticut men under the command of Captain Thomas Knowlton and several wagons loaded with entrenching tools.

When the Americans arrived on the peninsula (probably around 10:00 PM), a long and contentious conference took place between Prescott, Gridley, and, in all probability, Israel Putnam. The decision to fortify Breed's rather than Bunker's Hill posed some logistical challenges. It would increase by fivefold the area of the peninsula the patriots would have to defend, and it placed the main force five times as far from the supplies and reinforcements that would come across the

Neck. It also left the Americans appallingly vulnerable to being cut off by British landings in their rear. But it did do one thing that probably reflects the persuasive powers of the highly aggressive, some might say hotheaded Putnam. Being only 500 yards from the Charles, emplacements on Breed's threatened Boston and the shipping in the harbor in an immediate way Bunker's could not. It was, to use a modernism, "in the face" of the British, and if the strategic goal was to draw the British away from Dorchester Heights, it achieved that goal admirably.

Gridley plotted out the redoubt on Breed's summit. Private Peter Brown recorded that they "made a fort of about ten rod [1 rod = 16½ feet] and eight wide [an oblong box 165 feet long by 132 feet wide, with the long sides facing north and south] with a breastwork of 8 [132 feet] more."[4] In the side facing Charlestown (that is, facing almost due south) Gridley planned a redan, a triangular "spur" jutting from the wall face, that would enable defenders to enfilade attackers. In the back wall (facing north, back toward Bunker's Hill) was a narrow opening that was the only entry or exit to the redoubt. The earth walls were approximately six feet high with a firing step but without an adequate platform for guns or embrasures through which they could fire—a surprising lapse of Gridley's considering his military background. The breastwork Peter Brown mentioned ran at right angles from the east corner of the back wall of the redoubt, northeast down the hill to a track. Two hundred yards back from where the breastwork ended, Thomas Knowlton constructed a barricade of fence rails stuffed with hay. The gap between the end of the breastwork and the beginning of Knowlton's fence, which would become a crucial weak spot in the American defensive system, was covered only by three flèche, V-shaped barricades made of fencing. The construction of the redoubt and breastwork started at midnight.

When did the British get wind of the Americans' preparations? Henry Clinton's account would have us believe that he was the only one prescient enough to be prowling around sometime after midnight and heard the construction work across the waters. He then, again by his own account, roused Gage and Howe, and urged them to organize for a daybreak attack. Neither Gage, Howe, nor Burgoyne made any mention of this in their subsequent accounts of the battle. Burgoyne, for example, says that the first the British command knew of events on Charlestown was when the twenty-gun HMS *Lively* opened fire on the redoubt at dawn (about 5:00 AM). Of course, if Clinton's account is correct, it puts the other British generals, and Gage in particular, in a very bad light, but Clinton (although one of the ablest tacticians of the war) was an isolated, somewhat paranoid character who may not have been above "adjusting" the record to show himself to advantage. Clinton's account is grist for the mill for many historians because it conveniently feeds into the stereotype of the British high command as terminally lethargic and self-indulgent.[5] Howe, in his account, concedes that British sentries indeed heard the work in progress but did not report it until morning (and then only in casual conversation), and that the first Gage knew of it was when he was awoken by the *Lively*'s gunfire. Burgoyne also says the British knew nothing until the "dawn of day."

Wherever the truth lies, the British command made their assessment of the situation early in the morning of the seventeenth. Clinton offered to the council of war what seems to be the obvious suggestion: land troops close to the Neck and cut off the redoubt. Gage rejected it because it would put the British landing party between two enemy forces, and he was probably supported by Howe, who thought a frontal attack on Breed's would be "open and of easy assent and in short would be easily carried." Clinton, characteristically, ascribed Gage's rejection of his plan to the fact that Gage resented Clinton's experience (During the Seven Years' War Clinton had served in Germany, which was considered the premier league in contrast to America, where Gage had played out most of his career.)

Howe, as the ranking major general of the Clinton-Burgoyne-Howe triumvirate, took command of the operation. Nothing could be done until the tide turned. (Most historians put high tide at 3:00 PM, but according to the Loyalist Peter Oliver in his *The Origin & Progress of the American Rebellion,* "It was high Water about one o'clock after noon."[6] Peter Thacher, stationed in the redoubt, stated that the British left Boston "between 12 and 1 o'clock." And, in fact, Oliver's and Thacher's timing would make sense, as the British left Boston for the peninsula around 1:00 PM.) Around noon Howe assembled the first wave of ten light companies (under Lieutenant Colonel George Clark) and ten grenadier companies (under Lieutenant Colonel James Abercrombie) to be embarked from Long Wharf. They were joined by the battalion men of the 5th and 38th Foot. The remaining grenadier and light companies together with the battalion men of the 52nd and 43rd Foot assembled at the North Battery. These units, totaling about 1,500 men, were the first to be ferried over to Moulton's Point (sometimes called Morton's Point) at the southeast tip of Charlestown Peninsula. The 47th Foot and the 1st Battalion of marines under Major John Pitcairn waited at the North Battery for the boats to return and take them over.

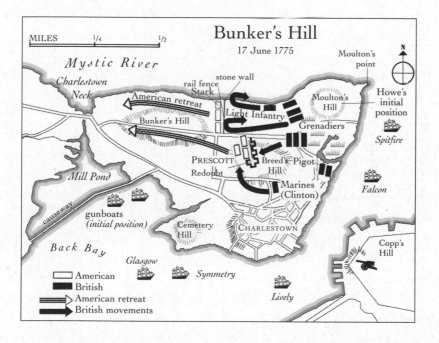

By 2:00 PM the first wave of British troops had landed[7] under the covering fire of the big twenty-five-pound guns on Copp's Hill at Boston, as well as the *Falcon* and *Lively* in the harbor and the *Glasgow* and *Symmetry*—two floating batteries up near the Neck on the Charles

River side of the peninsula. Artemus Ward, back in Cambridge, had been reluctant to reinforce Prescott. First, he must have been dismayed that orders to fortify Bunker's Hill had been ignored, and would have been unwilling to send more men into what he must have considered a disaster waiting to happen. He was not the only one who was horrified at the precarious position Prescott and Putnam had chosen; many men on Breed's thought they had been staked out as sacrificial scapegoats, perhaps by treachery. Second, he was ever mindful that his main responsibility was to husband the main army at Cambridge. But once it was clear that the British had entirely missed the opportunity to outflank the men on Breed's and were intent on a frontal attack, Ward committed reinforcements, including John Stark's New Hampshire men camped across the Mystic River, east of the Charlestown Peninsula at Medford.[8] Stark and his men would be crucial.

Howe and his second-in-command, Brigadier General Robert Pigot, reviewed their options from the top of Moulton's Hill while they waited for their artillery to come up. Lord Rawdon reported, "We had halted for some time till our cannon came up,"[9] which is a much more convincing explanation for Howe's delay than the popular jibe that Howe, ever indolent and stupid, stopped for a nice long lunch. The delay certainly gave the patriots precious time to complete their defenses. When Stark had arrived at the redoubt, he had immediately appreciated the threat to the American left flank posed by the beach along the Mystic. He had his men scramble to build a rough stone barricade as an extension of Knowlton's fence. It would prove to be the keystone to the American defense. And so, as the British waited, the only door that might have let them in behind the patriot lines closed in their face.

Howe could see the redoubt, the breastwork, Knowlton's fence, and many men milling around on Bunker's Hill, and he sent back to Boston for his reserves. But the British, in an extraordinary lapse that reflects badly on Gage and Howe in particular, had not reconnoitered the Charlestown Peninsula in the previous weeks and months. They knew this peninsula, so close to the city, could be vital and yet still failed to carry out the obvious precaution of surveying the ground. Howe's ignorance of the particular characteristics of the terrain, critically the network of fences and walls over which his men would struggle and the swampy ground at the foot of Breed's that would mire his artillery, would have disastrous consequences for British arms while favoring the patriots. "The ground on the peninsula," wrote an English officer, "is the strongest I can conceive for the kind of defence the rebels made, which is exactly like that of the Indians, viz. small inclosures with narrow lanes, bounded by stone fences, small heights which command the passes, proper trees to fire from, and very rough and marshy ground for troops to get over. The rebels defended this ground well, and inch by inch."[10]

In any event, Howe decided on a flanking attack up the beach by his light infantry while sending in his grenadiers, supported by the 5th and 52nd Foot, against Knowlton's fence. Pigot was to take three companies each of light infantrymen and grenadiers plus the battalion men of the 38th, 43rd, 47th Foot, as well as the 1st Battalion of the marines, and conduct a holding attack against the breastwork and the southern wall of the redoubt while Howe turned the patriot left and came in through the back door.

It was now approximately 3:00 PM, and the first to go were Clark's light companies up the beach with the 23rd (Welch Fusiliers) at the head of a column of about 400, four abreast. Timing was critical to eighteenth-century warfare. The time it took defenders to reload a musket or rifle gave the attackers a crucial window of opportunity to overwhelm the defense at the point of the bayonet. Stark had three ranks of men behind the wall and had staked firing markers fifty yards out. Fire discipline would be critical to the defenders. If they all fired together or fired too soon (as inexperienced and frightened men could be expected to), they would be overrun. It is a testament to Stark's extraordinary leadership as well as his men's courage and discipline that they shattered their attackers with almost continuous volleys. On a day when impressive fire discipline would be one of the characteristics of the American troops generally, these New Hampshire militiamen were outstanding. Their will broken, the British light infantrymen were withdrawn, leaving ninety-six of their companions crumpled on the beach. The beach attack has all the hallmarks of an ad hoc tactical decision. Even the support of one light 3-pounder "grasshopper" artillery piece could have made a massive difference. (The 6-pounder, of which Howe had six, would probably have been too heavy in the soft ground of the shoreline.)

As Pigot, on the left, headed up Breed's Hill, and the grenadiers and battalion companies of the 5th and 52nd led by Howe advanced against Knowlton's fence, the British fieldpieces (six 6-pounders went forward while four 12-pounders were left firing from Moulton's Hill) became bogged down in the low-lying marshy ground at the foot of the rise. The men serving the guns sweated to get them up onto the firmer ground of a road. At some point it was discovered that twelve-pound balls had been issued to some of the six-pound guns. Much has been made of this snafu, and it seems to some historians only to underscore the almost laughable ineptitude of British command. Entertaining, if fanciful, theories have been put forward to account for it. For example, one theory suggests that Colonel Samuel Cleaveland, the head of ordnance back in Boston, had been seduced by the lovely daughter of a patriot family and had somehow been persuaded to sabotage the ammunition supply, and so on. What is more likely, though less picturesque, is that boxes of twelve-pound balls were mistakenly sent from Moulton's Hill. Once the mistake was discovered, it was relatively easy to rectify, and it certainly did not involve sending back to Boston for the correct caliber balls. In the meantime Howe temporarily switched to canister, which was largely ineffective against defenders protected by walls and earth breastworks.

Howe, by now committed to his attack, had no choice but to press on. The lines became disorganized as men climbed walls and fences. The ground was rougher than it had looked from Moulton's Hill, which further disrupted what had been meant to be an ordered attack with the bayonet. Men, disregarding their orders, began to stop and fire. Cohesion dissolved under withering fire from Knowlton's fence, and the attack failed, as did Pigot's for the same reasons. Pigot was also being "galled" by sniper fire from Charlestown, and so the order was sent back to Boston to fire on the town, which was done by lobbing "carcases" (hollow shells filled with an incendiary mixture of powder and pitch) into the mainly timber buildings.

Howe regrouped and after fifteen minutes launched the second attack. The light infantry were sent against Knowlton's fence, while Howe and Pigot moved against the breastwork and redoubt. Disciplined fire at short range from fence, breastwork, and redoubt cut the British down, and Howe was left looking into a black hole that seemed as though it would swallow his career and honor. It was, he wrote, "a moment that I have never felt before."

The third attack had a different tactical slant. Rather than approaching in line, the British moved as quickly as possible in column. Timing, again, was an important factor. The faster the attackers could cover ground, the less time they would spend in the target zone. An American in the redoubt described how the British "advanced in open order, the men often twelve feet apart in the front, but very close after one another in extraordinary deep or long files."[11] On the British right wing the guns were brought up the roadway running up the east side of the redoubt and turned to fire into the rear of the breastwork, forcing its occupants to evacuate.

Clinton, who had been watching the unfolding debacle from Copp's Hill on the Boston side, had himself rowed over and landed, under fire, on the beach near Charlestown where the 63rd Foot and the 2nd Battalion of the marines were aimlessly milling around. Clinton took charge: "I then collected all the Guards and such wounded men as would follow—which to their honour were many—and advanced in column with as much parade as possible to impose on the enemy. When I joined Sir William Howe he told me that I had saved him, for his left was gone." It was during this last attack that Major John Pitcairn, of Lexington and Concord fame, was mortally wounded.

Although the men in the redoubt poured a blistering fire into the attackers, they were running out of powder and ball and it was only a matter of time before the British began to swarm over the walls and converge on the defenders. Lieutenant John Waller, of the 1st Battalion of the marines, described the carnage: "I cannot pretend to describe the Horror of the Scene within the Redoubt when we enter'd it, 'twas streaming with Blood & strew'd with dead & dying Men, the Soldiers stabbing some and dashing out the Brains of others was a sight too dreadful for me to dwell any longer on." Prescott wrote that at this last stage of the battle he had "perhaps 150 Men in the Fort" and about 30 of them were killed within the earth walls or while desperately trying to fight their way out through the narrow exit. Among the fallen was the revered Dr. Joseph Warren, president of the Massachusetts Provincial Congress, shot in the back of the head and then bayoneted. Prescott himself fought his way out, parrying bayonet thrusts with his sword.

Although most of the American casualties were probably sustained during their retreat, it was certainly no rout. As men streamed toward the Neck, Colonel Moses Little's regiment crossed the Neck and plunged into the melee to form a defensive firing line that provided a shield for the fleeing Americans. It was heroic and costly. One of Little's companies, twenty-three men commanded by Captain Nathaniel Warner, took seventeen casualties. Stark and Knowlton made orderly fighting retreats. Lord Rawdon described how the patriots "continued a running fight from one fence, or wall, to another." Burgoyne, observing from Copp's Hill, concurred; it was "no flight: it was even covered with bravery and military skill." The whole battle had taken ninety minutes.

After the Americans had run the gauntlet of the Neck, which was being raked by the guns of the *Glasgow,* and the peninsula was finally cleared, the British were in no shape to pursue them onto the mainland. The cost of their victory had been punitive. Of a British force of about 2,500, 1,144 (45 percent) became casualties. Two hundred and twenty-six rank and file were killed and 828 wounded. Twenty-seven officers were killed and 63 wounded. "A dear bought victory, another such would have ruined us" said Clinton. Howe was shattered: "I freely confess to you, when I look on the consequences of it . . . I do it with horror." Of about 3,500 Americans on the peninsula, 138 were killed and 276 wounded (about 12 percent). A further 31 were wounded and captured, of whom almost two-thirds subsequently died.

Recriminations on both sides were plentiful. Ward was accused of staying snugly in his house in Cambridge rather than more actively supporting his troops on the peninsula. Prescott roundly lambasted Putnam for not personally leading the men on Bunker's Hill down to the front lines, rather than merely trying to persuade them. The performance of the American artillery was lamentable and resulted in court-martials and cashierings. There was cowardice cheek by jowl with courage; inspired command and the craven surrender of leadership. On the British side, an officer giving testimony to the parliamentary inquiry following the battle hit each nail of command failure squarely on its head.

> *Too great a confidence in ourselves, which is always dangerous, occasioned this dreadful loss . . . We went to battle without even reconnoitering the position of the enemy . . . Had we wanted to drive them from their ground, without the loss of a man, the Cymetry [sic] transport, which drew little water and mounted 18 nine-pounders, could have been towed up Mystic Channel and brought to within musket shot of their left flank, which was quite naked, and she could have lain water-borne at the lowest ebb-tide . . . Had we intended to have taken the whole rebel army prisoners, we needed only have landed in their rear and occupied the high ground above Bunker's hill . . . In advancing, not a shot should have been fired, as it retarded the troops, whose movement should have been as rapid as possible. They should not have been brought up in line, but in columns with light infantry in the intervals, to keep up a smart fire against the top of the breastwork.[12]*

He ended his list with the most damning indictment of all: "We are all wrong at the head . . . The brave men's lives were wantonly thrown away. Our conductor as much murdered them as if he had cut their throats himself on Boston Common."[13]

A Vaunting Ambition

QUEBEC, 31 DECEMBER 1775

I t started very well. "We were all in high spirits intending to endure with fortitude all the fatigues and hardships we might meet with," wrote Abner Stocking on 13 September 1775 as he and 1,100 others set off on the great adventure. They had been gathered by Colonel Benedict Arnold, under the express orders of the newly appointed commander in chief, to be the right hook to Brigadier General Richard Montgomery's left jab. Montgomery had set off up Lake Champlain with about 2,000 men headed for Montreal. Arnold's force was bound for Quebec. They were ill equipped, undermanned, mainly innocent of the realities of warfare, and they were going to conquer Canada.

Canada was crucial to both sides. For the Americans it was a dagger pointed at their heart. As John Adams described the threat, "In the hands of our Enemies it would enable them to inflame all the Indians upon the Continent." And, one might add, since the Quebec Act of 1774 had extended Canada's westward boundaries, it had put a crimp in the ambitions of American land speculators of whom George Washington, for example, was an enthusiastic member. The British, under Major General Sir Guy Carleton, had only about 800 regulars, supplemented by a Canadian militia. (This militia was mainly drawn from the French middle and upper classes and supported by the clergy. The laboring-class *habitants* cannily kept both British and patriot at arm's length unless there was a profit to be turned.) Faced first with Montgomery's invasion force, Carleton had been obliged to put most of his regular troops in forward positions—primarily the forts at Chambly and St. John, both on the Richelieu River which connected Lake Champlain and the St. Lawrence, with a few left in the crumbling and totally inadequate defenses of Montreal, where he stationed himself. Quebec seemed ripe for the picking, as Washington was thrilled to discover through an intercepted letter. He wrote to Philip Schuyler on 4 October, "If Quebec should be attacked before Carleton can throw himself into it, there will be a surrender without firing a Shot."

Montgomery had set off from Fort Ticonderoga on 25 August with about 1,700 (maybe as many as 2,000) men, mainly Connecticut and New York militias. The British force of about 600 at St. John hid in its shell like a terrified turtle (despite having access to a couple of armed schooners that could well have scuppered Montgomery's fleet of bateaux), and so Montgomery prudently bypassed it and headed up to Chambly, which, after a purely perfunctory resistance (exactly thirty-six hours), threw in the towel on 18 October and delivered up six tons of powder, 6,500 musket cartridges, 125 muskets, and a good quantity of artillery, including two 24-pounders. Thus fortified, Montgomery retraced his steps and laid siege to St. John, which fell on 2 November. Over on the right wing, however, things were not progressing quite as smoothly.

Arnold had elected (or perhaps more accurately was forced) to get to Canada the hard way. (The Royal Navy might not have looked kindly on his merely sailing up the coast and entering the St. Lawrence.) He had to lug his heavy boats and equipment up the Kennebec against the flow, get onto the swampy headwater plateau, and navigate the Chaudière—going with the flow, thankfully—until it broke out into the mighty St. Lawrence. It was a trip of about 300 miles. Arnold, ever flamboyant, claimed it was 600 and a march "not to be paralleled in history." Although mistaken about the mileage, he was not far wrong about the march. It was six weeks of sheer hell, not helped by setting off as winter came on (as it happened, one of the most vicious in living memory). It was a confounding command decision—Washington's, presumably—that crippled the expedition; more than half the men would never make it to Canada.

They dragged their bateaux through the icy waters; they humped crucifying loads across innumerable portages; they froze; they starved; they ate their moccasins, shaving soap, hair grease, and a dog, all of which quickened their bowels; then they were reduced to a flour-and-water paste, which made them intractably constipated. By the time they stumbled out onto the south bank of the St. Lawrence on 9 November, crossed-eyed with misery, their number had been whittled down to 600. On the pitch-black night of 13 November Arnold's ragged-assed and emaciated little army of sansculottes got across the river and established itself on the Heights of Abraham. ("We much resembled the animals which inhabit New Spain, called the Ourang-Outang," recorded Abner Stocking.)[1] Although the "confusion in Quebec was very great," it could not be exploited: "An examination of our arms, ammunition, &c. . . . found [the former] much deficient in numbers, much in disorder. No bayonets, no field pieces and upon an average of the ammunition there amounted only to about four rounds per man."[2] Fearful of being overwhelmed by a swift sortie from the Quebec garrison, Arnold pulled back to Point aux Trembles, about twenty-five miles away, until he could be reinforced by Montgomery, who, on 10 November, had unceremoniously forced Carleton out of Montreal. In the spirit of this whole campaign (with its derring-do and oversized characters, it seems ripe for Hollywood-does-History), Sir Guy, a resourceful man, dressed himself as a lowly *habitant* to effect his escape down the St. Lawrence; which he did with 100 soldiers—all of whom were captured, except Sir Guy.

When Montgomery and Arnold pooled their resources on 5 December, they came up with only 900 men. (Carleton, now safely back in Quebec, could call on about 1,800—mainly militia, but with smattering of British regulars and sailors.) Montgomery contributed only a little over 300, having started out with close to 2,000 back in September. Some had fallen in battle (very few); some had fallen to illness (considerably more); some had been left in garrison at St. John,

Chambly, and Montreal; and some had simply gone home. But the bottom line of this uncomfortable accounting was that there were simply too few men and a pitifully inadequate weight of artillery to undertake a siege of Quebec. (When the patriot bombardment opened up on 10 December, a defender scornfully recorded that it killed one boy, wounded one man, and broke the leg of one turkey.)[3] Carleton, on the other hand, could call on a "monstrous force of . . . 32s and 42s," which smashed the American batteries.[4] Montgomery was forced to go for the high-risk tactic: an all-out assault.

As with so many encounters during the war, the principal protagonists were connected by "six degrees of separation": their entwined histories as part of the British military establishment. Montgomery and Carleton were both Irish (Montgomery was born in the aptly named village of Swords, County Dublin) and wellborn. Carleton had been a relatively late starter, entering the army at twenty-one years old; Montgomery (fourteen years younger than his adversary) had entered at eighteen. They had served together in the siege of Havana in 1762.[5] Carleton had been wounded leading the grenadiers during James Wolfe's capture of Quebec in 1759. Montgomery had served with Jeffery Amherst at Ticonderoga, Crown Point, and Montreal in 1759. They had been brothers in arms.

Although Montgomery's options were painfully limited, he affected an almost cocky optimism: "The works of Quebeck are extremely extensive," he wrote to General Schuyler on 5 December, "and very incapable of being defended." He would assault them even though he feared that his victory would bring on the "melancholy consequences" that a victorious assault traditionally visited on the vanquished. His men, on the other hand, noticed that he was "extremely anxious, as if anticipating the fatal catastrophe," recorded Dr. Isaac Senter.[6] Perhaps he was concerned that the tour of duty of a good number of his troops expired at the year's end. As it was, he had too few for the job in hand, and they were falling victim to a range of "pulmonic complaints" as well as smallpox. Perhaps he knew his own limitations as a tactician. As Samuel Mott, who accompanied Montgomery, wrote to Governor Jonathan Trumbull of Connecticut on 6 October 1775, "I have no great opinion of Montgomery's generalship although I believe him to be a man of courage; that is but a small (though essential) qualification of a general."[7]

Weather conditions seem to have played a more important role than imminent expirations in determining the date and time of attack.[8] Montgomery wanted "a night by its darkness more favorable" and a stormy one that might cloak his preparations. On the night of the last day of the year "the storm was outrageous," recorded Private John Henry, a Pennsylvanian with Arnold's contingent. It was time to attack.

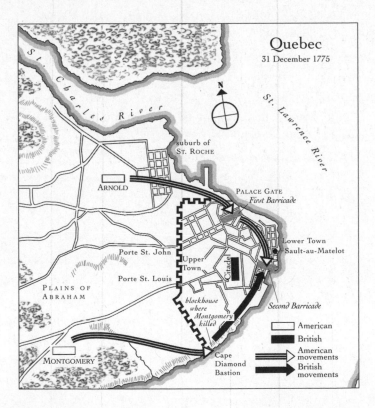

Montgomery divided his small army. Arnold would skirt around the northern perimeter of the walled city with 600 men, Montgomery around to the south with 300. They would meet up on the quayside of the St. Lawrence where the Sault-au-Matelot, a winding, narrow thoroughfare, led steeply up into the Lower Town, and force their way in through the back door, as it were. At 4:00 AM on 1 January Montgomery moved off down to Wolfe's Cove and along the ice-strewn margin of the St. Lawrence, where he gingerly approached an old brewery building now converted into a blockhouse and armed with four 3-pounders, about fifty militia, a Royal Artillery sergeant, and eight seamen under the command of Captain Barnsfare, a merchant ship's commander. Montgomery himself led a "forlorn hope" of about a dozen men toward the blockhouse, and Barnsfare held his fire until he could be "sure of doing execution." And execution it was. As the British fired down the narrow approach, the air exploded with canister and musket balls. Montgomery was hit in the head and killed instantly, and only one or two of his party (including the future vice president, Aaron Burr) escaped with their lives. Thomas Ainslie, a collector of customs duties, who was inside the blockhouse, described how "[British] musketry and guns continued to sweep the avenue leading to the battery for some minutes. When the smoke cleared there was not a soul to be seen."[9] Montgomery's second-in-command, "Col. Campbell, a Scotchman . . . very profane," took one look and removed the rest of Montgomery's force from the battle.

As Montgomery was meeting his fate, Arnold, having run a gauntlet of musketry from the defenders on the parapets of the northeastern walls, had come around to the southern side of the city, where he faced two barricades, one behind the other, blocking his route up into the Lower Town. Arnold poured his men into the narrow confines of the Sault-au-Matelot, where they had the tactical options of a fish in a barrel. ("Confined in a narrow street, hardly more than twenty feet wide, and on the lower ground, scarcely a ball well aimed or otherwise, but must take effect on us" was how Private Henry put it.[10] Arnold was almost immediately wounded in the left leg. The ball passed between the tibia and fibula, traveled down the calf, and ended up "at the rise of the tendon Achilles." As he was helped away, "it was observable among the soldiery . . . that the colonel's retiring damped their spirits," recorded Private Henry.[11] Captain Daniel Morgan (an imposing and pugnacious leader of a Pennsylvania company of frontier riflemen who would go on to play a key role against Burgoyne at Saratoga as well as in the war in the South) by an instantaneous and instinctive battlefield vote of confidence was "elected" Arnold's replacement. The first barrier was stormed again, and this time the critical phasing between the defenders' firing and reloading worked in the attackers' favor. They got through the embrasures (openings for cannon) immediately after the first discharge. Morgan scaled the barrier, slamming his back onto one of the cannon, which, he reported, "hurt me very much."

The severe weather (the winter of 1775–76 was one of the worst in living memory),[12] which was meant to help the attackers, had in fact dampened their spirits by soaking their muskets: "The guns," said John Henry, a Pennsylvania private, "were useless . . . The snow, which lodged in our fleecy coats, was melted by the warmth of our bodies."[13] Thomas Ainslie reported that the "weather is severe indeed. No man, having been exposed to the air about ten minutes, could handle his arms to do execution. One's senses are benumbed."[14] The patriots snatched up British arms and began to infiltrate the houses that lined the street.

Having got over the first hurdle, Morgan, shaken and in pain from his back injury, gave in to the great temptation of command under intense stress—spread the responsibility—and convened a committee. It would be a fatal error. He had wanted to make a frontal assault on the second barrier even though, as Private Henry recorded, it was "so strongly constructed that nothing but artillery could effectuate its destruction." Morgan's later recollection is ambiguous. On the one hand, he says he was "overruled by sound judgment and good reasoning." On the other, he regretted the "sound judgment": "I gave up my own opinion, and lost the town."[15] In this he was supported by Major Henry Caldwell, his adversary across the barricade, who recognized how desperately close it had been: "Had they acted with more spirit, they might have pushed in at first and possessed themselves of the whole of Lower Town."[16] What Morgan had given up, it transpired, was command itself.

The resourceful Major Caldwell, however, had spotted the potential danger in the Sault-au-Matelot and had organized an aggressive defense. To counter the American infiltration, he sent a detachment of Highlander Loyalists under Captain Nairne into the houses on his side of the barrier. Working from room to room, they drove out the Americans at the point of the bayonet. At the same time Carleton sent 500 men under Captain George Laws to loop around, come up behind Morgan, and bottle him up. Although Morgan urged a counterattack and breakout against the British troops now blocking his only escape, he was overruled by his fellow officers and "Arnold's subordinates agreed to surrender."[17] Morgan, cornered and snarling in defiance, refused to surrender to the British and would offer his sword only to a French priest he spotted nearby. The number of Americans captured (426), the high ratio of American dead to wounded (62:42),[18] and the small loss of the defenders (about five dead) were sanguinary reminders of the heavy price brave men paid for a commander's tactical naïveté.

Although reinforced periodically through the early months of 1776, the American besiegers were ravaged by smallpox and undermined by lack of money. Carleton, on the other hand, kept himself snugly and safely ensconced behind his stout walls and bastions as the morale of his enemy disintegrated. As the ice ceded the St. Lawrence to spring, British warships, carrying 200 men of the 29th Foot and a detachment of marines, made their way to the city, and almost immediately a strong sortie from Quebec had the patriots in a full-fledged and panicked retreat westward down the St. Lawrence and then over the river to the relative safety of Sorel.

"We Expect Bloody Work"

BROOKLYN, 22–29 AUGUST 1776

Charles Lee declared it was hopeless. Washington's bumptiously self-regarding second-in-command had been sent south from Boston to oversee the defenses of New York and immediately declared to Washington that the city was indefensible because it was "so circled with deep navigable water." Some others, like John Jay, the New York delegate to the Continental Congress, shared Lee's general pessimism and offered a very radical solution indeed. Burn New York City, scorch Long Island, and establish the army in the natural fortress of the Hudson Highlands. Washington, and Congress, saw it differently. The city was, in Washington's estimation, "a post of infinite importance both to them and us." Commanding the mouth of the Hudson, it was a springboard for naval actions north and south and therefore of immense strategic potential. Washington was determined to do his damnedest to deny it to the British.

For a man who saw the importance of the "deep navigable water" surrounding Manhattan, Lee was peculiarly blind to the possibilities of at least attempting to deny the British navy easy access to it. During February and March 1776 he busied himself with strengthening the defenses within the city. For example, all roads leading to the Hudson River were barricaded, batteries were installed down on the lower east side, and the fortress of St. George at Battery Point was reinforced. He built a string of redoubts across the island at Grand Street, protected the two bridges—Kingsbridge and Freebridge—which connected the mainland at the northern end of Manhattan, and, most important, fortified Brooklyn Heights, which dominated the city from the Long Island side. In order to deny the British access to the East River from the Sound, batteries were established on Montresors Island (today's Randalls Island). To guard the southern entrance to the East River, Horn's Hook, a small peninsula just south of Brooklyn Heights, and Governors Island, which sat practically in the mouth of the East River, were fortified.

Lee and a sequence of other American commanders (Lord Stirling, Nathanael Greene, and Israel Putnam) who succeeded him when he was sent down to South Carolina to take charge of the defense of Charleston neglected, despite congressional resolutions, to place batteries on the two choke points through which shipping had to pass to get into New York Harbor. The first was Sandy Hook, a slender peninsula jutting up from the Jersey shore that commanded the sandbars that lay across the threshold to the Lower Bay. At high tide shipping could pass over the bars. The second was the Narrows, a mile-wide channel between the western tip of Long Island and the eastern edge of Staten Island that led into the Upper Bay. Shore batteries placed here would have forced the British to run a gauntlet that could have proved costly indeed. (Colonel William Douglas, with somewhat more foresight, predicted, "It will be in vain for us to exspet [sic] to Keep the Shiping out of the North [Hudson] River unless we can fortify at the Narrows.")[1] Nevertheless, Lord Stirling expressed a supreme confidence in the defensive network that had been created during the spring and summer, writing to Washington that there was "little to fear from General Howe, should he attempt anything in this quarter," and adding for good measure, "I could wish General Howe would come here in preference to any other spot in America."[2]

When the British did come, they did so with a force of unprecedented size: the largest Britain had ever assembled for an overseas expedition. On 25 June Howe and 9,000 troops in the first contingent arrived. His 130-ship convoy from Halifax moved through the Narrows: "Some of the ships [were] within 7 or 800 yards of Long Island," commented a British officer. "We observed a good many of the Rebels in motion on Shore. They fired musquetry at the nearest ships without effect . . . Luckily for us the Rebels had no cannon here or we must have suffered a good deal."[3] They landed on Staten Island untroubled by any American resistance. General Howe was followed, over the next two weeks, by his older brother, Admiral William "Black Dick" Howe, with 13,000 men and 150 ships; Sir Peter Parker returned with another 3,000 men from South Carolina, and Commodore William Hotham from England (transporting 1,000 Guards and 7,800 German mercenaries under Philip von Heister). When it was all assembled, the British expeditionary force numbered 32,000 soldiers, 10,000 sailors, 2,000 marines, and 400 transports.

Washington, by comparison, had initially brought somewhere in the region of 10,000 Continentals. In May Congress had voted to mobilize 23,000 militia, of whom 14,000 were earmarked for the New York defenses. (General Howe wildly overestimated the American strength at 35,000, a miscalculation that may well have had a bearing on the eventual outcome of the battle.) But as with many of these grandiose plans, the result did not quite match the intention. For one thing, the squalor within the city that resulted from too many men crammed into too small an area resulted in debilitating disease. The city's water supply quickly became contaminated, and the "camp disorder" (typhoid fever) and "bloody flux" (dysentery) took their toll. One of the victims was Nathanael Greene, who was rendered hors de combat for the whole of the forthcoming battle—a grievous blow to Washington's command structure.

Where would Howe strike? The diversity of the patriot preparations, aimed at countering possible attacks on the lower west side of Manhattan, or higher up around what is now 125th Street, or perhaps coming down the Sound to land at Kingsbridge on the Harlem River, gives some idea of Washington's dilemma and Howe's opportunity. Henry Clinton, as he had at Bunker's Hill, advocated a sweeping outflanking maneuver, landing midway up on Manhattan and bagging

Washington and most of his army. Howe, though, held fast to his belief that the capture of Brooklyn Heights was the first priority. Although, given his enormous superiority in men, ships, and armament, it does not seem entirely unrealistic to wonder why an attack on the Heights coordinated with the broadly outflanking movement advocated by Clinton was ruled out. Washington's own agonized indecision, even after the British had landed on Long Island, reflects that he at least believed a multipronged attack a possibility and, given the vulnerability of his defensive position, a sound strategy.

With or without an outflanking strategy Howe had good reasons for attacking Brooklyn Heights. First, the city itself was heavily defended. Washington's adjutant general, Joseph Reed, wrote to his wife, "The city is now so strong that in the present temper of our men, the enemy would lose half their army in attempting to take it."[4] Second, he would have no security in New York without having captured the Heights. Third, Long Island was a highly productive agricultural area that would reduce his dependency on supplies shipped in from Britain. Fourth, it was a Loyalist stronghold. Fifth, it offered a long, proximate, and, as it turned out, undefended coastline. Sixth, he had a highly effective intelligence network of Loyalist contacts, and Clinton (unlike Washington, Putnam, and Sullivan, the primary American commanders in the battle) knew the terrain well. He had, after all, been born and raised in New York City when his father was governor, and had carried out detailed reconnaissances relatively recently.

At 9:00 AM on Thursday, 22 August, Howe landed an advance corps of 4,000 light infantry and grenadiers under the command of Henry Clinton and Charles Cornwallis at Gravesend Bay on the southwestern tip of Long Island, about seven miles south of the main American fortifications on the Brooklyn peninsula. Specialized landing craft (ramp-fronted, not unlike the LCTs used in World War II) deposited them on the beach and then returned to an armada of transports already anchored off Gravesend to pick up 11,000 more troops, cannon, wagons, and cavalry. By noon on the twenty-second Howe had approximately 15,000 men ashore—with more to follow.

The American defenses that Howe would have to reduce were

planned as a "collapsible buffer." The inner fortifications consisted of a necklace of redoubts and entrenchments that ran north-to-south across the neck of the Brooklyn peninsula, from Wallabout Bay in the north sweeping down to the marshes of the Gowanus Creek. Beyond them, farther inland, was the steep escarpment of the Gowanus Ridge (sometimes called the Heights of Guam or Prospect Heights), which ran west-to-east from Gravesend across to Jamaica, roughly parallel with the coast on which the British had landed. It was heavily wooded, perfectly tailored for the irregular, Indian-style tactics suited to American troops, and intersected by only a few passes that could be effectively defended by relatively few men. A contemporary wrote, "To an enemy advancing from below [from the south where the British were camped on the coastal flats], it presented a continuous barrier, a huge natural abattis [sic], impassable to artillery, where with proportionate numbers a successful defence could be sustained."[5]

From the point of view of the British, sitting down at Gravesend, there were five options. On their left was the coastal road leading up to Gowanus Creek and the Brooklyn fortifications; the next most westerly pass was Martense Lane Pass; then came Flatbush and Bedford passes; and then, far over at the eastern extremity of the Gowanus Ridge was Jamaica Pass, six miles away. Sullivan (who had taken over from the incapacitated Greene) was in charge of the outer defenses on the ridge. He placed 800 men in each of the first three passes but had not defended Jamaica Pass. It seemed so far away and such a long shot for the British that he ruled it out as a threat. However, he did send a five-man mounted scouting party out toward Jamaica Pass—at his own expense, as he later bitterly complained—but it was to prove a useless precaution. (Some weeks earlier Washington had rejected the services of 400 Connecticut cavalry because he could not feed their mounts—a costly mistake, as it turned out.)

Henry Clinton may have been a world-class whiner, but he was also a first-class strategist. (It was a combination that must have given Howe as much pleasure as a sharp stone in his shoe, or as Clinton admitted, with deadpan understatement, "My zeal may perhaps on these occasions have carried me so far as to be at times thought troublesome.") Between

22 and 26 August Clinton concocted a plan of exceptional boldness: a huge flanking sweep right around the eastern extremity of the Gowanus Heights, which, by use of the Jamaica Pass, would bring him up behind Sullivan on the Heights and cut Sullivan off from retreat to the main American fortifications at Brooklyn.

While the outflanking maneuver was in motion, Clinton further proposed that Major General James Grant would move up the coastal road and engage the American right wing. Simultaneously, the Hessians would move against the American center at Flatbush. It was important that these two supporting moves did not push too far. They were to be, initially, holding maneuvers designed to divert and engage American forces while Clinton, Cornwallis, and Howe completed their encirclement. Grant and the Hessians were to be the anvil to Clinton's hammer. Once the flanking force was in position, it would fire two guns to signal Grant and the Hessians that everything was set for the battle proper.

Brooklyn
22–29 August 1776

It was a brilliantly audacious plan, and Clinton was sure, given his experiences with Howe at Bunker's Hill and the earlier rejection of his plan to land a force on mid-Manhattan, that Howe would reject it: "In all the opinions he [Howe] ever gave me, [he] did not expect any good from the move."[6] Sir William Erskine offered to take Clinton's proposal to Howe (itself a reflection of the strained relationship). The Loyalist Oliver De Lancey, who also knew the terrain intimately, supported Clinton's plan, and Howe, teeth clenched no doubt, accepted it.

On Monday, 26 August, at 9:00 PM, with the tents of the main camp left standing and cooking fires burning, Clinton led the vanguard of the British flanking force, consisting of dragoons and light infantry, followed later by Cornwallis, Lord Percy (of Concord fame), and Howe, up the Kings Highway toward New Lots and the Jamaica Pass. In all there were 10,000 men in a column two miles long moving with the strictest regard for silence, first on the highway but soon on a much less conspicuous cart track. Units of the light infantry flanked the column with orders to detain anyone who might give the game away. However, it is almost impossible to move so large a force completely undetected. Colonel Daniel Brodhead complained that the American high command was detached and unresponsive: "Gen'ls Putnam and Sullivan and others came to our camp which was to the left of all the other posts and proceeded to reconnoiter the enemie's lines to the right, when from the movements of the enemy they might plainly discover they were advancing towards Jamaica, and extending their lines to the left so as to march round us."[7]

Colonel Samuel Miles, stationed between the Bedford and Jamaica passes with two battalions of riflemen, later insisted that he had forewarned Sullivan of Howe's intentions: "I was convinced when the [British] army moved that Gen'l Howe would fall into the Jamaica road, and I hoped there were troops there to watch them. Notwithstanding this information, which indeed he might have obtained from his own observation, if he had attended to his duty as a general ought to have done, no steps were taken."[8] The only troops that were watching, the five-man vedette so expensively procured by Sullivan, were captured easily and noiselessly and by 2:00 AM the whole British column was in place at the entrance to Jamaica Pass. Miles would be captured after a desperate but futile engagement with the rear of the British column.

Back on the coastal road, Grant began his move just before midnight, having allowed Howe's column a three-hour start. With 5,000 infantry and 2,000 marines he marched quietly up the Shore Road and about 1:00 AM engaged the picket of Colonel Atlee's Pennsylvania musketeers. By 3:00 AM Israel Putnam, Washington's designated overall commander on Long Island, made a rare appearance outside the confines of the interior lines at Brooklyn to order Brigadier General Lord Stirling of Sullivan's division south to counter Grant's advance, which was generally taken to be the main British thrust: "I fully expected, as did most of my officers, that the strength of the British army was advancing in this quarter to our lines," wrote Stirling later.

Available to Stirling were two regiments that would fight themselves into legend, here and throughout the war: William Smallwood's Marylanders (the "Dandy Fifth") and John Haslet's Delawares. Both regiments were at this time commanded by their majors, Mordecai Gist and Thomas McDonough respectively, the colonels being detained in New York City on court-martial duty, an astonishing piece of bureaucratic nonsense, given the circumstances. Washington too was enmeshed in the spider's web of administration that prevented his full involvement in the battle.

Stirling took six companies of the Marylanders and Delawares, together with another 400 men sent over from the center by Sullivan, and hurried down to reinforce the embattled Atlee. Stirling's total strength was now about 2,000 against Grant's 9,000. He drew his main force up on Blockje's Hill, facing the British, with Atlee on his right nearest the coast and a detachment of Marylanders pushed forward on his left. It was a classic formation: the inverted *V*, or *kettle* as Frederick the Great had termed it. The arms of the *V* opened toward the enemy in a warm welcome, a Venus flytrap, into which, hopefully, the British would enter. There was a problem, however. The British did not enter. In fact, it was they who glued Stirling to their flypaper. Standing off on a facing rise, Grant bombarded Stirling's men, who stood stoically in rank in the "true English taste," as one of them ironically recorded, and took a pounding from very early in the morning to midday: "Both the balls and shells flew very fast, now and then taking off a head. Our men stood it amazingly well, not even one showed a disposition to shrink."[9] The Americans could not advance against a numerically superior Grant; nor could they retreat until overwhelmed. Grant did not overwhelm. He knew what he had to do: keep Stirling in place.[10]

Around 7:00 AM the Highlanders and Hessians under Philip von Heister and Carl von Donop engaged Sullivan's men at the edge of the wooded escarpment at Flatbush. Again, it was a holding action rather than an all-out drive.

At 8:00 AM Howe's force, now safely through Jamaica Pass and at Bedford village, fired their signal cannon.[11] Relieved of any need for secrecy, Cornwallis and his grenadiers continued on to sweep down on Stirling's rear. Clinton's light troops and dragoons peeled off south down the road to Flatbush, the light infantry fanning out into the woods on either side of the road to descend on Sullivan's men, who were preoccupied with the Hessians and Highlanders on their front. The hammer was about to strike the anvil.

The Hessian jägers—expert riflemen deployed as skirmishers— and uniform ranks of the Scots and German infantry, "with martial music sounding and colors flying," broke through Flatbush Pass and began to move up through the woods, flushing out the enemy at the point of the bayonet. It was, in the French term, a *battue*: the driving of game to the guns. As the Americans fled out of the woods in a desperate attempt to reach their interior lines, they had to cross an intervening plain, and it was here that the British dragoons swept down while the light infantry, combining with the Hessians and Scots, systematically corralled and destroyed groups of patriots in the woods. An officer of Fraser's Highlanders (71st Foot) reveled in the slaughter.

Rejoice . . . that we have given the Rebels a d. . . . d crush . . . The Hessians and our brave Highlanders gave no quarter, and it was a fine sight to see with what alacrity they dispatched the Rebels with their bayonets after we had surrounded them so they could not resist . . . It was a glorious achievement . . . and will immortalize

*us and crush the rebel colonies. Our loss was nothing. We took care
to tell the Hessians that the rebels had resolved to give no quarter—
to them in particular—which made them fight desperately, and
put all to death that fell into their hands."*[12]

But not all British officers were as brutish. One wrote that the
"Americans fought bravely and (to do them justice) could not be broken
till greatly outnumbered and taken flank, front and rear. We were greatly
shocked by the massacres made by the Hessians and Highlanders after
victory was decided."[13] As the American center imploded, Cornwallis
crashed in behind Stirling's left shoulder while Grant pressed at his
front and von Heister his left, leaving the only route of escape back over
the marshes and Gowanus Creek, eighty yards wide and running with
a strong incoming tide.

William Alexander, Lord Stirling, was not, at first sight, a man
cut from the cloth of heroes. He was "overweight, rheumatic, vain,
pompous, gluttonous, inebriate," according to one of his biographers.[14]
There was something comical about his lordly pretensions. His claim
to the Scottish earldom of Stirling had been rejected by the House of
Lords. Married into the powerful Livingston clan, he soon dissipated
a considerable fortune because of his lavish spending, and just prior to
the outbreak of the war he lost his fine mansion and its contents to his
creditors.[15] Now pressed by a different sort of creditor, Stirling decided
to make a preemptive strike with 400 of his best troops, the Marylanders,
against Cornwallis at the Vechte-Cortelyou House (sometimes referred
to as "the Old Stone House") in order to buy time for the rest of his
command to make their escape over the creek. It was a forlorn hope, a
suicide mission, and magnificent. The Marylanders made seven attacks
against 2,000 British. Stirling, according to Cornwallis, "fought like a
wolf." And, when the action was over, 256 Marylanders would share a
grave. Another 100 were wounded or captured—an astounding casualty
rate of almost 90 percent. Stirling himself eventually surrendered his
sword to von Heister.[16]

By 11:00 AM Howe "lay before the rebel lines." He had prevented his
grenadiers from attacking the fortifications, writing to Lord Germain,

"It required repeated orders to prevail upon them to desist from the attempt."[17] Why? Howe later told a parliamentary inquiry that he was not prepared to "risk the loss that might have been sustained in the assault," believing he could take the position at "a very cheap rate."[18] The received opinion is that his traumatic experience at Breed's Hill had made him gun-shy. Like most received opinions, it can benefit from a little examination.

First, there were occasions both before and after the battle of Brooklyn when Howe attacked or planned to attack strongly defended high ground. For instance, just eight months after the debacle at Breed's, Howe, now in overall command at Boston, had planned a frontal attack against the Americans up on Dorchester Heights, a position infinitely better prepared and defended than the one he'd faced at Breed's Hill. It was only a massive storm that forced him to recall the boatloads of troops already on their way to carry out what would have been a very costly assault. After Brooklyn, Howe would not hesitate to attack Chatterton's Hill during the battle of White Plains, or assault the defended heights of Fort Washington (thought by its commander to be pretty well impregnable). Second, there were good reasons to hold off a final assault on the twenty-seventh. His men, having marched all through the night and fought for half the day, were exhausted. "Howe's army was not the coherent, well-supplied, well fed and fresh force it had been the night before . . . Muskets were fouled and ammunition consumed. The troops had not eaten that day and were burdened by over a thousand prisoners."[19]

Washington, on the other hand, had been sending over reinforcements steadily as soon as he concluded the landings of the twenty-second were not a feint. There were in the region of 9,500 troops behind the Brooklyn defenses, and the possibility of an American counterattack must have figured in Howe's decision. In addition, the Hessians warned him, having interrogated John Sullivan, whom they had captured, that a large American force was marching from Hell Gate (the confluence of the Harlem and East Rivers), or might land on the north shore of Long Island. One British officer had no doubt that an immediate assault would be disastrous.

The [American] lines could not be taken by assault; but by approaches. We had no fascines [bundles of wood used much like sandbags in later wars] to fill ditches, no axes to cut abatis [entanglements of wood, functioning like the concertina wire of later wars], and no scaling ladders to assault so respectable a work. The lines were a mile and a half in extent, including angles, cannon-proof, with a chain of five redoubts, or rather fortresses with ditches, as had the lines that formed the intervals; the whole surmount with a most formidable abatis, finished in every part.[20]

Some commanders are lucky, and if the eventual outcome goes in their favor, history forgives what might have been reckless, even potentially disastrous, decisions. The losers are left wearing posterity's dunce's cap. George Washington needed all the luck he could get at Brooklyn on the twenty-seventh. He had repeatedly reinforced a precarious position which, if captured, would have had the most shattering, perhaps fatal, impact on the patriot cause. Even if he had been supremely confident of his defenses to withstand frontal assault, he must have known that if the British navy worked its way up the East River it would have made a fine omelet of all the eggs Washington had placed in that one basket. His choices of old Putnam as commander at Brooklyn and of Sullivan, a man about whom Washington had already expressed deep reservations, were perverse. Putnam's performance at Brooklyn replicated the puzzling lack of involvement he had shown on Bunker's Hill that had so much enraged William Prescott.

The firepower of the British navy was formidable. Of the two capital ships on station, HMS *Asia* alone, for example, could deliver 1,024 pounds of shot in one broadside while "the strongest fort in Brooklyn could muster only around 288lbs per volley."[21] On the twenty-sixth and twenty-seventh, British ships attempted to get up the river, but, recorded the *New England Chronicle*, "the wind at both times entirely obstructed them." On the twenty-eighth and twenty-ninth a northeasterly roiled over New York and Brooklyn to further frustrate the Royal Navy.

On the afternoon of the twenty-ninth Washington convened a council of war that comprehensively enumerated all the reasons to quit the Brooklyn fortifications.

> *1st. Because our advance party had met with defeat . . . 2nd. The great loss sustained in the death or captivity of several valuable officers . . . 3rd. The heavy rain which fell two days and two nights [28th and 29th] without intermission, had injured the arms, and spoiled a great part of the ammunition . . . 4th. . . . several large ships had endeavored to get up, as supposed into the East River, to cut off our communications (by which the whole army would have been destroyed,) but, the wind being N.E., [they] could not effect it . . . 5th. [ditto] . . . 6th. Though our lines were fortified by some strong redoubts, yet a great part of them are weak . . . 7th. . . . the troops had become dispirited by their incessant duty and watching . . . 8th. Because the enemy had sent several ships of war into the Sound . . . there was reason to apprehend they meant to pass over land, and form an encampment above Kingsbridge in order to cut off and prevent all communication between our army and the country beyond them.* [22]

The withdrawal began at 8:00 PM on the twenty-ninth. Luckily, one of the units on hand was John Glover's Marblehead Regiment, composed mainly of Massachusetts fishermen. It would play a critical part in ferrying the garrison back to Manhattan.

If Washington had made grievous mistakes, he was now about to redeem them with one of the most masterful evacuations in military history: at night, across difficult water in small boats, under the noses of a numerically superior enemy and its powerful fleet. A carping critic might say that he had no choice; eventually, imminently perhaps, the British fleet would have penetrated the East River, and General Howe would have stormed the patriot lines. But the resolution, the sheer nerve, the cool head, as well as the tactical skill needed to bring it off, were nothing short of brilliant.

Throughout the night the small boats plied back and forth between

Brooklyn and Manhattan (essentially where the Brooklyn Bridge is now). "We were strictly enjoined not to speak, or even cough." recorded Private Martin. There were scares and near misses, but Washington's luck held. At 11:00 PM the wind dropped, but the British fleet did not exploit its opportunity. Washington got away with not only every last man, but also much of his stores. And at dawn a mercifully thick fog, "so very dense . . . that I could scarcely discern a man at six yards' distance," wrote Colonel Benjamin Tallmadge, rose to shroud them—a lagniappe to Washington from the gods of war.

The dogs of war, however, are not in the business of handing out tidbits and were soon snapping at Howe's heels. Not surprisingly, Henry Clinton, true to form, was deeply disappointed in his chief, and Commodore Sir George Collier, aboard the *Rainbow*, raged in his diary: "To my inexpressible astonishment and concern the rebel army have all escapd across the [East] River to New York! How this has happened is surprising, for had our troops followd them close up, they must have thrown down their arms and surrenderd; or had our ships attackd the batteries, which we have been in constant expectation of being orderd to do, not a man could have escapd from Long Island."[23]

The estimates of losses for both sides vary widely. Howe claimed a whopping 3,200 American casualties and prisoners, while the American returns for 8 October show 1,012 casualties and prisoners, of whom it was claimed that only about 200 were killed, and the rest captured and wounded. Douglas Southall Freeman, Washington's great biographer, estimated 312 American dead with 1,095 wounded and captured. Sir John Fortescue put British and Hessian dead at 63 with 314 wounded.[24]

If Howe had thought he had achieved the destruction of the patriot army by his "indirect" method, and cheaply to boot, who could have blamed him? He could certainly have been forgiven for claiming so. Washington was close to despair. For example, within a few days of the evacuation of Brooklyn the 8,000-strong Connecticut militia shrank to 2,000, and the commander in chief wrote to Congress that "the militia . . . are dismayed, intractable and impatient to return. Great numbers of them have gone off; in some instances by whole

Regiments . . . their example has infected another part of the Army."[25] It went farther than the army. Nathanael Greene noted that "the country is struck with a panic."[26]

Washington's army on Manhattan numbered in the region of 20,000 (of whom 16,000 were fit for duty). It was divided into three areas of responsibility. Putnam's division was in the city at the far southern tip; General William Heath's two brigades were sent up to Kingsbridge, on the Westchester side of the Harlem River. Greene, now recovered from his "bloody flux," occupied the center with a largely militia division. His area of responsibility included Kip's Bay on the western shore of the East River facing Howe's army over in Brooklyn. For Washington it was not a happy strategic position. As the ever-sensible though spelling-challenged Colonel William Douglas put it: "I Don't like the Chance we run in being beet in our Communications with the Cuntry, and I fear this Island of NYork will cost Amarica too much. If we were once on the Main[land] they could not support their Land forces with their Shiping, but at this Place they may guard three sides of us."[27] And during the ten days following the American evacuation of Brooklyn, Howe exploited Washington's anxiety by moving men from Red Hook in the south to Hell Gate in the north. At the same time Howe the admiral tried unsuccessfully to conclude some kind of peace agreement with a patriot delegation that included John Adams and Benjamin Franklin. When those talks broke down, Howe the general went to work.

Early on Sunday, 15 September, British warships began to bombard the flimsy American defenses at Kips Bay, predictably against the advice of Henry Clinton, who had advocated a landing higher up in the Bronx to take control of Kingsbridge. "Had this been done without loss of time," he wrote, "while the rebel army lay broken in separate corps between New York and that place [Kingsbridge], it must have suddenly crossed the North River or each part of it fallen into our power one after the other."[28] The dangerous waters of Hell Gate and the American battery at Horn's Hook that would have threatened the transports as they moved upriver have been advanced as reasons that influenced Howe's decision to reject Clinton's proposal. Neither is entirely persuasive. Hell Gate was navigated a few weeks later, albeit with difficulty, and the firepower of the British navy could have neutralized the guns on Horn's Hook. It is more likely that Howe knew of the weakness of Washington's defenses at Kips Bay. Private Martin, for example, characterized "the lines" at Kips Bay as "nothing more than a ditch dug along the bank of the river, with the dirt thrown out towards the water." He also described the low morale of the militia manning them: "The demons of fear and disorder seemed to take full possession of all and every thing." In addition, Washington had come to the same reasonable conclusion as Clinton and expected the British landing in Harlem and had, in anticipation, placed some of his best troops there.

When the 4,000 British and Hessian vanguard hit the beaches of Kips Bay at around 10:00 AM, they were unopposed. Under bombardment ("The breastwork was blown to pieces in a few minutes," wrote Lord Rawdon), the militia ran away, to Washington's utter disgust. In a fit of almost suicidal despair he sat immobile on his horse, "so vexed at the infamous conduct of the troops, that he sought death rather than life," wrote Nathanael Greene.[29] An aide-de-camp had to take the bridle and lead his despondent chief to safety.

Washington and Greene, however, had been the architects of their own predicament. The militia failure was a failure of command. First, unseasoned troops should never have been put in the position in which the men at Kips Bay found themselves. Second, the physical defenses were entirely inadequate. Third, there was a grievous absence of leadership at all levels. Martin excoriated his officers: "The men were confused, being without officers to command them;—I do not recollect of seeing a commissioned officer from the time I left the lines on the banks of the East river, in the morning, until I met . . . one in the evening. How could the men fight without officers?"[30] "The men were blamed for retreating, and even flying," wrote Benjamin Trumbull, chaplain to the 1st Connecticut, "but I imagine the fault was principally in the general officers in not disposing of things so as to give the men a rational prospect of defence and a safe retreat should they engage the enemy."[31]

Howe's primary objective for the fifteenth, as stated in his order book, was the Inclenberg Heights (modern Murray Hill). Even before the landing, he had written into Clinton's orders that he was to halt there to allow time for his second wave (another 9,000 troops) to come over from Brooklyn and be consolidated with his advance force. However, the precise timing that had been so impressive during the landings on Long Island was not repeated on Manhattan, and the second wave did not come up until 5:00 PM. In the meantime, Clinton, who led the vanguard, was constrained by his orders and prevented from advancing across the island to trap Putnam's division, then retreating up the Hudson River shore road. The legend of Mrs. Murray—she who plied the indolent Howe with Madeira wine in order to buy time for Putnam's escape—was just that: a legend (although one dutifully trotted out in most popular histories). There were good military reasons to consolidate on the Heights, but there is no doubt that an opportunity to bag Putnam's command was missed, and we do not need the fairy tale of Mrs. Mary Murray to explain why.[32] Certainly one question raises its awkward head: where was the British cavalry, those dragoons that had done such execution at the battle of Brooklyn? In a scouting role they would have been invaluable at this juncture, yet they seem to have been entirely absent.

With about 10,000 Americans now safely gathered into the fortified lines of the Harlem Heights, which stretched across the neck of the Harlem Peninsula from the Hudson to the Harlem River, and another 6,000 with General Heath at Kingsbridge, Howe sent part of his force to take possession of the city and moved the bulk of his army up through McGowans Pass (modern-day 100th Street and Fifth Avenue) to take up a position across from the Harlem lines, the two armies separated by a valleylike depression known as the Hollow Way. At dawn the next day Washington sent Thomas Knowlton (of Breed's Hill renown) across the no-man's-land of the Hollow Way to probe the British positions. It was like poking a stick into a hornet's nest. Elements of the British light infantry stormed out of their positions, forcing Knowlton to retreat. Full of confidence, the British pursued—into a classic entrapment. And soon the hunters became the hunted as other American units hit their exposed flanks. British reinforcements were brought up pell-mell, and Washington prudently withdrew his attack before it too became overextended. It may have been only a skirmish (the Americans lost about thirty killed, including the valiant Knowlton; the British fourteen), but it did wonders for the battered American morale: "This Affair I am in hopes will be attended with many salutary consequences as it seems to have greatly inspired the whole of our Troops," wrote Washington.[33] It also seems to have given Howe the jitters, for he now took over three weeks to consolidate his positions before attempting a flanking attack via Throgs Neck over on the Sound on the 12 October.

The Throgs Neck landing was a humiliating failure in which initially twenty-five Pennsylvania riflemen under Colonel Edward Hand (later reinforced by William Prescott's Massachusetts Continentals, among other units) held up 4,000 British at a choke point leading off the Neck until Howe was forced to seek the more propitious landing place of Pells Point, farther up the coast. Howe landed there on the same day (18 October) that Washington began to march all of his now 13,000-strong army (leaving a garrison of about 2,000 at Fort Washington) to White Plains. To have stayed on Manhattan would have invited encirclement. At White Plains there was a supply depot as well as an escape route into the Hudson Highlands.

An army on the march is horribly exposed to flank attack, and Howe was technically in a strong position to exploit that vulnerability. Once ashore, Clinton and Cornwallis quickly (Clinton later recorded that at first his proposal "did not seem to be much relished") moved 4,000 British and Hessians to Eastchester, just over a mile from Washington's column, but John Glover and his Marbleheads, about 750 men, executed a classic hit-and-retreat defense, brilliantly utilizing the network of stone walls and narrow lanes. It bought Washington invaluable time.

On 19 October Howe moved farther up the coast to New Rochelle and, ever cautious, waited three days for 8,000 Hessians under Wilhelm von Knyphausen, recently arrived from Europe, to join him. The delay allowed another opportunity to slip by, as Israel Putnam noted: "The stupidity of the British general, in that he did not early on the morning of the 20th send a detachment and take possession of the post and stores at Whiteplains."[34] Washington arrived at White Plains on the twenty-second, but it was not until the twenty-eighth that Howe moved his 14,000 men to confront the Americans dug in on high ground on a line from the Bronx River on their right flank, with the left anchored in the hills. Across the Bronx River, about a mile and a half from the right flank of Washington's main force, was Chatterton's Hill, a north-south ridge rising steeply 180 feet above the river. Only belatedly had Washington realized the threat it could pose, and sent Colonel Rufus Putnam to begin a hasty entrenchment of the hilltop, supported by two regiments of militia plus another 1,000 men under General Alexander McDougall. (It would later be further reinforced by Haslet's Delaware Continentals, Smallwood's Marylanders, and Webb's Connecticut regiment and would become the key to the battle.)

On the whole, the American main force was in a strong position, as Clinton reported: "[I] could not from what I saw recommend a direct attack . . . their flanks were safe and their retreat practicable when they pleased." Howe, always on the qui vive for a flanking opportunity, detached 4,000 men and twelve guns to attack Chatterton's Hill. Alexander Leslie and Carl von Donop came up the east slope, at first in line, but after having taken a beating from the defenders above, they switched to column to reduce their target profile (a lesson hard learned in the third attack on Breed's Hill). Colonel Johann Rall's Hessians stormed up the southern slope. The militia "fled in confusion, without more than a random, scattering fire . . . The rest of General McDougall's brigade never came up . . . Part of the first three Delaware Companies also retreated in disorder," wrote Haslet, until he too was forced to give way. It rained heavily on the twenty-ninth, which prevented immediate pursuit, and when, on the following day, Howe prepared to attack Washington's main body, it had gone.

Fort Washington remained in Howe's rear like the last ripe plum on the tree, and on 5 November he moved south to pick it. Washington had always had his doubts about the wisdom of maintaining the fort at all.

It was, insisted Congress, essential for guarding the chevaux-de-frise (a "chain" of sunken obstacles that stretched across the Hudson). But the barrier had already proved useless in preventing British warships from moving up the Hudson. As Washington noted to Nathanael Green (a staunch advocate for holding the fort): "The late passage of the 3 Vessels up the North River . . . is so plain a Proof of the Inefficiency of all the Obstructions we have thrown into it. . . . If we cannot prevent Vessels passing up, and the Enemy are possessed of the surrounding Country, what valuable purpose can it answer to attempt to hold a Post of which the expected Benefit cannot be had . . . but as you are on the Spot I leave it to you to give such Orders as to evacuating Mount Washington as you judge best."[35] It was an abdication of command which, years later, he would ascribe to "that warfare in [his] mind" caused by the contradictory pull of his own judgment (which was correct) and those of Congress and Greene.

Fort Washington was a disaster waiting to happen. It had been designed by Rufus Putnam, who, although the army's chief engineer, by his own admission had no "knowledge of the art." In addition, as Captain Alexander Graydon pointed out, "There were no barracks, or casemates, or fuel, or water within the body of the place."[36] Water had to be hauled up from the Hudson, hundreds of feet below. There was no magazine in which to store powder. Because the topsoil was so thin, there could be no ditch, and what entrenchments there were had to be made of earth hauled up from the foot of the hill. And, to cap it all, William Demont, the adjutant to Colonel Robert Magaw, the fort's commander, deserted to the British on 2 November and sold them plans and troop dispositions—the whole kit and caboodle—for a paltry £60.[37]

On 15 November Howe hit the fort from three sides. Knyphausen and Rall's Hessians clambered up the "Excessive Thick Wood" of the steep northern slope under galling fire (in fact the Hessians would take the lion's share of casualties that day) from Colonel Moses Rawling's 250 Maryland and Virginia riflemen. In time the rifles became impossibly fouled, and the lack of a bayonet began to tell as the Hessians closed. After "very great difficulties and hard labors," the Hessians gained the summit, "which as soon as the rebels saw they ran away towards the fort with great precipitation."

On the southern approaches, at the Harlem Heights lines, Lord Percy engaged Lieutenant Colonel Lambert Cadwalader's 800 Pennsylvanians. Down on Laurel Hill at the foot of the eastern slope of Mount Washington, Colonel William Baxter's militia were assaulted from across the Harlem River by light infantry led by Major General Edward Mathew as well as Cornwallis with the Guards, the grenadiers, and the 33rd Foot. Baxter was killed, and the militia fled to the fort. Under intense pressure from his front and the threat of being cut off behind his left flank, Cadwalader managed to get his men back into the fort by a whisker.

Two thousand eight hundred patriots were now cooped up in what would become a slaughterhouse if the British guns were turned on them. At 4:00 PM Magaw surrendered and the tattered garrison marched out: "A great many of them were lads under fifteen and old men, and few had the appearance of soldiers," wrote Lieutenant Frederick MacKenzie. And despite assurances to the contrary, they were soon stripped of whatever meager possessions they had. Greene wrote to his friend Henry Knox, "I feel mad, vexed, sick, and sorry . . . This is a most terrible event; its consequences are justly to be dreaded."

Howe, quick to exploit his victory, sent Cornwallis with 4,000 men across the Hudson on 19 November to take Fort Lee (via a carelessly unguarded route up the steep face of the Palisades from the Hudson), forcing Greene to evacuate with unseemly haste. It was deeply humiliating for one who only a few days previously not only had assured his commander in chief that the situation at Fort Washington was under control but also had felt confident enough to disregard his commander's advice to move precious stores from Fort Lee. No wonder Washington allowed himself a moment of despair: "I am wearied almost to death with the retrogade motion of things, and solemnly protest that a pecuniary record of £20,000 a year would not induce me to undergo what I do."

Fire and Ice

TRENTON I, 25–26 DECEMBER 1776;
TRENTON II, 30 DECEMBER 1776;
AND PRINCETON, 3 JANUARY 1777

For Washington the retreat through northeastern New Jersey was a desperate thing. His army was evaporating before his eyes. On 30 November, for example, 2,000 Maryland and New Jersey militia, whose term of service had expired, quit. At Hackensack, Washington ruefully noted there were "not above 3,000 men and they much broken and dispirited." Charles Lee, who remained in Westchester, saw his original 7,000 dwindle to 4,000 effectives by early December. General Heath, in the Hudson Highlands, could muster approximately another 2,000.

Washington's pursuer, Lord Charles Cornwallis, although perhaps the most aggressive of the British field commanders, seemed to arrive with uncanny regularity just as Washington was putting out the lights and exiting through the back door. He had not, for instance, pursued Greene after the fall of Fort Lee, advising the Hessian captain Johann Ewald, "Let them go, my dear Ewald, and stay here. . . . One jäger is worth more than ten rebels."[1] Ewald was stunned: "Now I perceived what was afoot. We wanted to spare the King's subjects and hoped to terminate the war amicably, in which assumption I was strengthened the next day by several English officers."[2] The Loyalist Charles Stedman was equally convinced of Howe's bad faith: "General Howe appeared to have calculated with the greatest accuracy the exact time necessary for the enemy to make his escape." The facts, however, are like square wheels on the otherwise sleek chariot of conspiracy.

Howe's overarching strategy after his victories of the previous months was to consolidate an army that had been campaigning for four months. As winter drew on he desperately needed forage for his horses and food supplies for his men. The Jerseys, with a significant population of Loyalists, were fertile and untouched by war. When Cornwallis reached Brunswick on the Raritan River on 1 December, it seemed (to his and Howe's critics) an act of perversity, stupidity, sloth—the list is extensive—not to hop over and destroy what Joseph Reed, Washington's odious adjutant general, described as the "wretched remains of a broken army." One of the problems, and not an insubstantial one as later engagements would bear out, was that the "broken army" still had formidable artillery that vigorously engaged the British and Hessians from the west bank of the Raritan. Perhaps more important, Cornwallis's force was exhausted, and its lines of communication were dangerously stretched. As Cornwallis saw it: "I could not have pursued the enemy from Brunswick with any material advantage, or without greatly distressing the troops under my command . . . But had I seen that I could have struck a material stroke by moving forward, I should certainly have taken it upon me to have done it."[3]

Passing through Princeton on his way to the Delaware, Washington left Lord Stirling in the college town with a rear guard of 1,400 men that left at 3:00 PM on the seventh. Cornwallis arrived one hour later. The whole of Washington's part of the army was now down to about 2,600, and Charles Lee, despite the ever more desperate entreaties of his chief to join him, dragged his feet while enjoying the schadenfreude of Washington's predicament. (Lee's pleasure was cut short when he was ignominiously captured, "in his slippers . . . his shirt very much soiled from several days' use," by a detachment of dragoons under the young Banastre Tarleton, a firebrand cavalry officer who would make an even greater mark later in the war.)

The patriot army crossed the Delaware between the second and seventh. On the eighth Howe and Cornwallis entered Trenton and were promptly bombarded by thirty-seven guns from the far bank. The British commanders, as was expected of gentlemen-officers, sat their horses with a nonchalant disregard of danger. A Hessian officer in attendance wrote, "Wherever we turned the cannon balls hit the ground, and I can hardly understand, even now, why all five of us were not killed."[4]

Fortune, it seemed, was smiling on the British commander. He could look back with some satisfaction on the achievements of the campaigning season. He had outgeneraled, outfought (and outnumbered) his enemy. He had all but destroyed Washington's ability to fight. He controlled New York City, all of Canada, and large swaths of eastern New Jersey, where 300 to 400 men a day were declaring their allegiance to the Crown. Congress had been panicked out of Philadelphia. He knew that the reverses he had inflicted on the patriots encouraged desertion and sapped the will to reenlist. (On 20 December Washington wrote to John Hancock, "Ten more days will put an end to the existence of our army." And to his cousin Lund Washington, "I think the game is pretty near up.") He had dispatched to Rhode Island the irksome Clinton, who successfully secured an all-season port for the big ships of his brother, Admiral Howe. He had an army of 10,000 compared with perhaps 5,000 under Washington. And his domestic arrangements with his mistress, Mrs. Loring (the wife of his accommodating commissary of prisoners), were, to put it delicately, pleasing.

Howe's army, however, was stretched and vulnerable to guerrilla attack. ("Every foraging party was attacked . . . they could be called nothing more than mere skirmishes, but hundreds of them happened in the course of the winter.")[5] He needed winter cantonments to secure the territory he had taken, rather than undertake a potentially perilous crossing of a major river in midwinter and make an amphibious landing against an entrenched enemy enjoying significant artillery support. The irony was that Washington looked across the Delaware and felt nothing but dread and vulnerability: "Happy should I be if I could see the means of preventing them. At present I confess I do not."[6] Hindsight is an unforgiving critic, and, of course, there were many (Clinton and Lord George Germain among them) who later deplored his decision.

Howe himself was not entirely happy with the arc of posts that ran from Burlington and Bordentown in the south, up through Trenton and on to Princeton and Brunswick (the location of a major supply dump as well as the army's £70,000 war chest), and back to Amboy. On 20 December he wrote to Germain, "The chain, I own, is rather too extensive, but . . . trusting to the general submission of the country to the Southward of this chain, and to the strength of the corps placed in the advanced posts, I conclude the troops will be in perfect security."[7] Von Donop, stationed at Bordentown (some eight miles south of Trenton), had overall command of the Hessian troops in the sector. At Trenton was Colonel Johann Rall with approximately 1,500 men of the Lossberg, Knyphausen, and Rall regiments, together with a small detachment of British dragoons.

Rall had played a major role at Chatterton's Hill during the battle of White Plains and again in taking Fort Washington, but history has been harsh to him, as it so often is to commanders who not only lose battles but also contrive to get themselves killed in the process. The conventional wisdom is that he was a drunk, a crude loudmouthed slob—"conceited and insolent"[8]—who got his comeuppance. Much has been made of his "failure" to entrench, which is usually portrayed as an arrogant contempt for his enemy. It is true that he held the patriot "peasant army" in low esteem, as indeed did most of the British and Hessian officer corps. It was as much a social as a military prejudice. But there were other reasons. As Rall saw it, it was pointless trying to entrench against an enemy that could hit him almost at will and from any direction. He was in "Indian" country and being driven half crazy with anxiety: "I have not made any redoubts or any kind of fortifications because I have the enemy in all directions." When he was urged by some of his officers to dig in, his exasperation expressed itself trenchantly. "*Scheiszer bey Scheisz!* [Shit on shit] Let them come . . . we will go at them with the bayonet."[9] It was a response fueled by desperation rather than complacency.

Many histories imply that the situation at Trenton was static, with Rall and his Hessians lolling around. The opposite was true. Under constant guerrilla attack and well aware of his vulnerability, Rall called often to von Donop at Bordentown, General Grant at Brunswick, and General Leslie at Princeton for reinforcements that never materialized. Far from being complacent, he kept his men on an exhausting regimen of patrolling and round-the-clock surveillance. They slept dressed for action.

Washington too was in a desperate situation. He knew that unless he made some kind of demonstration against the enemy the rebellion was likely to wither away by the New Year. On 31 December the service agreements of large numbers of men would expire. He was driven to action, come what may, while he still had at least the semblance of an army. Joseph Reed underscored the predicament, writing on 22 December: "We are all of the opinion my dear general that something must be attempted . . . even a failure cannot be more fatal than to remain in our present situation. In short some enterprise must be undertaken in our present Circumstances or we must give up the cause."[10] After taking counsel, Washington fixed on 26 December for an attack on Trenton, where, he knew, the garrison was close to breaking point. ("We have not slept one night in peace since we came to this place. The troops can endure it no longer" a Hessian officer wrote.)

British intelligence quickly picked up the scent of the impending attack on Trenton, even though it did not know the exact day. On the evening of the twenty-fourth Grant warned Rall of a possible attack, and the Hessian responded energetically by increasing his surveillance. The story of the unheeded warning given to a drunken Rall on Christmas night by a Loyalist farmer is in most popular histories but makes no sense given that Rall was already all too painfully aware of the danger he was in, not only from Grant's informer but from various other sources.

The aura of composure, rationality, almost detachment, that surrounds Washington like a slightly saintly penumbra can be misleading when it comes to Washington the general. His natural bent as a commander, though often disguised by the conventions of his class and time, and occasionally tempered by strategic necessity, was highly aggressive—sometimes to the point of recklessness. Although many of his contemporaries labeled him a modern-day Fabian (the ancient Roman who preferred to run away in order to fight another day), he was by nature a fighter and a opportunist and, when cornered by circumstances as he now was, would almost always elect to roll the dice.

Washington also had a fondness (weakness, some might say) for complicated battle plans. His attack on Trenton would require three forces, separated by considerable distances, in the dead of night, in the depth of winter, in (as it turned out) a screaming storm, to coordinate their efforts. He would lead the most northerly group of 2,400 men in two divisions across the 800-yard-wide Delaware, now choked with ice floes.[11] One division, under Nathanael Greene, was to head inland and approach Trenton from the northeast, while the other, under John Sullivan, was to move down the river road and come at the Hessians from the west. These maneuvers in themselves required complicated coordination.

In the center, almost opposite the town, James Ewing would cross with his 800 men and secure the bridge over the Assunpink Creek, the only escape route for the garrison. Farther south, Colonel John Cadwalader (whose brother Lambert had been with Washington at the battle of Brooklyn) with 1,200 Philadelphia Associators (militia) and 600 New England Continentals under Colonel Daniel Hitchcock would cross and engage the Hessians at Burlington to prevent them from coming to the rescue of their compatriots at Trenton.

Washington's troops were late at their assembly points, and by the time he had crossed with the vanguard he was three hours behind schedule, which, he said, "made me despair of surprising the Town, as I well knew we could not reach it before the day was broke."[12] By the time the army got under way it was 4:00 AM and the plan was running four hours late. But Washington was always lucky with the weather. (One thinks immediately of the saving storm and fog at Brooklyn.) The raging nor'easter that slammed into the transports around 11:00 PM as they crossed the Delaware was a wicked trial for the ill-clad soldiery, but it brought its own benefits, lasting throughout the night and early morning to mask their approach and offer them the surprise that was key to Washington's gamble.

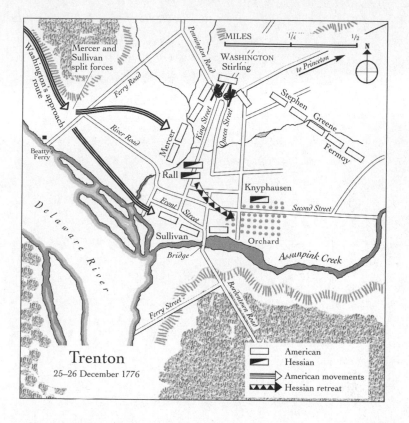

Trenton
25–26 December 1776

All gamblers are grateful for the occasional ace, and the eighteen cannons under Henry Knox ensured a massive supremacy of firepower over the six the Hessians could muster. It also meant, on that sodden night, that the artillery, which, unlike the musketry, could to some extent be weatherproofed, could be relied on to perform. In addition, seven batteries on the Pennsylvania side would also engage the Hessians.[13]

Farther south, Cadwalader could make only a partial landing, which he had to recall, while Ewing failed to make it across at all. Although some historians have suggested that both these commanders were, to put it euphemistically, deficient in determination, Washington knew very well that the river at their crossing points was even more treacherously clogged with ice than where he had crossed at McConkey's Ferry, and was forgiving of their failure.

Rall had been thorough in covering the approaches to the town, but at 8:00 AM the outposts on the northeastern and northwestern sides were driven in. (After the battle, Washington would commend them on their orderly retreat.) The alarm was sounded, and the Hessians poured out of their quarters, dressed and fully prepared. Certainly this was no drunken rabble, as John Greenwood, an American soldier, testified: "I am willing to go upon oath, that I did not see even a solitary drunken soldier belonging to the enemy,—and you will find . . . that I had an opportunity to be as good a judge as any person there."[14]

Nine of Knox's cannons, firing down the main north-south thoroughfares of the town—King and Queen streets—with canister and ball, knocked out the two Hessian guns facing them, exposing the Hessian foot to a murderous fire that was further augmented by the musketry of Sullivan's men pushing through the westerly side streets and alleys to hit the Hessians on their left flank. Rall had no option but to pull his men out of the town center and move toward the east in an attempt to turn the American left wing—a maneuver that Washington adroitly forestalled by extending his left to threaten the Hessian right flank.

For the Hessian commander it was a critical moment. He might have retreated across the bridge and taken up a defensive position on the high ground on the southern side of the Assunpink (as Washington would do very successfully when he returned to Trenton a few days later). If, by doing so, he bought time for reinforcements to reach him, it could have been disastrous for the patriot cause, trapped up against the Delaware and far from their transports. But history remains supremely indifferent to ifs and maybes, and Rall did not retreat (partly because he was under the misapprehension that the bridge was already impassable). In fact, he did the opposite: driving back into the town center in a heroic but futile attempt to recapture his guns and regimental standards. The Hessians were hit on three sides simultaneously by General Mercer ahead, firing from the west; Lord Stirling and the artillery firing into their right flank from the north; and General St. Clair firing into their left flank from the south. Amazingly, the Hessians recovered their two cannons, but Rall received two mortal wounds to his side. Demoralized, the Rall and Lossberg regiments surrendered, followed shortly after by the Knyphausen regiment, which had tried but failed to cross the Assunpink. It was all over by 9:30 AM.

Given the ferocity of the street fighting (Sergeant Joseph White recalled that his "blood chill'd to see such horror and distress, blood mingling together, the dying groans . . . The sight was too much to bear") and the overwhelming numerical superiority of artillery and muskets available to Washington, it is a dramatic illustration of the nature of the warfare of the period and the effectiveness of its weaponry that out of 1,500 Hessians, only 22 were killed and 83 wounded (834 were captured). The Americans losses were astonishingly small: two men dead of exposure and two wounded: Captain William Washington, a distant cousin of the commander in chief, shot through both hands; and a future president, Lieutenant James Monroe, who survived an almost fatal shoulder wound.

On 26 and 27 December Washington led his triumphant, exhausted, and drunk army—Joseph Reed reported that "the soldiers drank too freely to admit of Discipline or Defence," even though Washington had ordered forty hogsheads of rum destroyed—with its prisoners and loot back to Pennsylvania. They had crossed not only the Delaware but also a psychological barrier: the frontier that separates those who believe they are condemned to be beaten and those who are convinced they are destined to win. The battle may have been only a large-scale raid, but its impact was massive. A Loyalist, Nicholas Cresswell, reported on the sea change in his patriot neighbors: "A few days ago they had given up the cause for lost. Their late successes have turned the scale and now they are all liberty made again."[15] The raggedy-assed soldiery had done something extraordinary, and they knew it. Equally extraordinary, they would be back in Trenton in only three days.

The second battle of Trenton, although given fairly short shrift in most histories, is full of interest not only for the student of Washington's command style but also as a fine example of what might be called the "organic" evolution of strategy: the role of happenstance as an important factor in warfare (much to the chagrin of those historians who prefer to ascribe intention and planning that only hindsight lends to what in reality is the chaotic ricochet of events).

First, Washington had no clearly thought-out strategy after Trenton.[16] What happened next was taken, in his characteristically aggressive and opportunistic manner, on the fly. What triggered Washington's return to Trenton was, ironically, Cadwalader's eventual successful landing in New Jersey on 27 December. There they now were, 1,800 men stuck out on their own limb. After the usual war council Washington decided to support Cadwalader with everything he had: "a considerable force," by Washington's own estimation.

During the twenty-ninth and the thirtieth Washington sent over 6,800 men and a formidable artillery train to Trenton, ranging them on the high ground on the southern bank of the Assunpink.[17] In retrospect it all seemed perfectly logical. Lure Cornwallis to attack a strongly held defensive position and then, while he banged at the front door, slip out the back and capture the enemy's rear bases (including that succulent £70,000).

It is an appealing theory (seemingly sanctified by events), except that almost everything points in the opposite direction. Washington had placed his men in the gravest peril and risked everything to chance. It is difficult to understand how the strategic benefit could ever have outweighed the massive risk. He had deployed his army with its back to the Delaware without what in modern jargon would be called "an exit strategy." In fact, many in his army were panicked by the situation in which they now found themselves. A Virginian, Ensign Robert Beale, in the American center, recognized the danger: "This was the most awful crisis, no possible chance of crossing the River; ice as large as houses floating down, and no retreat to the mountains, the British between us and them."[18] Captain Stephen Olney concurred: "It appeared to me then that our army was in the most desperate situation I had ever known it; we had no boats to carry us across the Delaware, and if we had, so powerful an enemy would certainly destroy the better half before we could embark."[19] One of Washington's generals, Arthur St. Clair, referred to the "probability of defeat."

Before he sallied out of Princeton to do battle with Washington, Cornwallis had been urged by Colonel von Donop, who knew the country behind the American lines, to take a route that could be used to flank the American right. Cornwallis, with the airy self-confidence of the blinkered, decided to stick with his frontal strategy. It was on just such arbitrary turns of the cards that Washington's fate hung.

The British and Hessian column was delayed by a hit-and-retreat maneuver, carried out brilliantly by Colonel Edward Hand and Colonel Nicholas Haussegger, which bought valuable time for Washington to consolidate his positions on the Assunpink. Once in Trenton, inspiration seems to have abandoned Cornwallis, who accepted battle entirely on Washington's terms. He ordered attack after attack—all inspiringly heroic, all bloodily sterile—on the bridge and one or two other main choke points across the Assunpink, which, of course, Washington had taken very particular care to defend in depth with his most experienced troops, interspersed with the more fragile militia units. The fighting at the bridge was fierce, and Washington's own conduct there is an exemplar of the detachment that was so highly prized by officers of his social class. With his horse's chest pressed against the bridge rail he made a flamboyant gesture of nonchalance in the face of danger. "The firm, composed, and majestic countenance of the General inspired confidence and assurance in a moment so important and critical," remembered Private John Howland.[20]

American fire tactics at the bridge provide a valuable insight into the tendency to shoot high during the stress of battle, perhaps caused by involuntary flinching at the moment of firing. (The same tendency had plagued the British attackers at Breed's Hill and would again at the battle of Princeton.) To counter it, Colonel Charles Scott instructed his Virginians:

Now I want to tell you one thing. You're all in the habit of shooting too high. You waste your powder and lead, and I have cursed you about it a hundred times. Now I tell you what it is, nothing must be wasted, every crack must count. For that reason boys, whenever you see them fellows first begin to put their feet upon this bridge do you shin 'em. Take care now and fire low. Bring down your pieces, fire at their legs, one man Wounded in the leg is better a dead one for it takes two more to carry him off and there is three gone."[21]

As brilliant as Washington's defense was that evening of 2 January, one question begged to be answered: what now? Cornwallis's army of some 6,000, despite a few bloody noses, was not going away and almost certainly would eventually have opened up the patriot right flank. Even with his advantage in artillery (forty compared with Cornwallis's twenty-eight) Washington was hardly likely to take the offensive. To cap it, there were British reserves at Maidenhead and Princeton which, momentarily, would be mobilized.

Quitting while ahead must have been a particularly appealing notion to Washington at that point, but there is no evidence he had a plan to extricate the army from the predicament in which he had placed it. Washington, in council that evening, expressed his own pessimistic forecast, as recorded by Major James Wilkinson on the American right wing: "The situation of the two armies were known to all; a battle was certain, if he kept his ground until the morning, and in case of an action a defeat was to be apprehended; a retreat by the only route thought of, down river, would be difficult and precarious." He called for advice. General Arthur St. Clair pointed out the cross-country back-road route that would lead into the undefended southeastern perimeter of Princeton six miles away to the east. Brilliantly, it offered the prospect of escape and a chance to attack. Not only that; it offered the only hope, and Washington grasped it with both thankful hands.

The night was exceedingly dark, and, true to form, the weather turned in Washington's favor. A hard freeze came down, transforming the glutinous mud of the previous few days into a good solid base—"as hard as pavement," noted a soldier—for carts, artillery, and infantry. Although Washington employed the usual ruses—campfires kept burning, some men left to make conspicuous noise of entrenching, and so on—his departure was certainly noticed by the British, who, taking it for a flanking attack, made the entirely understandable but completely inappropriate dispositions. Most histories have Cornwallis curling his disdainful aristocratic lip while making a complete ass of himself by pronouncing, "We will bag the fox tomorrow." It makes a good story, but, sadly, there is no evidence he used the phrase (although foxy Washington certainly was).

Early on the morning of 3 January Washington's column, marching undetected up the Quaker Road that approached Princeton from the southeast, came to the Stony Brook. Here Washington detached Greene's division to follow the road that ran alongside the brook, north to the main Princeton-Trenton road, while he and Sullivan with the main force (approximately 5,000) carried on toward the town. Greene was to destroy the bridge where the main road crossed the brook, which would hinder any attempt by Cornwallis to double back to Princeton. At this point events began to unfold with the randomness that is more characteristic of battle than chess-game tidiness.

The commander of the Princeton garrison, Lieutenant Colonel Charles Mawhood, had been instructed by Cornwallis to bring part of his garrison across to Trenton, and at 5:00 AM on the third he duly set off down the main road with the 17th and 55th Foot (700 or so men in total) and eight guns, leaving the 40th Foot to hold the town. Not far into their journey, the British column spotted Sullivan and Washington over to the east. Greene's force, 1,500 strong and working its way up the ravine of Stony Brook, remained undetected, though it soon became aware of the redcoats on the main road.

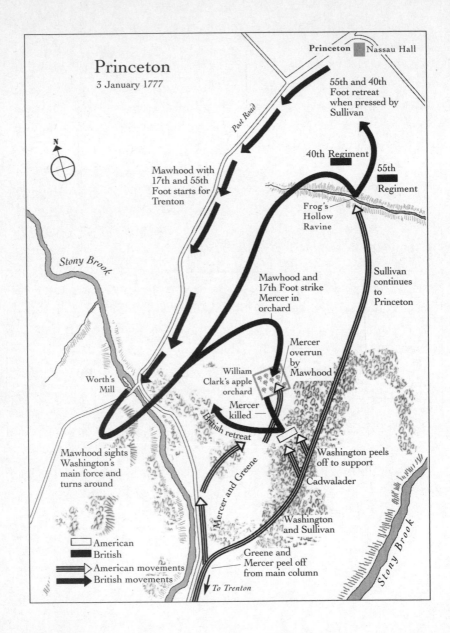

Princeton

3 January 1777

N

Princeton · Nassau Hall

Post Road

55th and 40th
Foot retreat
when pressed by
Sullivan

40th Regiment

55th
Regiment

Mawhood with
17th and 55th
Foot starts for
Trenton

Frog's
Hollow
Ravine

Stony Brook

Sullivan
continues
to
Princeton

Mawhood and
17th Foot strike
Mercer in
orchard

Mercer
overrun
by
Mawhood

Worth's
Mill

William
Clark's apple
orchard

Mercer
killed

British retreat

Mawhood sights
Washington's
main force and
turns around

Washington peels
off to support

Cadwalader

Washington
and Sullivan

Mercer and Greene

Greene and
Mercer peel off
from main column

Stony Brook

☐ American
■ British
⊳ American movements
➤ British movements

To Trenton

Mawhood immediately decided to attack Sullivan. He sent the 55th and most of the artillery to occupy a small hill to the south of the road, had his remaining infantry (supported by two fieldpieces), about 450 men, drop their packs and head off east in interception. At this point, Greene's vanguard (about 500 strong), commanded by General Hugh Mercer, debouched from the ravine and moved to engage Mawhood, who had no option but to turn his attention to this totally unexpected threat. They clashed in an orchard on William Clark's farm, and a brisk firefight and artillery exchange developed.

What next transpired was a classic illustration of the vulnerability of the rifle in a volley-fire situation. Although the British fired high ("Their first shot passed over our heads"), it took much longer for the patriot riflemen to reload, and this gave a critical advantage to the British musketmen, armed with their murderous fourteen-inch bayonets. Cohesion, unity, and discipline were key to eighteenth-century battles where big, solid punches of concentrated fire counted far more than the dissipated effect of numerous jabs. At this critical juncture the American ranks began to lose their coherence despite attempts to pull it back together. With a formality that would not have been out of place at Minden or Malplaquet, Captain John Fleming instructed his First Virginians, "Gentlemen, dress before you make ready." It was a classic command of eighteenth-century combat, intended to compact the troops and intensify their volley, but Fleming was shot down as soon as the words were out of his mouth, and the cohesion he was trying to create fell apart. Mawhood, although considerably outnumbered, took his opportunity, and the British moved in quickly. Mercer was bayoneted multiple times (and would die some hours later), and his men broke, streaming back into and destabilizing Cadwalader's Pennsylvania Associators as they moved up in support. It was the critical point of the battle.

Washington's personal intervention at the moment of crisis sheds another light on his complex personality. All too often he is depicted as a masterful civil servant, the great moderator, the inspired military bureaucrat—all Eisenhower and no Patton. But here he showed those characteristics so admired by his age: heroism and physical courage displayed with imperturbability—the studied insouciance of the ancient Roman. It was an inspiration, if also an agony, for those who watched him only thirty yards or so from the British line with, as he later recalled, "a thousand deaths flying around him," rallying Cadwalader's militia and bringing his reinforcements into play. The tactical center, which had disintegrated under the centrifugal force of panic, he now drew back together again, tightened and unified it, and sent it forward to deliver, as Washington ordered, "a heavy platoon fire on the march."[22] Overwhelming numbers did the rest, and Mawhood's corps broke and ran. In the pursuit, Washington's mask slipped for a while and he joined in exuberantly, unrestrainedly, thoroughly enjoying himself—until drear duty brought him back.

Mawhood's stand had cost him dearly. For example, the 17th Foot took 45 percent casualties (101 killed and wounded out of 224); the British artillery lost 50 percent of its men. But it cost Washington more. Princeton was taken, but the patriot army had been drained by the battle, and there was no time to move on to Brunswick. As Washington would later report to Congress, "In my judgement Six or eight hundred fresh Troops upon a forced March would have destroyed all their Stores and Magazines: taken (as we have learnt) their Military Chest containing 70,000 pounds and put an end to the war."[23] But Cornwallis was now closing in fast, and Washington's only option was to head in to the high lands around Morristown and hunker down for the winter.

17

The Philadelphia Campaign

BRANDYWINE, 11 SEPTEMBER 1777;
GERMANTOWN, 4 OCTOBER 1777;
AND MONMOUTH COURTHOUSE,
28 JUNE 1778

Most historians seem to agree—that it is impossible to agree. Of all the campaigns of the war, none is as confusing, when it comes to intention and execution, as that undertaken by General William Howe in the late summer and early fall of 1777, which led to his capture of Philadelphia (the seat of Congress and the unofficial capital of the new United States).

One line of argument holds that Philadelphia was not important strategically (as events would prove when the British eventually abandoned it in 1778); nor, among a loose confederation of often contentious states, did it have the symbolic potency of a national capital. Howe, so the argument goes, was simply deluded, or stupid, or both, to target it. But if he was, he was joined by George Washington among others, who certainly thought the city worth the contention. The political pressure that would compel the patriot commander to defend the city was, from Howe's point of view, a double blessing. It prevented Washington from interdicting General John Burgoyne, as he descended from Canada,[1] and it offered the chance of bringing the patriot army to battle.[2] The most common accusation against Howe has been his supposed predilection for the classic eighteenth-century notion of maneuvering for territorial possession while avoiding battle, and many have highlighted this as the main reason for Howe's failure during his years in command of the Royal Army in North America. A counterargument can be made that Howe sought to bring Washington to a decisive battle (Howe's strategy in New Jersey in June is a prime example), and on those occassions when annihilation of the enemy seemed likely (for example, at Brooklyn, and the chase across the Jerseys in 1776), the British were foiled more through the exhaustion of their troops and tactical considerations than a lack of intent or the commander's venality and sloth.

Another arena of debate concerns the method Howe chose to get within striking distance of the city and the risks and benefits involved. Following his setbacks at the two battles of Trenton and at Princeton, he had drawn in his horns in New Jersey and relinquished most of the territory he had once held. From his main bases at Amboy and Brunswick he tried unsuccessfully during June 1777 to bring the patriot army, now reduced to about 4,000, to battle. But Washington, ensconced in solid defensive positions at Morristown and Bound Brook, was not inclined to take the bait. By 30 June Howe gave up on the cat-and-mouse game in New Jersey and pulled his army back to New York (and in consequence delivered his Loyalist supporters to the tender mercies of their patriot neighbors).

Although many of his officers disagreed, the British commander in chief ruled out an overland approach across the Jerseys to Philadelphia. No sane commander, he argued, would want an army strung out over ninety miles with a highly aggresive enemy chewing on its flanks. So Howe went back to an earlier plan: an approach to Philadelphia by sea and via the Chesapeake (which was only later amended to the Delaware River route). On 9 July he embarked about 13,000 men and left them as Howe's officers complained "pent up in the hottest season of the year in the holds of the vessels"[3] while he waited for news of General John Burgoyne's satisfactory progress down from Canada en route to Albany (which he received on the fifteenth) and Henry Clinton's arrival from

England (perhaps less satisfactory from Howe's point of view). On 23 July 1777 the vast armada of 267 ships (the largest fleet ever to sail in North American waters) set sail from Sandy Hook.

The opposing commanders shared the same fear: that the other would go north. On 12 July Washington had confidently predicted, "His designs I think are most unquestionably against the Highlands."[4] And later, "There is the strongest reason to believe that the North River [the Hudson] is their object." It seemed obvious to Washington that Howe would go to support Burgoyne, and he was "puzzled . . . being unable to account upon any plausible Plan . . . Why he should go to the southward rather than cooperate with Mr Burgoyne."[5] As late as 21 August Washington remained convinced that Howe was making a feint southward, so strongly did he believe that the most profitable strategic move for the British would be in the north, bisecting the rebellious states down the line of the Hudson. Not knowing where Howe would fetch up—the Carolinas? Philadelphia? Boston? the Hudson?—Washington was forced to march and countermarch, driven by the winds of shifting intelligence: "compelled to wander about the country," said Nathanael Greene, "like Arabs."

Howe reached the Delaware Capes on 29 July and rendezvoused with Captain Sir Andrew Snape Hamond of the *Roebuck*, which had been on station in the Delaware for some time. At this point events become as murky as the waters of the turbid river. Most popular histories have Hamond warning Howe off a landing farther up the river. Or as Lieutenant William John Hall put it, Hamond went on board Howe's flagship and "produced a chimerical draught of fortifications."[6] Hamond, in fact, had done no such thing and was himself flabbergasted when the Howes rejected his advice to land at Reedy Island, about thirty-five miles south of Philadelphia.[7]

Hamond did, however, mislead the Howes about the whereabouts of Washington's army, informing them that it was at Wilmington, south of Philadelphia, when in fact it was at Coryell's Ferry in New Jersey, 100 miles to the northeast, and would not arrive at Philadelphia until 24 August.[8] Perhaps it was this crucial piece of misinformation that persuaded the British commanders to take the long route to the

Chesapeake. To have tried an amphibious landing at, say, New Castle, just south of Wilmington, might have courted disaster if opposed by Washington's supposedly proximate army. To have gone farther up to Chester would have been to commit the fleet to such narrow navigable channels that it would have been strung out for four miles[9] and exposed to shore batteries. Hardly a propitious springboard for an amphibious landing. Perhaps Howe was now convinced that Washington would stay in defense of Philadelphia, and a landing at Head of Elk at the most northerly reaches of the Chesapeake would give enough time for the British commander to prepare his ship-weary troops for the pitched battle he sought.[10]

After another wretched twenty-five days at sea (during which over 300 dead and dying horses were thrown overboard), Howe's exhausted army finally fetched up in the Chesapeake. After a short period of recuperation it headed up toward Philadelphia, while Washington on 9 September moved his army (by now some 11,000 strong and predominantly Continentals) into a blocking position at Chad's Ford where the main road crossed the Brandywine creek.

The Brandywine offered a solid defensive position, deep enough to force the British and Hessians to cross via fords which would act as choke points and very effective killing zones if adequately defended. On its east side (the American positions) there was good high ground and abundant wooded cover. The terrain, in other words, would have been excellent if the American command had been thoroughly conversant with it. However, as events would ruthlessly expose, it was not.

Washington's dispositions from south (downstream below Chad's Ford) to north were as follows: The most southerly (the American left wing) was anchored by Brigadier General John Armstrong's 1,000 Pennsylvania militia at Pyle's Ford, a particularly strong defensive position that Washington hoped would offer his militia maximum protection. A half mile upstream was Chad's Ford, the center of the patriot defense under Nathanael Greene's division with Anthony Wayne's brigade and Thomas Procter's artillery attached. A little behind them stood Adam Stephen's division in support. The right wing, under the overall command of John Sullivan, had Sullivan's own division at the

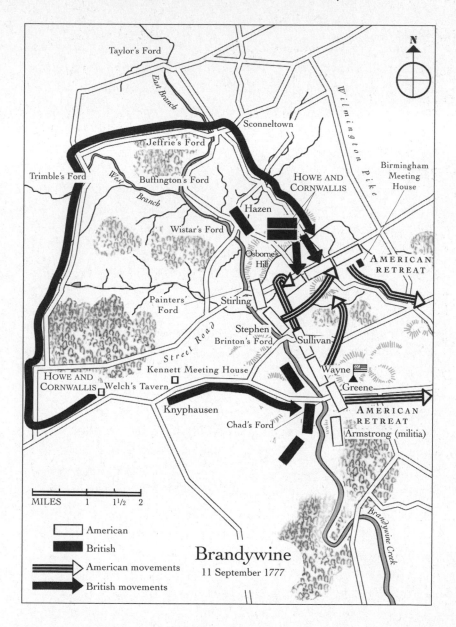

Taylor's Ford

East Branch

Sconneltown

N

Wilmington Pike

Jeffrie's Ford

Trimble's Ford

West Branch

Buffington's Ford

Wistar's Ford

HOWE AND
CORNWALLIS

Birmingham
Meeting
House

Hazen

Osborne's
Hill

AMERICAN
RETREAT

Painters'
Ford

Stirling

Stephen
Brinton's Ford

Sullivan

Wayne

HOWE AND
CORNWALLIS

Kennett Meeting House

Greene

Welch's Tavern

Knyphausen

Chad's Ford

AMERICAN
RETREAT

Armstrong (militia)

Brandywine Creek

MILES   1   1½   2

American

British

American movements

British movements

Brandywine
11 September 1777

next ford upstream, Brinton's. Lord Stirling's division was in support.
The next upstream ford, Painter's, one mile from Brinton's, was held
by the Delaware regiment of Sullivan's division; at the next, Wistar's,
a further mile from Painter's, a battalion of Moses Hazen's regiment

stood guard. The last ford controlled by the American's, Buffington's, a mile above Wistar's and close to the fork of the east and west branches of the Brandywine, was held by Hazen's second battalion. Across the river from Painter's Ford Sullivan had the army's lone cavalry detachment. Theodorick Bland's 1st Dragoons, which was both understrength and weakly led. Also across on the west side (opposite Buffington's) was a militia detachment under Major James Spear. The army's total strength was about 11,000.

Concerned about his vulnerability on his far right, Sullivan inquired on 10 September about other fords that might be even farther upstream of Buffington's. Washington's headquarters informed him that locals had completely ruled out any that might be a threat.[11] One mile above Buffington's was Jeffrie's Ford, but it was dismissed as being too far north (about ten miles from Chad's) and approached from the west by a difficult road.

At 5:00 AM on Thursday, 11 September Howe and Cornwallis led a column of twelve British and three Hessian regiments totaling approximately 6,500 men, and headed up the Great Valley Road on the eastern side of the Brandywine toward the unguarded Jeffrie's Ford. A half hour later, a second column of around 6,000, commanded by General Wilhelm von Knyphausen, set off down the main road (today's Route 1) to Chad's Ford. Howe's battle plan was to be a replay of the battle of Brooklyn: Knyphausen was to hold the American center in place while Howe and Cornwallis worked a huge detour around the American flank. By 10:30 AM Knyphausen had driven Brigadier General William Maxwell's light infantry skirmishers back across the Brandywine and was in position.

Moving over 6,000 men without being spotted is not easy, and, as at Brooklyn, the British flanking column was sighted. The first alarm came from Hazen at Painter's Ford, who passed it along to Sullivan and from him to Washington. It was around 11:00 AM. Washington, sensibly, ordered Sullivan to have Bland's cavalry detachment investigate. Meanwhile, another sighting came in from Lieutenant Colonel James Ross, scouting out by the Great Valley Road. It was now 11:25 AM, and Washington's instinct for aggressive action was piqued. If Howe had split his force, he had also opened himself to defeat in detail. Washington ordered Sullivan, Greene, Maxwell, and Armstrong to cross the creek and attack Knyphausen. Stephen's and Stirling's divisions would move northward up the Brandywine to screen the American right flank.

Even as units of Sullivan's division clashed with the 4th Foot across from Brinton's Ford, Washington received yet another message that threw his whole strategy into disarray. At 1:30 PM Sullivan passed on Major Spear's observation that he had traversed most of the Great Valley Road opposite the American positions and had seen nothing of a British column above the forks of the Brandywine. Perhaps, concluded Washington, what the earlier reports had seen was merely a feint, and the whole British force was about to attack at Chad's and Brinton's. He immediately ordered Stirling and Stephen to halt and stay put and canceled all plans for the attack.

No sooner were the orders issued than two more reports threw Washington into further confusion. A farmer, Thomas Cheyney, insisted to a doubting Washington that he had seen the British massing on the eastern side of Jeffrie's Ford. And at about 2:10 PM Washington received a report from Sullivan forwarding Bland's account of seeing the British "to the rear of [Bland's] right about two miles, coming down," in an eerie repetition of the passage through Jamaica Pass during the battle of Brooklyn. The British had slid through a narrow defile on the eastern side of Jeffrie's that could have easily been denied them with "a hundred men," thought the Hessian Captain Johann Ewald. It was, one of them remembered, "left wide open for us." Washington had been blindsided.[12]

Now Stephen and Stirling were hurried to block the British at Birmingham Meeting House. Sullivan was ordered to join them there, and Greene's division was pulled back from Chad's Ford into a reserve position from which it could equally support the positions either at Chad's or at Birmingham Meeting House. Stirling and Stephen established their divisions on the north-facing slope of Plowed Hill, facing the British on Osborne's Hill. The Americans had taken up a formidable defensive position; the British chief of engineers, John Montresor, acknowledged it was "remarkably strong, having a large body advanced, small bodies still further advanced and their rear covered by a wood wherein their main body was posted." Sullivan, however, was having a hard time joining them. He simply did not know the way: "I neither knew where the enemy were, nor what route the other two divisions were to take, and of course could not determine where I should make a junction with them."[13]

The British deployed in three frontline assault groups. On their right, the Guards; in the center, British grenadiers; on the left, British light infantry and German jägers. Hessian grenadiers backed up the Guards, while the 4th brigade supported the center and left. James Agnew's 3rd Brigade was held in reserve. The British attacked at 4:00 PM. The light infantry bit into the Frenchman Prudhomme de Borré's division on the far left, which crumbled and ran (de Borré leading the way, for which disgrace he would resign a few days later). The Guards and grenadiers hit the patriot center and Sullivan's still-forming division on the American left. Sullivan's men wilted under the bayonet attack and fell back in confusion. The American center, well served by its artillery, held on until attacked by grenadiers who, an observer noted, "ran furiously at the rebels" without stopping to fire (further evidence that heavy packs were discarded before battle). What followed was fifty minutes of an eighteenth-century slugfest. The New Jersey Continental Joseph Clark recorded in his diary, "The firing, while the action lasted, was the warmest, I believe, that has been in America since the war began."[14] Stephen Jarvis, serving with the Loyalist Queen's Rangers, watched the American army from an eminence and reported, "We saw our brave comrades cutting them up in great style."[15] His safe distance may lend a certain charm to the battle that those at the sharp end found hard to appreciate, as an anonymous British officer who was in the thick of it described.

> *What excessive fatigue. A rapid march from four o'clock in the morning till four in the eve, when we engaged. Till dark we fought. Describe the battle . . . Thou hast seen Le Brun's paintings and the tapestry [depicting Marlborough's victories] perhaps at Blenheim. Are these natural resemblances? Pshaw! . . . There is a most infernal fire of cannon and musquetry. Most incessant*

shouting, "Incline to the right! Incline to the left! Halt! Charge!"
etc. The balls plowing up the ground. The trees cracking over one's
head. The branches riven by the artillery. The leaves falling as in
autumn by the grapeshot.[16]

In another echo of Brooklyn, Washington remained detached from the crucible of the battle, as though mesmerized—in "a daze," as his biographer Freeman put it—by the developments on two fronts. At around 5:00 PM he ordered Greene's reserve to shore up the collapsing front to the north. He would also go there; except the commander in chief did not know the way to his own battle and had to be led northward by a local, Curtis Lewis. In all, it had been an inept performance that could well have cost him his army had it not been for the magnificent rear-guard actions of George Weedon's and Peter Muhlenberg's brigades of Greene's division, and the tenacious defense of Wayne's division at Chad's Ford.

As evening drew on, elements of Cornwallis's and Howe's right wing made contact with Knyphausen's left, and the whole of Washington's army began streaming back toward Chester. The British had been on the go since 5:00 AM with only one brief halt. Howe's column, for example, had marched seventeen miles before going into a battle that lasted three hours. They were in no shape to pursue, and Howe must have bitterly regretted his lack of cavalry.

For Washington it was a low point as commander in chief, and the overwhelming exhaustion of the defeated engulfed him. The usually punctilious correspondent could only fall into bed with the final order of the day: "Congress must be written to, gentlemen, and one of you must do it, for I am too sleepy."[17] Later he would report to Congress that his losses were "not . . . considerable." Nathanael Greene estimated them at 800 to 900 killed and wounded with an additional 400 captured.[18] (Howard Packham's *Toll of Independence* puts American losses much more modestly at 200 killed, 400 wounded, and 400 captured.) Howe's losses were 90 killed, 448 wounded. (All but 40 of his killed and wounded were British.) Indeed, considering the numbers involved and the intensity of the firing, these figures represent low percentages for each side, as a young local who visited the battlefield immediately after the cessation noted: As "awful was the scene to behold," there were "a few dead but a small proportion of them, considering the immense quantity of powder and ball that had been discharged."[19]

In the days following the battle Washington sought to keep his army between Howe and Philadelphia. Despite defeat, the morale was surprisingly high, and the commander in chief was filled "with the firm intent of giving the Enemy Battle."[20] And he would have done so on 16 September but for a stupendous rainstorm that ruined 400,000 cartridges. On 20–21 September Anthony Wayne's brigade of 1,500 men, which had been ordered to harass the British rear guard, was completely surprised at Paoli by a midnight bayonet attack conducted by the 40th, 42nd, 44th, and 55th Foot; the 2nd Battalion of light infantry, and some elements of the 16th Dragoons, under Major General Charles Grey. Major John André described the action from the British point of view.

> No soldier . . . was suffered to load; those who could not draw [unload] their pieces took out their flints. . . . It was represented to the men that firing discovered us to the enemy, hid them from us [because of the considerable smoke caused at discharge] killed our friends and produced a confusion favourable to the escape of the Rebels. . . . On the other hand, by not firing we knew the foe to be wherever fire appeared and a charge ensured his destruction.[21]

It was certainly a onc-sided affair, and there were exaggerated claims of huge numbers of American dead—200, even 300—which led to a charge of massacre.[22] Other authorities put the patriot dead at just over 50 and point out that the American wounded, far from being mercilessly bayoneted, were taken from the field to British hospitals.[23]

With Washington on the east side of the Schuylkill, Howe now marched north up the west side, threatening the important American military depot at Reading Furnace. When Washington responded by also marching north, Howe doubled back during the night and crossed his whole army over to the east bank of the Schuylkill. He was now between Washington and Philadelphia, and the patriot army, which had marched 140 miles in eleven days and had insufficient clothing and food, was in no condition to contest the prize. On 26 September Cornwallis entered the city, while the main part of the army encamped at Germantown.

One commander invariably preferred a straightforward approach: fix the enemy's front in place and, while he is preoccupied, work around a flank. Perhaps the relatively straightforward strategy reflected a man of a blunt, none-too-sophisticated disposition. The other commander had a predilection for the complicated, multipronged attack that required highly developed military skills in his men, and particularly in his officers. Paradoxically, the more experienced commander usually opted for simplicity, like a master chef who cooks a simple omelet. The inexperienced commander, like many amateur cooks, enjoyed doing complicated things with foie gras and truffles.

Washington had a taste for the fancy when it came to planning attacks, and the upcoming battle at Germantown, five miles north of Philadelphia, would be a classic example. Howe had stationed the bulk of his army here, with a subsidiary garrison in Philadelphia under Cornwallis. At the end of September Washington learned that the British commander in chief had sent off a sizable contingent for duties elsewhere, reducing his numbers at Germantown to approximately 9,000. In contrast, the patriot army had grown through various reinforcements and levies to 8,000 Continentals and 3,000 militia. Many of them were green, but Washington saw an opportunity and constructed a plan of impressive intricacy that depended on a degree of coordination that would have been remarkable for even the most experienced and trained troops.

As at Trenton, Washington would rely on separate columns converging on target more or less simultaneously. At Germantown he wanted a grand pincer movement involving four assault elements. Column one would consist of himself, Sullivan, and Wayne (with

Stirling's division in support) and would barrel down the Skippack
Road (the main approach from the west) with 3,000 men to engage
the British center head-on. Meanwhile, column two, the American
left wing, under Nathanael Greene, with General Adam Stephen's
division and Brigadier General Alexander McDougall's brigade,
would come in from the northeast with about 6,000 men (two-thirds
of the total army), down the Limekiln Road, and flank the British
right wing.

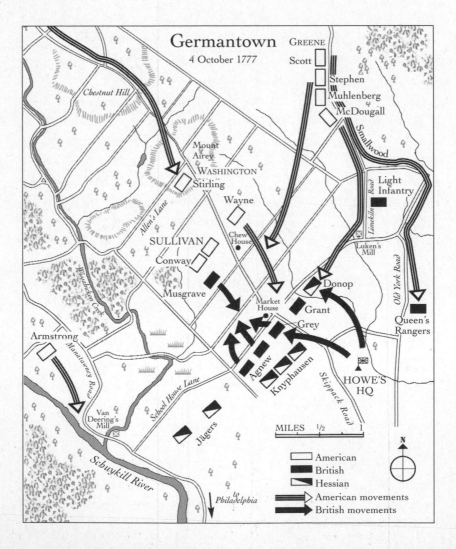

About a mile farther east, off to Greene's left, column three, composed of Brigadier General William Smallwood's 2,000 Marylanders and New Jersey militia, was meant to sweep down the Old York Road, around the British right wing, and come in from the rear. Finally, way over on the American right wing (about eight miles away from Smallwood) Brigadier General John Armstrong was to lead his Pennsylvania militia down the Ridge Road (parallel to the Skuylkill River on Armstrong's right) against the Hessians under Knyphausen, who were in a commanding position on a rise on the southeastern bank of the Wissahickon Creek as it runs into the Skuylkill.

If these dispositions were not complicated enough, Washington's timetable for attack (to be carried out at night over unfamiliar ground) was, to put it mildly, optimistic: "Each column to make their dispositions so as to attack the pickets in their respective fronts precisely at five oClock with charged bayonets, the columns to move on the attack as soon as possible. The columns to endeavor to get within two miles of the enemy's pickets on their respective fronts by two oClock and halt till four and make the disposition for attacking the pickets at the time mentioned."[24]

No sooner had Washington's intricately designed machine moved off (at 7:00 PM on 3 October) than the wheels began to wobble. Greene's group became lost and fell behind the timetable, and the consequence would have a significant impact on the outcome. At 5:00 AM, the hour for the concerted attack, Washington became uneasy. At 6:00 AM on the fourth Washington's column made contact with the British pickets on Mount Airy and began to drive them back down the main road toward Germantown. The 2nd Battalion of the British light infantry came up in support, but they too were forced back, which, as an officer recorded, did not come easily: "We charged them twice till the battalion was so reduced by killed and wounded that the bugle was sounded to retreat. Two columns of the enemy had nearly got round our flank. But this was the first time we had ever retreated from the Americans and it was with great difficulty that we could get the men to obey our orders."[25]

As the early morning fog grew steadily denser, Lieutenant Colonel Thomas Musgrave brought up his 40th Foot (he had become their commanding officer following the death of Lieutenant Colonel

James Grant at the battle of Brooklyn), which carried on a skillful fighting retreat until Musgrave managed to get six companies, about 120 men, inside Cliveden (known in the battle as the "Chew House"), the handsome and stoutly constructed home of the Loyalist justice Benjamin Chew. For much of the remainder of the battle the Musgrave garrison heroically held off repeated American attacks, and fifty-six patriots would die in that action alone.

During Greene's approach down the Limekiln Road, Stephen had veered off to his right, perhaps drawn toward the intense gunfire at the Chew House. In the swirling fog his command blundered into the left rear flank of Anthony Wayne's division, which was part of the Washington-Sullivan column and at that time closely engaged with the British right-center near the Market Square, the keystone of the British positions. In the confusion American fired on American, and the now decidedly wobbly wheels of Washington's battle plan started to fall off. Wayne's men disengaged and began to withdraw, leaving Sullivan's open to attack by the 5th and 55th Foot. Panic began to spread, exacerbated by the arrival of General Agnew's division, which was able to be released from the British left wing only because Armstrong's pincer had failed to exert any pressure whatsoever in that quarter. Sullivan's men were also running out of ammunition, and men peeled off to the rear holding open their empty cartridge boxes to show any officer who might doubt their motives. Although Agnew was killed almost as soon as he arrived on the scene, his men pushed on, fanning the wildfire panic that was now racing through the regiments of the Washington-Sullivan contingent.

Greene, unaware of Wayne and Sullivan's withdrawal, successfully penetrated deep into the British right-center, where one of his units, the 9th Virginia of Muhlenberg's brigade, broke into the Market Square and began plundering the British camp. While preoccupied with their loot they were counterattacked by the Guards and the 27th and 28th Foot and killed or captured to a man. Greene, now isolated, began an orderly retreat, as Tom Paine, who was in Greene's division, recorded: "The retreat was extraordinary. Nobody hurried themselves. Everybody marched his own pace. The enemy kept a civil distance behind."[26] Others, however, remembered a mad panic, "passed the powers of description, sadness and consternation expressed in every countenance."[27]

Smallwood and Armstrong, out on their respective far-flung flanks, were left high and dry, their only option to join in the general retreat. Back they all went the way they had come, all the way back twenty-four miles to their original camp at Pennypacker's Mill. In twenty-four hours they had marched over forty miles and fought a pitched battle for about four hours. What should have been a demoralized rabble was in fact extraordinarily buoyant—"high in spirits and appear to wish ardently for another engagement," wrote an unidentified officer.[28] Far from falling from grace, Washington's stock rose. Adam Stephen, on the other hand, was cashiered for being drunk and incompetent. He had been an adversary of Washington's, in matters military, political, and commercial, long before the outbreak of war, and his disgrace would be one of the few satisfactions enjoyed by Washington that day.

Although Douglas Southall Freeman, Washington's revered biographer, dismissed the importance of Musgrave's stand at the Chew House in influencing the outcome, it does seem to have had an important impact in several ways. It sucked in a good number of Washington's reserves that otherwise would have been engaged in the main battle; it drew off Stephen's division, which, in turn, triggered the friendly fire incident with Wayne's division and the escalating panic that followed. Some of Sullivan's and Wayne's men were also spooked by the rumor that the firing behind them indicated that they were surrounded. Washington, although he had come close to victory, was robbed by a series of mishaps and miscommunications that in their accumulation cost him the battle. He was left with that bitterest of dreg of consolation—"if only . . ."

The British defeat at Saratoga in October 1777 changed everything. It convinced the French that the odds had shortened on a patriot victory, and they laid down a big bet in the shape of a formal alliance, ratified by Congress on 4 May 1778. Now Britain would be involved in a war far outreaching the importance of that being waged against the thirteen colonies. At stake would be not only its empire but also the very real possibility of an invasion of Britain itself. The West Indies were economically crucial to the British economy, and to defend them against French incursions would involve moving troops and ships from North America. Sir Henry Clinton, who had succeeded Sir William Howe as commander in chief in May 1778, was soon called on to send 5,000 men for an expedition against St. Lucia and, consequently, would be forced to abandon Philadelphia.

On 16 June 1778 the evacuation of the city began, and by the eighteenth all 10,000 British and Hessian troops had crossed the Delaware to begin the long hot slog back to New York. Danger lay all along the route, of course, but for Clinton the overland option seemed far preferable to having his whole army intercepted on the high seas by a French fleet. In any event, the thousands of Loyalists who desperately fought for berths in the limited transports available (a scramble reminiscent of the evacuation of Saigon in 1975) would have made it impossible for the British commander to have shipped his army and its huge baggage and artillery trains back to New York.

Washington, still in his encampment at Valley Forge, was faced with something of a dilemma. To take on the retreating British column offered the prospect of a victory that, on the heels of Burgoyne's humiliation at Saratoga, almost certainly would have ended the war *instanta*. On the other hand, he could simply allow the British to return unmolested to New York, where they could be bottled up and allowed to wither without any risk to his own force. Washington, by nature, invariably favored the more aggressive approach. And although some historians have claimed that a resounding American defeat at the hands of Clinton in New Jersey could have jeopardized the whole war, it is difficult not to agree with Washington that at this stage of the game the benefits of victory far outweighed the risks.

The battle of Monmouth Courthouse that would take place in a three-mile-long by one-mile-wide corridor of "morasses" and ridges on

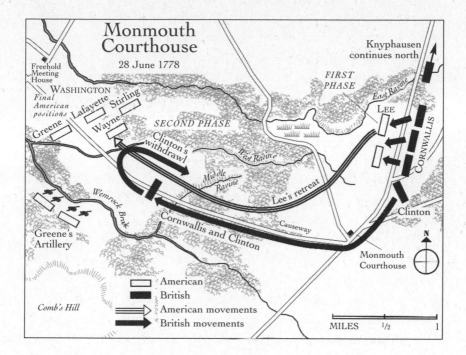

Monmouth Courthouse
28 June 1778

Freehold Meeting House

WASHINGTON
Final American positions

Lafayette
Stirling
Wayne
Greene

SECOND PHASE

Clinton's withdrawl

West Ravine

Middle Ravine

Lee's retreat

Wemrock Brook

Greene's Artillery

Cornwallis and Clinton

Causeway

Comb's Hill

Knyphausen continues north

FIRST PHASE

East Ravine

LEE

CORNWALLIS

Clinton

Monmouth Courthouse

N

American
British
American movements
British movements

MILES   ½   1

28 June 1778 would be not only the longest of the whole war but also the most tactically confused at the individual-unit level. Yet its overall architecture and dynamics are quite clear.

Clinton's column, stretched out over twelve miles, lumbered through the scorching heat. Temperatures would top 100 degrees Fahrenheit (37 degrees Celsius) on the day of battle. At its head was the division commanded by the Hessian general Knyphausen. In the rear was the largest element, commanded by Lord Charles Cornwallis; and sandwiched between both were the baggage and artillery, dictating the pace. The column moved slowly through dry sand roads that dragged on the wheels as effectively as mud. The army was like a great shaggy bear heading off to its lair with its shanks being constantly nipped and worried by patriot skirmishers that the beast would occasionally and ineffectively try to swat.

On 12 June Washington could report his main army strength at over 13,000, and despite the generally anemic response of most of his military council to the prospect of a general engagement (Charles Lee, for example, who had been exchanged from captivity in March, was particularly outspoken against what he saw as a disastrously overambitious attack, calling it "criminal"), Washington felt emboldened to test the British positions with a strong advance probe. He initially offered this assignment to Lee as the highest-ranking general under the commander in chief. Lee, perhaps bridling at what he saw as Washington's favoritism of the young Lafayette, refused it with a patronizing nod to his youthful rival. This refusal of the command is interesting because it is usually seen as the crabby elder soldier's condescension to a younger contender as being "a more proper business for a young volunteering general." But perhaps Lee had grounds for his umbrage.

Lafayette has been described as "the logical candidate."[29] Washington was now placing his trust in him, but because Lafayette was only twenty years old in June 1778 and had very little combat or command experience, one has to doubt his credentials in the face of a slew of other contenders. And although a biographer, Harlow Giles Unger, describes Lafayette before the battle of Monmouth as "the next-highest-ranking combat officer" (after Lee),[30] Washington's favoring of the young man and his clumsy handling of Lee show a leader whose command effectiveness had been compromised by equivocation.

Washington held Lee in awe, as did many in the patriot officer corps, and his reward was to be the butt of Lee's sneering. When Washington was struggling for survival during the retreat across the Jerseys in 1776, Lee almost contemptuously disregarded his pleas for reinforcement. When Lee was exchanged from captivity in March 1778, Washington put on a lavish show of welcome that Lee treated ungraciously. Washington would have had to be a saint not to have harbored some resentment of the Englishman's hubris. Whereas Charles Lee was a difficult man to like, unkempt, foulmouthed, and generally bizarre, Lafayette was charming and amiable, and showed an almost filial affection toward his commander in chief. Whatever the psychological shoals and riptides, Washington's vacillations and maladroit attempts to "manage" the two men would be disastrous.

When the size of the advance guard became a substantial 4,000-plus, Lee about-faced and asserted his right by seniority to its command (he would be "disgraced" otherwise, he said), and Washington acquiesced, putting aside the uncomfortable fact that Lee vehemently opposed the commander in chief's whole strategy. Reflecting the confusion he had created, as well perhaps as his reluctance to tackle Lee head-on, Washington's orders were vague and contradictory. Lee was to bring on "an engagement or attack the enemy as soon as possible" but yet not allow himself to become embroiled.[31] As to the tactical means, that was left up to Lee. The problem was Lee also had no idea what he intended to do, telling his commanders at 5:00 PM on the twenty-seventh that he could make no plan because he was ignorant of the terrain and the enemy's strength.

Although Lee had been ordered to reconnoiter during the early hours of the twenty-eighth, he neglected to do so until around 6:00 AM, by which time Clinton had already sent off Knyphausen and the baggage on the northeast road toward Middletown, and shortly thereafter followed him with Cornwallis's division. A rear guard was left at Monmouth Meeting House, and it was this that whetted Lee's appetite. Lee's disposition of his forces, however, was chaotic. They were "shifted about in kaleidoscopic arrangements and rearrangements"[32] and served only to alert Clinton and Cornwallis, who, like an enraged bear, swung around and turned on their attackers. It was a very big bear. Cornwallis's command (which Clinton accompanied) was composed of the brigade of Guards; both battalions of British grenadiers; all the Hessian grenadiers; both battalions of the British light infantry; the 3rd, 4th, and 5th infantry brigades; the 16th Light Dragoons; and John Simcoe's Queen's Rangers, the premier Loyalist regiment.

It was little wonder that Lee's command began to fall apart, and one of the ironies, given the complications of the Washington-Lee-Lafayette triangle, was that it was probably Lafayette's shifting of position (unauthorized by Lee) which was mistakenly interpreted as a retreat and triggered a panicked response in other regiments.[33] In no time Lee found himself swept up in a full-scale retreat, his men flowing back due west toward their starting positions to seek the refuge of the rest of the army some five miles back. A battle that Washington had always intended (despite his more pusillanimous advisers) to be highly aggressive had been turned on its head. The hunters were now the hunted.

With chivalric heroism, legend has it, Washington rode among the retreating and defeated, astride his great white charger, rallying and returning them to the fight by the sheer magnetism of his personality. To add to the irony of Lafayette's contribution to Lee's discombobulation, one of the primary contributors to the Washington legend (some might even say deification) was none other than Lafayette himself. In his *Mémoires,* published in 1837, three years after his death, he remembered: "General Washington was never greater in battle than in this action. His presence stopped the retreat; his strategy secured the victory. His stately appearance on horseback, his calm, dignified courage, tinged only slightly by the anger caused by the unfortunate incident in the morning, provoked a wave of enthusiasm among the troops."[34] Lee was stopped by Washington and given, by popular account, a scorching public tongue-lashing, which reduced the little bombast to spluttering incoherence. (Washington would later deny using any "singular expressions.")

Heroic intervention may well have played its part, but as Washington created his defensive line on the high ground above the western "morass," it was something altogether less romantic that held the army together: von Steuben's long training sessions of formation drill and volley fire during the Valley Forge winter. (Alexander Hamilton would later record that he had never understood the point of military discipline until he saw von Steuben's "reformed" army at work that afternoon.) At first against Stirling on the American left wing, then against Greene on the right and Washington in the center, the British and Hessians threw in attack after attack, which, although executed with extraordinary commitment and sacrifice, withered under disciplined musketry and artillery fire. The patriot lines would not be broken.

As evening drew on, this, the longest battle of the war, literally burned itself out, both armies exhausted by the unrelenting heat. The British recorded "3 sergeants, 56 rank and file died with fatigue [sunstroke]." The Americans lost thirty-six to the same cause. A standoff artillery duel was all either side could muster. Washington, aggressive as always, was not prepared to accept the failure of his original intention and ordered Brigadier General William Woodford's Virginians and Clark's North Carolinians to counterattack on both British flanks. The fast-closing night, however, put an end to any further action, and at midnight the wily Clinton slipped away and rejoined Knyphausen. A week later he and his army were safely in New York.

The battle of Monmouth, if reduced to a description of unit movements, is confusing and frustrating because many of the sources are contradictory. But there is a fascinating, if conjectural, connective vein that runs though it: the idea of betrayal. If Lee had betrayed Washington's trust in the past (he may also have more literally betrayed the American cause while in British captivity), Washington certainly had his revenge by battle's end. A cynic might conclude that Washington's special relationship with Lafayette, his muddying of the command structure, his vague instructions, and Lafayette's own conduct during the first phase of the battle, "set up" the unfortunate and unsympathetic Lee for failure: a failure for which he would be, conveniently, the scapegoat.

Lee's own temperament could not have been more useful to Washington. Lee simply self-destructed under the disgrace, insulted the commander in chief in writing, and was thrown out of the army. One last irony, though, is that many of the officers—both American and British—present during the first phase attested that Lee's conduct in the face of an overwhelming British force was appropriate and carried out with military professionalism. For Washington it must have been an irony sweetly to be savored.

The Saratoga Campaign

FREEMAN'S FARM, 19 SEPTEMBER 1777; AND BEMIS HEIGHTS, 7 OCTOBER 1777

Major General John Burgoyne, commander in chief of the royal forces about to invade America from Canada, was, in his other life, a successful playwright. As he set off from St. Johns in June 1777 he little knew that he would unwittingly rehearse the whole of the British experience of the War of Independence in miniature. It would be a play in three acts: a triumphant beginning; a grim middle; and a pathetic finale. The story line had seemed strong enough at first, but someone rewrote it halfway through and then everything began to go wrong. The scenery proved problematic, the supporting cast decamped for a better-paying show, and the necessary props either fell apart or failed to appear on cue. The author, who was also the director and principal actor, remained steadfast that he'd written a smash, if only the backers had supported him as they had promised.

He was also, in the spirit of his age and class, a chancer, described by Arthur Lee, the special agent for Massachusetts in London, as "an abandoned & notorious gambler."[1] On the Christmas Day before he left to take command of the invading force in Canada, Burgoyne had wagered fifty guineas with his good friend (although a bitter opponent of the war) Charles James Fox that he would return victorious by the following Christmas.[2]

In prospect, the plan Burgoyne had proposed to Germain on 28 February 1777 was imaginative, ambitious, and bold. Certainly it seemed to be worth a bet.[3] His army would bisect the rebellion by thrusting down the linked waterways of the Richelieu River, Lake Champlain, and Lake George, and then make a short hop (approximately ten miles) overland from the southern tip of Lake George ("the most expeditious and most commodious route" as Burgoyne described it to Germain)[4] to connect with the Hudson, and so on down to Albany. "These ideas are formed," wrote Burgoyne, "on the supposition that it be the sole purpose of the Canada army to effect a junction with General Howe; or, after cooperating so far as to get possession of Albany and open the communication to New-York, to remain upon the Hudson's-River, and thereby to enable that general to act with his whole force to the southward."[5]

A secondary force of 800 British regulars and Canadian volunteers together with an equal number of Indians, all commanded by a veteran of the French and Indian War, Lieutenant Colonel Barry St. Leger, would land at Oswego on east Lake Ontario, come up the Oswego River, cross Lake Oneida, capture Fort Stanwix (present-day Rome, New York), and finally come down the Mohawk River to make the junction with Burgoyne at Albany.

Sir William Howe, however, had been sending conflicting signals. Back in October of 1776 he had told Germain that his primary strategic objective was to open "a communication with Canada" to control the Hudson and attack New England, the heartland of the revolt, for which he would need an additional 15,000 men. By 20 December, with Washington close to defeat and miserably huddled east of the Delaware, Howe changed his mind. An offensive in Pennsylvania would end the war: "The principal Army should act offensively on that side, where the enemy's chief strength will be certainly collected."[6]

Consequently, on 2 April Howe responded to Burgoyne's proposal by informing Sir Guy Carleton, commander in chief in Canada and technically still Burgoyne's immediate superior, that there could be little hope of Howe being able to help Burgoyne: "[I have] but little expectation that I shall be able, from the want of sufficient strength in this army, to detach a corps in the beginning of the [Burgoyne's] campaign to act up Hudson's River," and therefore Burgoyne's men "will, I fear, have little assistance from hence to facilitate their approach, and as I shall probably be in Pennsylvania when that corps is ready to advance into this province, it will not be in my power to communicate [that is, cooperate] with the officer commanding it so soon as I could wish; he must therefore pursue such measures as may from circumstances be judged most conducive to the advancement of his Majesty's service consistently with your Excellency's orders for his conduct."[7] Howe glibly supposed that Loyalist support would be so strong that Burgoyne would find it "no difficult task to reduce the more rebellious parts of the province" (the opposite proved to be the case) and signed off with a half promise of help later on: "I shall endeavour to have a corps upon the lower part of Hudson's River sufficient to open the communication for shipping thro' the Highlands."[8]

In response, Germain approved Howe's Pennsylvania expedition but stressed, "[It is to to] be executed in time for you to co-operate with the army ordered to proceed from Canada and put itself under your command." Whatever the niceties of language, there is no doubt that Burgoyne and his army expected a rendezvous with a British force at Albany. On 6 May, for example, Burgoyne repeated the objective laid out in his February plan. He wrote to General Simon Fraser, "The military operations, all directed to make a junction with Howe, are committed to me."[9]

Although in the end Howe would fulfill the letter of his "endeavour," he failed to support the spirit of Burgoyne's enterprise. A more charitable explanation might be that he felt he could serve Burgoyne's mission by keeping Washington tied up in Pennsylvania. That Burgoyne's enterprise had merit is reflected in George Washington's astonished disbelief and amazed relief that Howe refused to make a priority of linking with Burgoyne. With his keen strategic eye he saw the danger. On 2 July he wrote to Jonathan Trumbull: "If it [Burgoyne's expedition] is not merely a diversion, but a serious attack, of which it bears strongly the appearance, it is certainly the proof that the next step of General Howe's army will be towards Peekskill, and very suddenly, if possible to get possession of the passes in the Highlands before this army can have time to form a junction with the troops already there."[10] He would write to General Gates on 30 July, after the fall of Ticonderoga (seen at the time as a catastrophe for the cause), "General Howe's in a manner abandoning General Burgoyne [Howe had set sail for the Chesapeake on 23 July] is so unaccountable a matter, that, till I am fully assured it is so, I cannot help casting my eyes occasionally behind me."[11] He could not believe his luck.

Burgoyne's expedition embarked from St. Johns on the Richelieu on 17 June, and it was a fine sight. Almost 4,000 British regulars; just over 3,000 Brunswick and Hesse-Hanau troops, under Major General Baron Friedrich Adolphus von Riedesel; 400 Indians (Iroquois, Algonquin, Abenaki, and Ottawa, under the command of the seventy-year-old French Canadian Pierre St. Luc de Lacorne, whose attitude as to the best tactical use of his Indians—*"Il faut brutaliser les affaires"*—proves that even scalping sounds chic in French); 300 Canadian woodsmen; and 300 Loyalists set off in a fleet that included the twenty-four-gun *Royal George* and twenty-gun *Inflexible*, led off by the Indians in their birch canoes.

There was a sizable artillery train—too sizable for some critics back in England—consisting of 138 pieces, including twenty-six 6-pounders, seventeen 3-pounders, sixteen heavy and two light 24-pounders, and fifteen 12-pounders along with howitzers and mortars. Its size was justified, according to Burgoyne, because the fortress of Ticonderoga was expected to be a tough nut to crack and he knew from James Abercrombie's disastrous failure to take it from the French in 1758 that artillery would decide the issue. (It did, but only against a fraction of the resistance he had expected.) He would also need to leave armament in defense of various posts along the way and, of course, would want to have sufficient left over to garrison Albany on his triumphal arrival. As it turned out, after shedding artillery en route, Burgoyne had only forty-two pieces for his final battles.

Spirits were high, and Burgoyne felt moved to make a pontifical proclamation of the kind that seems so often to be the necessary preface for colonial invasions masquerading as "liberations."

The army embarks tomorrow, to approach the enemy. We are to contend for the King, and the constitution of Great Britain, to vindicate Law, and to relieve the oppressed—a cause in which his Majesty's Troops and those of the Princes his Allies, will feel equal excitement. The Services required of this particular expedition are critical and conspicuous. During our progress occasions may occur in which, nor difficulty, nor labour, nor life are to be regarded. This Army must not Retreat.[12]

"General Burgoyne shone forth in all the tinsel splendour of enlightened absurdity," wrote a perceptive commentator.[13]

Early and easy victories for an invading army can often resemble the first couple of drinks for an alcoholic. They tend to give a rosy glow to an impending and inevitable disaster. And so it was for Burgoyne. The capture of Fort Ticonderoga, practically on the first anniversary of the Declaration of Independence, was seen at the time as a devastating blow to the patriot cause. The fort, "a perfect Mouse Trap," declared General Arthur St. Clair, the hapless commander of the American defenders of Ticonderoga, was on a peninsula at the confluence of Lake Champlain, the outlet from Lake George, and the South Bay of Lake Champlain, and was woefully undermanned: "Had every man I had been disposed of in single file along the lines of defense, they would scarcely have been in reach of each other's voices," wrote St. Clair. In addition, the defenders "[are] miserably clad and armed . . . many are literally barefooted, and most of them ragged."[14] And once the British had established a battery on Sugar Loaf Hill (which had been left undefended, even though St. Clair had been warned of its potential danger if occupied by the British)[15] overlooking both the fort and an encampment on Mount Independence across South Bay, the game was well and truly up for St. Clair. He had no option but to abandon the works before he would have been completely overrun.

The shock was profound: "The abandonment of Ticonderoga and Mount Independence has occasioned the greatest surprise and alarm. No event could be more unexpected and more severely felt throughout our army and country. This disaster has given our cause a dark and gloomy aspect," wrote the surgeon James Thacher.[16] St. Clair and Philip Schuyler, his commanding officer, were both roundly reviled, even accused of treason, but St. Clair's action had preserved a force that would, over the following weeks, provide the essential core of an army that would turn and totally destroy its once-invincible tormentor. As Dr. Thacher presciently added. "It is predicted by some of our well-informed and respectable characters, that this event, apparently so calamitous, will ultimately prove advantageous, by drawing the British army into the heart of our country, and thereby place them more immediately within our power."[17]

With the patriot army of the North on the point of collapse ("The army is in danger of starving . . . The army have not a mouthful of fresh provisions . . . We may be utterly ruined for lack of wagons," bewailed General Schuyler from his headquarters at Fort Edward), the way seemed open for Burgoyne's triumphal procession down to Albany. However, one of those "well-informed and respectable characters"— George Washington—saw things differently and gave heart to Schuyler by emphasizing Burgoyne's strategic dilemma. On 22 July Washington wrote, "From your accounts he appears to be pursuing a line of conduct, which of all others is most favourable to us; I mean acting in Detachments." Two days later he underscored the point: "As they can never think of advancing, without securing their rear by leaving garrisons in the fortresses behind, the force with which they can come against you will be greatly reduced by the detachments necessary for the purpose."[18] The revised script Washington was describing would not be to John Burgoyne's liking, but it would become the new drama he would be forced to play out.

It was as if Burgoyne had been standing at Washington's elbow. Only six days after Washington had written Schuyler, Burgoyne was writing with foreboding, if not mounting panic, to Carleton back in Canada that without his help in garrisoning Ticonderoga his manpower would be dangerously reduced.

The construction your excellency puts upon the orders of the Secretary of State [Germain] is too full and decisive for me to presume to trouble you further upon the subject of a garrison for Ticonderoga from Canada, I must do as well as I can, but I am sure your Excellency, as a soldier, will think my situation a little difficult. A breach into my communication must either ruin my army entirely, or oblige me to return in force to restore, which might be the loss of the campaign. To prevent a breach, Ticonderoga and Fort George must be in very respectable strength, and I must besides have posts at Fort Edward and other carrying places. The drains added to common accidents and losses of service, will necessarily render me very inferior in point of numbers to the enemy.[19]

Carleton let Burgoyne stew.

Like the witches in *Macbeth*, three actions were the evil portents of what was in store for the Burgoyne expedition. On 7 July, during its retreat from Ticonderoga, the American rear guard, about 1,000 men under Seth Warner, was caught at breakfast "in a very unfit posture for battle" by units of Simon Fraser's advance corps. After the initial shock, the Americans managed to form a battle line on Zion Hill at Hubbardton. Accurate patriot fire began to take its toll. Major Robert Grant of the 24th Foot was killed, and the regiment took heavy casualties as it struggled uphill to engage Colonel Turbott Francis's 11th Massachusetts and the 2nd New Hampshire under Benjamin Titcomb. Fraser detached Major Lord Balcarres'ss light infantry and the grenadiers of the 9th, 20th, 29th, 34th, and 62nd Foot under Major John Dyke Acland to try a flanking move around the American left in the hope of cutting off any retreat down the road to Castle Town, where St. Clair and the main American force were resting. Fraser also sent out an urgent appeal for von Riedesel's Brunswickers to hurry to the support of the now-weakened British left, which was in danger of being surrounded. They duly came on (singing hymns to the accompaniment of their regimental band, a psychological tactic the Germans employed regularly) and delivered a bayonet charge against Titcomb's 2nd New Hampshires on the American right, driving them back.

For the Americans there were to be no reinforcements. St. Clair tried vainly to persuade militia units under Colonel Bellows to march immediately to the aid of the embattled rear guard, but "not a man made a move in the direction of the fighting. The troops simply refused to obey."[20] Francis, already wounded, was killed by a musket ball, and the patriot defense disintegrated. Although there is some debate on final casualties (*The Toll of Independence* states that 30 Americans were killed, 96 wounded, and 228 captured, compared with 200 British and German casualties, of whom 35 were killed), the British and Germans knew they had been in a desperately hard fight.[21]

The second witch turned up almost exactly one month later. Burgoyne's diversionary force under the newly promoted Brigadier General Barry St. Leger had left Montreal on 23 June 1777 with a force of 1,000 Iroquois under Chief Thayendanegea (also known as Joseph Brant), 380 Loyalists under Sir John Johnson, 80 Hesse-Hanau jägers, 250 British regulars from the 8th and 34th Foot, and about 400 Canadian bateaux-men for a total of approximately 2,000. By 2 August they had begun their siege of Fort Stanwix (later renamed Fort Schuyler), defended by Colonel Peter Gansevoort and about 550 men mainly from his own regiment, the 3rd New York.

Brigadier Nicholas Herkimer, the commander of militia in Tryon County, set out to relieve the besieged fort but was ambushed on the way by the Indians and Loyalists of Brant and Sir John Johnson. In brutal hand-to-hand combat Herkimer, although wounded in the leg (he would die ten days later from complications caused by its amputation), managed to hold his force together, fighting in two-man teams that ensured one was always loaded and able to protect the other from being rushed while reloading. A sortie from Fort Stanwix that ransacked the Indians' camp relieved the pressure, and Herkimer was able to effect a retreat.

Although hard-pressed for men himself, Schuyler sent Benedict Arnold with 950 Continentals to relieve the fort. Arnold used the Loyalist Hon Yost Schuyler (no relation to the general), "a proprietor of a handsome estate in the vicinity . . . taken as a spy,"[22] to spread word among Brant's Indians that a huge patriot force was on its way. The Indians, already disaffected by the sacking of their camp, in their turn ransacked the British lines, forcing St. Leger to break the siege and return to Canada. Triumphant, Arnold would be back with the main army by the first week of September.

After Ticonderoga Burgoyne had a tough decision to make. Should he stick to his original plan and transport his army by water across Lake George, or should he strike out overland from Skenesborough and cut across sixteen miles of virtual virgin woodland and swamp until he hit the Hudson? He feared that a "retrograde motion" to take his army back to Ticonderoga would encourage the enemy and dishearten his own men and the local Loyalists. The overland route would also, he argued (retrospectively), force the American garrison at Fort George (at the southern tip of Lake George) to quit the place.

Howe had written to Burgoyne on 17 July not only to congratulate him on taking Ticonderoga but also to disabuse him of any hope of an imminent linkup: "My intention is for Pennsylvania, where I expect to meet Washington, but if he goes to the northward contrary to my expectations, and you can keep him at bay, be assured that I shall soon be after him to relieve you. After your arrival at Albany, the movements of the enemy will guide yours." He ended his letter with a salutation that must have felt to Burgoyne about as sweet as sucking on a lemon: "Success be ever with you."[23]

Five days prior to receiving Howe's letter Burgoyne had ordered all excess baggage to be sent back to Ticonderoga. He was about to take the great gamble, notwithstanding Howe's warning, and abandon his supply route and rear bases. He would make a push for Albany, even though he knew full well that on the overland route "considerable difficulties may be expected." He was not the only one enthusiastically rolling the dice. Back in London Germain, whose choice Burgoyne had been to lead the invasion, anticipated a success against the odds: "I am sorry the Canada army will be disappointed in the junction they expect with Sir William Howe," he wrote to his undersecretary, William Knox, "but the more honour to Burgoyne if he does the business without any assistance from New York."[24] Neither was the army unenthusiastic. Lieutenant William Digby was particularly gung-ho: "In my opinion, this attempt showed a glorious spirit in our General, and worthy alone to be undertaken by British Troops as the eyes of all Europe, as well as Great Britain were fixed upon us."[25] The baggage, artillery, and ammunition were to be sent back and transported over Lake George. On 24 July the infantry began its overland journey, but Burgoyne's logistical situation was deteriorating quickly. Only half of the transport wagons were operational. (They had been knocked together of unseasoned wood back in Canada and were now falling to pieces.) He was short of draft animals and fresh meat. In fact he was advancing so quickly that he was simply outstripping his channels of supply from Canada.

General Schuyler, although often lambasted for being away from the front, and as a patrician forced to bear the mistrust of hoi polloi (he would be replaced by Horatio Gates on 14 August, against Washington's wishes), had done an enormously effective job denying Burgoyne forage and foodstuffs as he passed through the countryside. Like Marshal Mikhail Kutuzov facing Napoleon's invasion of Russia in 1812, Schuyler was no swashbuckler, but his scorched-earth policy and patient wearing-down of his enemy did as much as anything else to bring Burgoyne to his knees on the final battlefield. A Hessian officer described the efficacy of Schuyler's strategy.

> *I have called it a desert country, not only with reference to its natural sterility—and heaven knows it was sterile enough—but because of the pains which were taken, and unfortunately with all too great success, to sweep its few cultivated spots of all articles likely to benefit the invaders. In doing this the enemy showed no decency either to friend or foe. All the fields of standing corn were laid waste, the cattle were driven away, and every particle of grain, as well as morsel of grass, carefully removed; so that we could depend for subsistence, both for men and horses, only upon the magazines which we might ourselves establish.[26]*

The need for forage in an eighteenth-century army was as pressing as the need for oil in a mechanized one. (Even a tactical master like Erwin Rommel, after all, was defeated as much by the strangulation of his oil supplies as any other factor.) Burgoyne had intelligence that a happy confluence of forage, cattle, "wheel carriages," horses, and Loyalists could be found at Bennington. Better yet, the market town was defended only by a small contingent, perhaps 300, of militia. Burgoyne therefore dispatched Lieutenant Colonel Frederich Baum with 723 men, including three squadrons of Brunswick dragoons;[27] fifty British corps of marksmen under Captain Alexander Fraser; 150 local Loyalists under Francis Pfister, a retired British officer; and about 150 Indians, to make the raid to the east.

Baum was unlucky. Only a short time before he set off, John Stark and over 2,000 New Hampshire militia and Vermont Rangers had moved down to Bennington and now took up a position facing Baum across the Walloomsac River. Stark, who had the tactical instinct of a chain saw, decided to hit Baum from every direction possible, simultaneously. Baum's main force (primarily three companies of Dragoon Regiment Prinz Lüdwig) was on a hill overlooking the bridge across the river. It was a strong-enough position except that Baum's security was as porous as a sponge and patriots whom he mistook for Loyalists began to infiltrate his position. Hit from the front (east), north, and west, Baum's position imploded. A Hessian officer's description of the patriot tactics eerily resembles Communist attacks on isolated firebases during the Vietnam War.

> *The troops lining the breastworks replied to the fire of the Americans with extreme celerity and considerable effect. So close and destructive, indeed, was our first volley that the assailants recoiled before it, and would have retreated, in all probability, within the woods; but ere we could take advantage of the confusion produced, fresh attacks developed themselves, and we were warmly engaged on every side and from all quarters. It became evident that each of our detached posts were about to be assailed at the same instant.*[28]

They were overrun. As the historian Brendan Morrissey has pointed out: "Stark's men are often depicted as inexperienced farmers, in contrast to Baum's and Breymann's regulars. In fact, New Hampshire and Vermont were major recruiting areas in the French and Indian Wars and many men were former rangers or Provincials."[29] Once the defenses had been breached, hand-to-hand fighting followed: "For a few seconds... The bayonet, the butt of the rifle, the saber, the pike, were in full play, and men fell, as they rarely fall in modern war, under the direct blows of the war."[30] This was classic partisan warfare, so rarely encountered in the formal volley battles of European theaters of the eighteenth century. Baum was killed, and most of his command either became casualties or were captured.

Colonel Heinrich Breymann's relief column arrived just in time to be overwhelmed in its turn by Stark and Seth Warner's Green Mountain Boys. Breymann was wounded but managed to escape. In this one action Burgoyne had lost 15 percent of his total force (almost 700 captured and 200 killed). Horatio Gates, on the other hand, would enjoy not only the addition of Arnold's returned force and Morgan's elite riflemen sent up from Pennsylvania, but also a healthy influx of militia carried into his army on the crest of enthusiasm created by the great victory at Bennington. Burgoyne would ruefully report to Germain, "Wherever the King's forces point, militia to the amount of three or four thousand assemble in twenty-four hours."[31] After Hubbardton and Stanwix, Bennington was the most crushing blow to Burgoyne's ambitions, and one from which he never recovered. He could hear the third witch cackling offstage.

By the end of August Burgoyne was seriously rattled and already beginning to cast about for excuses in case of failure. Writing to Germain on 20 August, he lined up four. First, the number of insurgents: "The Hampshire Grants [Vermont]... a country unpeopled and almost unknown in the last war, now abounds in the most active and rebellious race of the continent." Second, the scorched-earth policy: "In all parts the industry and management in driving cattle and removing corn are indefatigable and certain." Third, the failure of Howe to support him: "Another most embarrassing circumstance is the want of communication with Sir William Howe... No operation, my Lord, has yet been undertaken in my favour." Fourth, the unreasonable restrictions of his orders: "Had I latitude in my orders, I should think it my duty to wait in this position, or perhaps go back as far as Fort Edward . . . but my orders being positive to 'force a junction with Sir William Howe,' I apprehend I am not at liberty to remain inactive." He ended, somewhat ominously, "Whatever may be my fate my Lord, I submit my actions to the breast of the King, and to the candid judgment of my profession."[32] It sounded like the last words of a condemned man.

As to Burgoyne's fourth point, no eighteenth-century commander 3,500 miles away could be held to the letter of his orders when it took months for communications to go back and forth to London. There was, of necessity, a great degree of autonomy, because it was impossible for London to micromanage the American war (although generals complained of micromanagement nevertheless). Germain, on receiving Burgoyne's letter, was understandably taken aback: "What alarms me most is that he thinks his orders to go to Albany to force a junction with Sir William Howe are so positive that he must attempt at all events the obeying them, tho' at the same time he acquaints me that Sir William Howe has sent him word that he has gone to Philadelphia, and indeed nothing that Sir William says could give him reason to hope that any effort would be made in his favour." As if to underscore the issue of time lag, Burgoyne's army would already be in the bag when Germain received the letter.

Horatio Gates chose his ground well. Having first established his headquarters at Stillwater, he moved three miles north onto the bluffs above the west bank of the Hudson. Stillwater lay in flat land that would lend itself advantageously to the massed volley and artillery tactics of the Anglo-German army. On the other hand, the heavily wooded terrain of his new location, slashed with deep ravines, would disrupt troop deployment in solid masses and favor the more irregular tactics of the patriot army. Perched high above the river and the river road to Albany, he enjoyed a commanding advantage that was further reinforced by the skillful fortification engineering of Polish-born and French-trained Thaddeus Kosciuszko.[33]

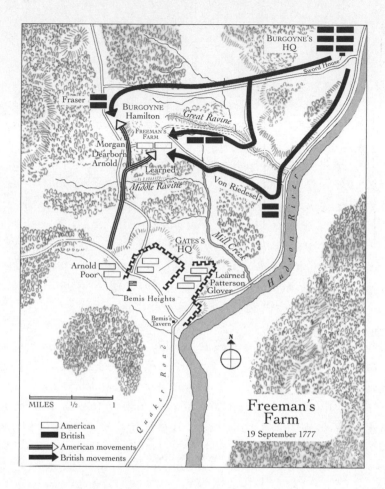

Freeman's
Farm

19 September 1777

MILES   ½   1

American
British
American movements
British movements

On 13 September Burgoyne crossed over to the western side of the Hudson. His choices were limited. There was no adequate road down the eastern shore, and even if there had been, he would eventually have been forced to make an opposed crossing of the river to get to Albany—not something to be relished. The Americans had to be tackled sooner or later; they could not simply be bypassed.[34] Gates had about 7,000 men, Burgoyne about 4,600 regulars plus another 1,000 in assorted Canadians, bateaux-men, and Indians; but one crucial factor separated them. Burgoyne's losses could not be replaced, whereas Gates was receiving a steady stream of reinforcements. The odds, as

Burgoyne was about to discover, were heavily stacked in favor of the house.

Lieutenant Colonel James Wilkinson, Gates's deputy adjutant general, described the strength of the American positions.

> *General Gates's right occupied the brow of a hill near the river, with which it was connected by a deep intrenchment . . . his front was covered from the right to the left of the centre, by a sharp ravine running parallel with his line and closely wooded; from thence to the knoll on the extreme left, the ground was level and had been partially cleared . . . beyond which in front of his left flank, and extending to the enemy's right, there were several small fields in very imperfect cultivation, the surface broken and obstructed with stumps and fallen timber . . . The right was almost impracticable [to be attacked]; the left difficult of approach.* [35]

On the right wing, Gates had Benjamin Lincoln commanding John Glover's brigade (approximately 1,600 men), John Nixon's brigade (approximately 1,300), and John Paterson's brigade (about 1,240). On the left, Gates had the formidable if impossible (from Gates's point of view) Benedict Arnold commanding Enoch Poor's 1,300 men. Ebenezer Learned's brigade was about 1,400 all told. In addition, Arnold had Daniel Morgan's 374 riflemen, certainly ranked among the finest light infantry of the day, and Henry Dearborn's 300 light infantry. Gates could also call on an artillery train of twenty-two pieces.

Burgoyne knew that Gates's defenses were too strong to allow him to blow open the river road. He had to probe inland and see if he could work his way around the patriot left flank, perhaps by exploiting the high ground to the west. He divided his force into three columns. The left wing (approximately 1,100 men) under von Riedesel would move down the river road and hold in place the American right wing. The right column (2,000 strong) under Simon Fraser went the farthest inland. It was made up of the crack units of Burgoyne's army: ten companies of light infantry under Major Lord Balcarres, and ten companies of grenadiers under the hard-drinking

Major John Dyke Acland. The battalion companies of the 24th Foot (Fraser's own regiment) were also attached, as were about 150 Loyalists, 50 Indians, and 100 or so British and German marksmen. The center column (about 1,100 men) was commanded by Lieutenant Colonel James Inglis Hamilton. He had the battalion companies of the 9th, 20th, 21st, and 62nd Foot as well as the Royal Artillery under Captain Thomas Jones. Burgoyne accompanied the center column. Separated by distance and rough terrain, Burgoyne's three columns would never be properly coordinated. In total there were about 4,200 British and German troops in the field. Gates could muster something close to 9,000 men.[36]

Hamilton entered the Great Ravine, which ran parallel to the American lines, and waited there until Fraser's force had reached the high ground farther west.

Back in American headquarters Arnold stormed and raged at Gates. Rather than sitting behind their defenses, Arnold urged his commander to use the wooded terrain to best advantage and make a preemptive strike. To wait would be to invite a flanking attack. Gates eventually agreed to release Morgan and Dearborn on the left.

At about 12:30 PM elements of Hamilton's advance guard (9th Foot) moved across the roughly cleared field of Freeman's Farm and were hit by Morgan's rifles firing from the cover of the woodland on the southern fringe of the clearing. The effect was devastating, "every officer being killed or wounded," recorded Lieutenant William Digby of the 53rd Foot. The surviving advance guard ran back to the safety of the main body. Morgan's men, unable to contain their elation, chased after them and were counterattacked by units from Fraser's light infantry, which advanced with disciplined and coordinated volley fire. For a while things quietened.

After a good deal of acrimonious wrangling between Gates and Arnold, Gates ordered 900 men from Enoch Poor's brigade (primarily Colonel Joseph Cilley's 1st New Hampshires and Colonel Alexander Scammell's 3rd New Hampshires) to support Morgan and Dearborn. The British lined the northern edge of the Freeman's Farm clearing with the 21st Foot on the right, the 62nd in the center, and the 20th on the left. The 9th was kept back in reserve. Burgoyne and his staff positioned themselves immediately behind the center troops.

It was about 1:30 PM when the center exploded. Arnold attempted to drive a wedge between Hamilton's and Fraser's forces but was beaten off with heavy losses. Arnold was in the thick of it, as a man from Dearborn's corps recorded, "riding in front of the lines, his eyes flashing . . . a voice that rang as clear as a trumpet."[37] Burgoyne too, in sharp contrast to Gates (whom Colonel Richard Varick said spent the battle "in Dr Pott's tent backbiting his neighbours"),[38] did not flinch when the bullets were flying. Sergeant Roger Lamb remembered: "General Burgoyne, during this conflict behaved with great personal bravery. He shunned no danger; his presence and conduct animated the troops, for they greatly loved the General. He delivered his orders with precision and coolness, and in the heat, danger, and fury of the fight, maintained the true characteristics of the soldier—serenity, fortitude, and undaunted intrepidity."

The battle for the center raged back and forth for over three hours. (Scammell called it "the hottest Fire of Canon and Musquetry" that he ever heard in his life.) "The British artillery," recorded Wilkinson, "fell into our possession at every charge but we could neither turn the pieces upon the enemy, not bring them off; the wood prevented the last, and the want of a match the first, as the lint stock was invariably carried off, and the rapidity of the transitions did not allow us time to provide one."[39] The 62nd Foot, fired on from its front and both flanks, took a terrible beating, losing 83 percent of its original complement of 350, including 35 officers and two lieutenant colonels.[40] Of the approximately 800 men engaged in the three center British regiments, about 350 (44 percent) became casualties.[41] The British gunners were also hammered. Lieutenant James Hadden was the only artillery officer serving with the center column not to be hit. Thirty-six out of 48 artillerymen (75 percent) became casualties. Fraser, out on the British right, had been all but neutralized by Morgan and Dearborn, and Burgoyne had not yet called in von Riedesel. Arnold was being sucked into the center, but Burgoyne could not, did not, organize his wings to move in and crush him.

Von Riedesel was an astute soldier. Realizing that the British center was perilously close to collapse, he organized a defensive force of 800 men under Brigadier General Johann Specht and set off with about 500 men of his own regiment plus two companies of the Regiment Rhetz, and two 6-pounders commanded by Captain Georg Pausch. It was now around 5:30 PM. Von Riedesel's intervention was critical, for when he arrived he found the battle "raging its fiercest. The Americans far superior in numbers . . . The three brave English regiments had been . . . thinned down to one-half, and now formed a small band surrounded by heaps of dead and wounded."[42] The Germans charged into Arnold's right flank "on the double quick," with Pausch opening up at close range with deadly volleys of canister. As at Hubbardton, von Riedesel had saved the British bacon.

Arnold again went to Gates for more men, and Gates (who never laid eyes on the front lines) reluctantly agreed to commit Learned's brigade to the critical battle for the center. One more push, pleaded Arnold, and the battered British regiments would cave. Learned's brigade, however, lost its way and wandered off to attack Fraser on the British right, where it was easily repulsed and failed to make any impact on the outcome.

Mutual exhaustion and encroaching darkness ended what would prove to be one of the pivotal battles of history. Technically the British had "won" because they remained in possession of the battlefield, but in reality they were finished. The British lieutenant Thomas Anburey viewed the carnage and the plight of the wounded: "Some of them begged they might lie and die, some upon the least movement were put in the most horrid tortures, and all had near a mile to be conveyed to the hospitals."[43] The Anglo-German forces had suffered 524 casualties, 160 killed and 364 wounded. The Americans suffered around 300 casualties, of whom 80 were killed in action.

Lieutenant Digby's journal gives some insight into the tactical problem the terrain posed the British: "From the situation of the ground, and their being perfectly acquainted with it, the whole of our troops could not be brought to engage together, which was a very material disadvantage."[44] For an army that was trained to lay down concentrations of volley fire, this was obviously a huge problem. For the patriots, on the other hand, the broken and wooded ground played to their strong suit. An American, Captain E. Wakefield, recalled that Morgan "astonished the English and Germans with the deadly fire of his rifles." After the battle, Burgoyne (perhaps remembering what he had seen at Bunker's Hill) stressed that the bayonet alone could break the tactical impasse: "The impetuosity and uncertain aim of the British Troops in giving their fire, and the mistake they are still under, in preferring it to the Baynotte, is much to be lamented."[45] The problem, however, was much better appreciated by the poor old infantryman who, in order to get within bayonet-point range, would have to receive fire without being able to return it, an act of self-sacrifice that took courage, discipline, and training of a very high order. It was not that these qualities were lacking, but the understandable temptation was to try to keep the enemy at a distance with volley fire, no matter how inefficient it turned out to be.

After Freeman's Farm Burgoyne's instinct was to attack the next morning. Perhaps he was dissuaded by Simon Fraser, who begged for time to rest the exhausted troops. Perhaps it was the message from Henry Clinton that arrived on the twenty-first, informing him of an imminent attack on the Hudson forts. Would it not be wise, given Gates's numerical superiority, to wait and see if Clinton's intervention might draw off a significant chunk of the patriot army? Burgoyne informed Clinton that he could hang on until 12 October, but no later. The school of "what ifs," on the other hand, points out that the Americans were also exhausted and in chaos after Freeman's Farm. A quick strike might have broken open the way to Albany. In any event, Burgoyne chose to dig in, and for over two weeks there was an uneasy standoff. The Americans constantly probed and harassed Burgoyne's exhausted army so that no "officer or soldier ever slept during the interval without his clothes," and no officer "passed a single night without being upon his legs occasionally at different hours."

Burgoyne's options were strictly limited, but desperation or his old cavalry officer's aggressive instinct led him on. Against von Riedesel's counsel (which favored a retreat back to the Batten Kill on the eastern shore of the Hudson, preparatory to a full-scale retreat), he decided to make a reconnaissance in force to again test the American left. But as von Riedesel feared, it was to be neither fish nor fowl. Burgoyne did not have the manpower to deal with the now considerable weight of Gates's army of perhaps 10,000 to 13,000 men against Burgoyne's 4,000 effectives. The 1,500-man force Burgoyne took out on 7 October was a forlorn hope.

Once more there were three columns. Fraser was again on the right with the light infantry (under Lord Balcarres) and his own 24th

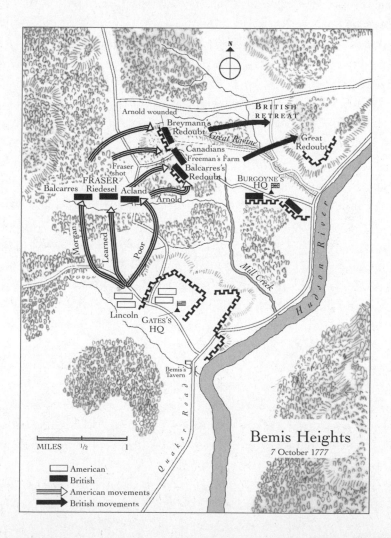

Foot. Von Riedesel was in the center, and on the left were Acland's grenadiers. At around 1:00 PM Fraser reached Barber's wheat field, a mile or so to the west of Freeman's Farm, and sent out a foraging party. Von Riedesel was not happy with the disposition. The flanks were in woods, which offered cover to attackers, and off on the British right flank was a hill, which would serve the patriots very nicely as a platform for enfilading fire.

Once more Arnold and Gates locked horns. If Gates did not move his complacent rump, Arnold railed, there was a real danger of his left wing being rolled up. Gates suggested Morgan and Dearborn make a modest excursion around the British right flank, and when Arnold lost his temper at what he thought was a pusillanimous response, Gates promptly sacked him. Benjamin Lincoln prevailed on Gates to unleash a coordinated attack with Poor targeting the British left (Acland's grenadiers) while Morgan circled around and came down off the hill to hit Fraser. Learned would take on von Riedesel in the center when both the flank attacks were in place.

At about 3:00 PM Poor's 1st, 2nd, and 3rd New Hampshire Continentals hit Acland's grenadiers like a wrecking ball. Acland was shot through both legs and was about to be finished off by an enthusiastic teenage patriot when he was saved by the timely arrival of James Wilkinson. (Acland would die, years later, from a chill caught while fighting a duel.) At the same time Morgan's riflemen "poured down like a torrent from the hill" and struck Balcarres's light infantry in its front and right flank. As Balcarres tried to maneuver to the right to fend off the attack, Dearborn's light infantry attacked his rear. Von Riedesel's misgivings were well-founded, as Lieutenant Digby described: "Our heavy guns began to play, but the wood around being thick, and their exact knowledge of our small force, caused them to advance in great numbers, pouring in a superiority of fire from Detachments ordered to hang about our flanks."[46] The British ranks broke and made a run to Balcarres's Redoubt, a substantial log-and-earth fort on the old Freeman Farm battlefield.

Arnold, perhaps "exhilarated" with a dipper or two of rum,[47] charged into the battle, behaving "more like a madman than a cool and discreet officer," as one of Gates's aides described him, swept up Learned's corps, and attacked von Riedesel's Brunswickers. The Germans held steady at first, but the collapse of Balcarres's light infantry exposed their right flank, and they too were forced to retreat. At this critical point Simon Fraser was hit (a gut wound, invariably fatal in those days), and the British battle line collapsed. As Burgoyne said later, losing Fraser "helped turn the fate of the day."

Arnold then turned his attention to the formidable Balcarres's Redoubt, but it was too well-defended and he was repulsed with heavy losses. Now he moved on to Breymann's Redoubt, which anchored the far right of the British entrenchments, galloping between the British and American lines, so legend has it, defying the fire of both. Morgan had already engaged the German defenders of the redoubt, but Arnold forced his way in through a rear sally port and the position fell. Breymann was killed (perhaps fragged), and Arnold took another wound in the same leg that had been hit at Quebec in 1775. (Typically, he stubbornly refused to have the leg lopped.)

For Burgoyne, his "reconnaissance" had been a disaster. In precisely fifty-two minutes he had lost more killed (278, of whom 94 were German) than at Bunker's Hill (226); 331 were wounded, and 285 taken prisoner—a 56 percent loss. The Americans lost 30 killed and 100 wounded. It was, said Wilkinson, "a scene of complicated horror."

How grotesque now was the comparison between Burgoyne's jauntiness at the beginning of the expedition and this squalid ending with the British pinned back in their main camp, where even the wounded had no respite. The British positions were untenable, and Burgoyne ordered a general retreat back to the Fishkill, a stream feeding into the Hudson on its western side. Hunkered down in two encampments on the northern side of the Fishkill, Burgoyne threw the dice one last time. He had laid out his remaining artillery—twenty-seven guns—in the hope that Gates might just walk into them. Gates almost obliged. In the morning fog of 11 October he ordered a general advance across the Fishkill. It was only the timely warning of a British deserter that halted Glover, Learned, and Nixon's divisions just as the morning mist burned off and the British opened fire. In a panicked scramble Gates managed to withdraw his force out of gun range. It was the last fling. Shelled from high ground to the west, cut off to the north by John Stark, and bombarded from the south, Burgoyne was, observed Lieutenant Anburey, "worn down by a series of incessant toils and stubborn actions."[48] Von Riedesel described the situation: "Even for the wounded, no spot could be found which could afford them a safe shelter—not even, indeed, for so long a time as might suffice for a surgeon to bind up their ghastly wounds."[49]

On 16 October Burgoyne capitulated on what he thought, with good reason, were extremely favorable terms. Under a "convention" agreed on with Gates the army was to be repatriated with the promise that it would not take any further part in the war. Congress would later renege on Gates's agreement.

Only two days before, Dr. Thacher reported Henry Clinton so close to Albany that "it must involve General Gates in inexpressible embarrassment and difficulty by placing him between two armies, and thereby extricating Burgoyne from his perilous situation."[50] But almost as Thacher was writing this, Clinton had pulled back down the Hudson. It had been close; but what good is a ten when you need an ace?

The War in the South

19

The Laurels of Victory, the Willows of Defeat

CAMDEN, 16 AUGUST 1780

During the two-year campaign to subdue the South the British were engaged in ten major battles. They "won" seven of them and lost the war. Defeat sometimes comes not in some apocalyptic bloodbath but through a thousand cuts that drain away the blood, the energy, the will. Bright-eyed, confident, and feisty at the beginning, by the end the British resembled a punch-drunk slugger, both eyes swollen closed, the legs gone, but still gamely, desperately, pathetically, willing to swing away.

The British brigadier general Charles O'Hara put his finger on it in a letter to the duke of Grafton on 1 November 1780, ten months after the first British landings.

> It is a fact beyond doubt that their own numbers are not materially reduced, for in all our Victories, where we are said to have cut them to pieces, they very wisely never staid long enough to expose themselves to those desperate extremities . . . how impossible must it prove to conquer a Country, where repeated success cannot ensure permanent advantages, and the most trifling check to our

Arms acts like Electric Fire, by rousing at the same moment every
Man upon the vast Continent to persevere upon the most distant
dawn of hope.[1]

It was, and is, the great colonial conundrum: the more vigorously the undergrowth of insurrection was hacked, the more robustly it grew back.

The opening moves of the campaign, however, augured well for the invaders. During the last days of 1778 the port of Savannah was taken by Lieutenant Colonel Archibald Campbell, who had come down from New York with 3,500 men. General Robert Howe, the American commander, outflanked and comprehensively routed, was forced to abandon the city. Nine months later the Swiss-born Major General Augustine Prevost and his 3,200 defenders beat off the combined attentions of Admiral Count Charles d'Estaing's 4,000 French infantry and General Benjamin Lincoln's 1,500 Continentals, militia, and cavalry. A classic eighteenth-century siege of the kind mounted against Savannah, complete with all the arcane business of approaches—first, second, and third parallels; lines of contravallation and circumvallation; saps; gabions; fascines, and siege guns—took time, and with the hurricane season approaching, d'Estaing was concerned for the safety of his fleet. A hurriedly planned and ineptly executed assault went badly wrong. The wonderfully named Major Beamsley Glazier led a successful counterattack, which left 173 French and American soldiers dead. (The final casualty count would be a staggering 800 by the time they were forced to abandon the siege.) Count Casimir Pulaski, with a fatal élan that seems to characterize Polish cavalrymen, was mortally wounded leading an attack against the fortifications, not something for which cavalry is best adapted.

Sir Henry Clinton was being given a chance to redeem the embarrassing failure of his 1776 attempt to capture Charleston (or Charles-Town, as it was more usually called in the eighteenth century). On 11 February

1780 his invasion force landed on Simmons (now Seabrook) Island, and the commander in chief took very special care to prepare his investment of the city. By 1 March he controlled James Island, and soon after he had crossed over to the mainland. After a month of spirited skirmishing, the British opened their siege lines. The Royal Navy, even under the command of the geriatric Admiral Marriot Arbuthnot, managed to sail past Fort Moultrie without injury and establish itself in the inner harbor (abetted by a spectacularly ignominious misuse of the American fleet by Commodore Abraham Whipple, who preferred to scuttle rather than fight).

If an eighteenth-century siege had an almost stately progression about it, the experience for besieger and besieged was anything but genteel. Captain Johann Ewald of the Hesse-Hanau jägers recorded the difficulties the besiegers faced.

> *The intolerable heat, the lack of good water, and the billions of sandflies and mosquitoes made up the worst nuisance. Moreover, since all our approaches were built in white, sandy soil, one could hardly open his eyes during the south wind because of the thick dust . . . I tried to protect the workers as much as possible, but there were at least one hundred sharpshooters . . . whose fire was so superior to mine that the jägers no longer dared fire a shot.*[2]

The defenders had to endure almost constant bombardment: "It appeared as if the stars were tumbling down," recorded General William Moultrie in his journal for 9 May 1780. "The fire was incessant almost the whole night; cannon-balls whizzing and shells hissing continually amongst the ammunition chests and temporary magazines blowing up; great guns bursting, and wounded men groaning along the lines. It was a dreadful night!"[3] The British victories at Moncks Corner (14 April) and Lenuds Ferry (6 May) snuffed out all hope of evacuation or reinforcement and also raised the profile of the young Lieutenant Colonel Banastre Tarleton, who, said Horace Walpole, "boasts of having butchered more men and lain with more women than anybody else in the Army"—proof that an Oxford education was not incompatible with licensed thuggery.

On 8 May General Benjamin Lincoln refused Clinton's invitation to surrender, but on 11 May the panicked civilian authorities of the city demanded Lincoln accept the best terms he could get. In the end, despite a gallant defense, there would be no "honors of war." The Americans (over 3,300 of them) were to march out with colors cased to ground their arms. The common soldiery was destined for prison hulks; the officers for better quarters and the hope of exchange. For Lincoln there would be a very sweet revenge. But he would have to wait until 18 October 1781, at Yorktown, Virginia.

When Clinton left the South on 5 June 1780 to head back to New York, he instructed Cornwallis not to jeopardize South Carolina and Georgia by any intemperate adventure into North Carolina. This was about as realistic, given Cornwallis's military instincts, as expecting restraint from a hungry rottweiler adjacent to a juicy steak. Charles, Lord Cornwallis knew one thing. If the South was to be held, and the patriot cause defeated, it would not be done by sitting on his defensive duff in Charleston. The enemy had to be aggressively rooted out and defeated. It would not be lack of testosterone that lost the South to the British.

Colonel Abraham Buford's 350 men of the 3rd Virginia Continentals and the mauled survivors of the brutal little actions at Moncks Corner and Lenuds Ferry, the only military presence of any standing in South Carolina following the fall of Charleston, were dragging their sorry way back to Hillsboro. On 27 May Tarleton, with 40 troopers of the 17th Light Dragoons, 130 horsemen of his Loyalist British Legion doubled up with 100 Legion infantrymen, set off in pursuit. After a characteristically punishing ride Tarleton caught up with Buford at 3:00 PM on 29 May 1780 and demanded his surrender, or else . . . Buford responded with the customary rhetoric of intending to resist "to the last extremity." It was an exchange within the context of the combat etiquette of the eighteenth century. Tarleton, however, was not a great one for etiquette.

Buford drew up his men in one line with the order not to fire until the enemy was within fifty yards—very effective on infantry but a disastrously inappropriate tactic against cavalry. (Tarleton later wrote his own tactical assessment: "Colonel Buford also committed a material error, in ordering the infantry to retain their fire . . . which when given had little effect either upon the minds or bodies of the assailants, in comparison with the execution that might be expected from the successive force of platoons . . . commenced at the distance of three or four hundred paces.").[4] Tarleton's horsemen smashed through the American line, and his infantry followed up with the bayonet.[5] Tarleton went down when his horse was shot from under him, and the fact that his men might have thought their leader killed he later used to explain the butchery that followed: an egotistical flourish completely in character with this proto-Flashman. His after-battle report to Cornwallis was tersely efficient: "I have cut a hundred and seventy officers and men to pieces." One hundred and thirteen Americans were killed and 203 captured, of whom 150 were wounded (a 65 percent casualty rate). Tarleton lost five killed and twelve wounded.[6]

It had not been George Washington's choice, but Congress had appointed Horatio Gates to the command of the southern theater. He was received by the troops there "with respectful ceremony," noted Otto Holland Williams, and immediately began to act with a pomposity compounded by profound ignorance. For example, he informed his cavalry commander, William Washington, that cavalry was irrelevant to warfare in the South, and this in some of the finest horse country in the land, and the only theater of the war in which cavalry played an important role. The army, although ill-prepared, was put on immediate alert, and against the advice of commanders with local knowledge Gates determined to take a direct route to Camden through countryside so barren that the army starved, and when it did eat, the results were dire: "Instead of rum we had a gill of molasses . . . which instead of livening our spirits served to purge us as well as if we had taken a jallop, for the men all the way as we went along were every moment obliged to fall out of the ranks to evacuate."[7]

Gates was not a fighter and probably had no intention of fighting. He had no faith in his militia, and yet his army was composed predominantly of militia.[8] He intended to do what he had done at Saratoga: take up defensive positions (to the north of Camden) and wait. Unfortunately for him, Cornwallis was a fighter. He set out in search of a battle, despite the fact that he believed Gates's approaching army numbered 5,000, whereas his own stood at 2,239. Around midnight on a "hot, humid, and moonless" 15 August the cavalry screens, like Matthew Arnold's "ignorant armies" that "clash by night," blundered into each other, and after a brief exchange both pulled back.

Now it was time for Gates to make a decision, and true to form he ducked his responsibility. At an awkwardly silent command council he simply defaulted to Brigadier General Edward Stevens's despairing "Gentlemen, is it not too late *now* to do anything but fight?" Otho Holland Williams, Gates's assistant adjutant general, recorded the extraordinary moment: "No other advice was offered, and the general desired the gentlemen would repair to their respective commands."[9]

The cream of the American army, the Continentals of Mordecai Gist's 2nd Maryland brigade under the overall command of Baron Johann de Kalb (who was appalled at the decision to give battle but had not spoken up and would pay a very high price for his silence), was placed on the right (west) of the Waxhaws Road, which, running north-south, divided a mile-wide battlefield bracketed by swamps. Smallwood's 1st Maryland brigade was posted in reserve to de Kalb's rear. The Continentals totaled approximately 900 muskets.

To the left of the road Gates posted all of his militia (around 2,500 in total). Anchored on the road were General Richard Caswell's 1,800 North Carolinians, and to their left Stevens's 700 Virginians. Colonel Charles Armand's 120 cavalry were positioned behind the militia but would play no part in the battle. Gates, as he had been at Saratoga, was safely in the rear, behind Smallwood's Marylanders and probably over 400 yards behind his front-line Continentals. Seven pieces of artillery were, according to Williams, "removed from the center of the brigades and placed in the center of the front line."[10]

Cornwallis's deployment was a mirror image of Gates's. On his right, facing the militia, were his best regulars: five companies of light infantry, the 23rd (Royal Welch Fusiliers), and Lieutenant Colonel James Webster's own Yorkshiremen of the 33rd Foot. (Webster was in overall command of the British right wing.) The left wing, under the twenty-six-year-old Irishman Lieutenant Colonel Lord Francis Rawdon, was composed mainly of Loyalist units: Rawdon's own Volunteers of Ireland, the infantry of the British Legion, Lieutenant Colonel John Hamilton's Royal North Carolina Regiment, and Colonel Morgan Bryan's North Carolina Volunteers. Five companies of the formidable 71st Foot (Fraser's Highlanders) were held in reserve behind Rawdon, and behind them was the cavalry of Tarleton's British Legion.

Gates's deployment has generally been criticized because he placed his weakest troops, the militia of the left wing, against the British right wing—the strongest that Cornwallis could muster. He should have predicted his adversary's dispositions, his many critics claim, because the right wing was traditionally "the primary post of honor" and therefore reserved for elite troops.[11] But in Gates's defense he could not have known, in the pitch-black early hours of 16 August, how Cornwallis would deploy. Nor was it an invariable rule that the strongest troops in the British army were always posted to the right wing. At Brandywine, for example, the British main thrust had come from the left, as it would later at Guilford Courthouse. And if the criticism can be leveled at Gates, what about Cornwallis? He too placed his weakest forces (Cornwallis had described their commander, Hamilton, as "a block-head") against the very best in the American army. The Marylanders were as skilled and disciplined as any of the British. The realities facing commanders at the moment of battle are not always appreciated from the comfortable perch of hindsight. Nathanael Greene, for one, declared Gates's dispositions rational: "You was unfortunate but not blameable," he later wrote.[12]

As at Saratoga, Gates seemed to be transfixed, paralyzed perhaps by the awful realization that he had neither the skill nor, more

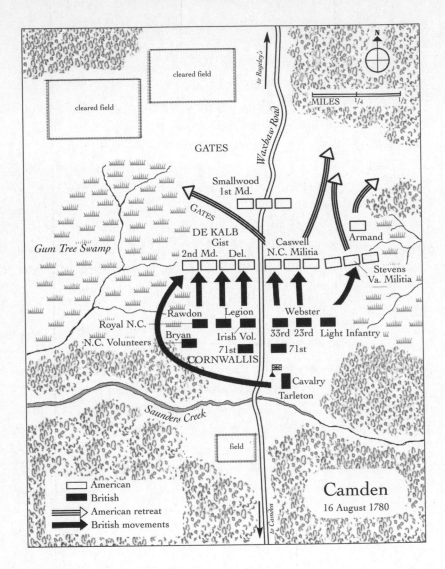

important, the nerve essential to field command. Just as Arnold had urged him to commit himself to action at the start of the Saratoga battles, so now Otho Williams begged him to set Stevens's militia in motion as the British were "displaying" (transitioning from column into their battle line) and therefore vulnerable. "The general seemed disposed to wait events—he gave no orders."[13] But finally, pressing his chief, Williams elicited from Gates a lame "Let it be done" (the last order Gates gave in the battle). But the moment had passed, and Webster and the British right wing moved "with great vigour" to close with the militia. Garrett Watts, a private in Caswell's North Carolina brigade, described the panic:

> *The militia were in front and in a feeble condition at that time. They were fatigued. The weather was warm excessively. They had been fed a short time previously on molasses entirely . . . we were close to the enemy, who appeared to maneuver in contempt of us, and I fired without thinking except that I might prevent the man opposite from killing me. The discharge and loud roar soon became general from one end of the lines to the other . . . I confess I was amongst the first that fled. The cause of that I cannot tell, except that everyone I saw was about to do the same. It was instantaneous. There was no effort to rally, no encouragement to fight. Officers and men joined in the flight.[14]*

As the American left wing evaporated, Webster, resisting the temptations of the chase, sheered to his left to roll up the left flank of de Kalb's Continentals, who, until then, had been more than holding their own against the Loyalists. "Being greatly outflanked," the 2nd Maryland and Delawares were forced to give ground. De Kalb called on Smallwood's 1st Maryland brigade to come up in support, but it was stopped by Webster and could not link up. De Kalb, unwilling to quit without orders from Gates (who by this time had fled the field), was hit multiple times and bayoneted. Tarleton now swept around to the rear of the 1st Maryland, and, as he put it, "rout and slaughter ensued in every direction," or, as Williams put it, "every corps was broken and dispersed."

Williams also says that Gates was borne away by the "torrent of unarmed militia," the implication being that Gates had no choice, but if he was stationed behind Smallwood rather than his militia, it would be interesting to understand how that could have happened. Perhaps the truth was, as Alexander Hamilton suspected, that Gates simply ran away, fast and far: "One hundred and eighty miles in three days and a half! It does admirable credit to the man at his age of life."

Of Gates's original 3,000, 2,000 militia had fled almost unscathed. Of the 1,000 Continentals (including about 250 North Carolina militia who had been closest to, and felt protected by, the bayonets of Robert Kirkwood's Delawares) left to make their final stand, estimates of casualties vary from 188 to 250 killed.[15] Most of the rest were captured and/or wounded. The kill rate (in the region of 20 percent) was extraordinarily high. Many "fine fellows lay on the field," including the stripped corpse of Johann de Kalb.

20

The Hunters Hunted

KINGS MOUNTAIN, 7 OCTOBER 1780; AND COWPENS, 17 JANUARY 1781

Cornwallis was not a commander with much faith in the static warfare of "posts." He had one goal: to find and destroy the patriot army. As he moved out northeast from Camden toward North Carolina, he would have to pacify the patriot bands that could threaten his left flank. So he sent Major Patrick Ferguson, a firebrand Scot and British army regular, with his Loyalist American Volunteers (around 70 men, plus another 900 or so Tory militia), to do the business.[1] As Ferguson (the only British soldier in the whole detachment) moved up toward the northwestern frontier bordering on the Appalachians, he triggered a response from the tough Scotch-Irish Over Mountain Men (mere "banditti," in Ferguson's estimation). Ferguson's threat to "hang their leaders, and lay their country to waste" did not cow a bunch of hard-fighting frontiersmen. They came after him and tracked him down to Kings Mountain (actually a ridge shaped like a footprint, about 60 feet high, 600 yards long, 200 feet wide at the ball of the foot, and 60 feet wide at the heel).

Ferguson was only thirty miles away from Cornwallis and had urged his chief to come to his support. Rather than risk being caught in the open attempting to get back to the main British army, Ferguson thought his ridgetop defense would be strong enough to hold out; as he declared to Cornwallis, "I arrived today [6 October 1780] at Kings Mountain & have taken a post where I do not think I can be forced by a stronger enemy than that against us."[2] This confidence might explain why he made no attempt to strengthen his position with breastworks or abatis.

The patriot strategy at Kings Mountain was starkly simple, as Colonel William Hill explained: "All that was required or expected was that every Officer & man should ascend the mountain so as to surround the enemy on all quarters which was promptly executed."[3] Like a cornered animal, Ferguson's force parried and counterattacked with the bayonet. The patriot Robert Henry, in the act of cocking his rifle, was charged by a Tory musketeer: "His bayonet was running along the barrel of my gun, and gave me a thrust through my hand and into my thigh . . . Wm Caldwell saw my condition and pulled the bayonet out of my thigh, but it hung to my hand."[4] The bayonet was ineffective in the fragmented encounters on the hillside. It was primarily a weapon dependent for its success on mass, and not only were the Loyalists unable to form in sufficient concentrations, but their enemy would not provide them with a consolidated target. As a result the defenders were run ragged and picked off.

The cover provided by the wooded slopes favored the patriot riflemen, while firing downhill tended to cause the Loyalists to fire high, as James Collins, a patriot, recorded: "Their great elevation above us proved their ruin: they overshot us altogether, scarce touching a man, except those on horseback, while every rifle from below seemed to have the desired effect."[5] As the Over Mountain Men closed in, Ferguson led one last rally but was shot out of his saddle, riddled by rifle bullets—an ironic end to an advocate of rifle warfare who on Kings Mountain had put his trust in the bayonet. Given that Ferguson's force was completely surrounded and his defeat overwhelming, it is surprising that the Tories suffered only 157 killed and 163 wounded out of a total force of about 900. (Patriot losses were light: 28 killed, 62 wounded.)[6] There were incidents of killings after surrender, but given the opportunity for a full-scale massacre, the rough-hewn frontiersmen acted with great restraint.

——— ≕⊹≍⊱ ⋅⋅—

On 14 October George Washington finally managed to have Nathanael Greene replace Gates as commander in chief in the southern theater, and on 2 December Gates handed over command at Charlotte, North Carolina, just east of the Catawba River. The army Greene inherited was dispirited by defeat and desertion. He could probably muster 1,100 or so Continentals, of whom approximately 800 were fit for service.[7] Contrary to received military wisdom, Greene split his force—as he said, "partly from choice, partly from necessity" (the parents of most military decisions)—and sent Daniel Morgan with "the cream" of the army off west of the Catawba.

Morgan, as Greene intended, was a threat to Cornwallis's left flank, and the important posts of Ninety Six and Augusta, that could not be ignored if an expedition against Greene (and indeed into North Carolina) was to be undertaken. Tarleton was dispatched to neutralize Morgan and set off with his usual hard-driving determination on the first day of 1781. His force consisted of 550 dragoons and light infantry of the British Legion, the 1st Battalion of the 71st Foot (200), a similar number of the 7th Foot (Royal Fusiliers), 50 horsemen of the 17th Dragoons, and a 50-man contingent of the Royal Artillery with two three-pound "grasshoppers": in all, just over 1,000 men.

After eluding Morgan's shadowing force (by employing the old fake bivouac fires trick, used both by Washington after Trenton and Howe after Brandywine), Tarleton doubled back and crossed the Pacolet River. He was now closing in on his quarry with alarming speed. Scrambling, Morgan had two choices. He could try to cross the Broad River, which ran east to west across his line of retreat, and find good ground in the Thicketty Mountains. His other option would be to make his stand south of the river. In any event, to be caught by Tarleton in the act of crossing the Broad invited disaster, and Morgan would have been mindful of Greene's mission directive: "Employ [your force] . . . either offensively or defensively as your own prudence and discretion may direct, acting with caution and avoiding surprises."[8]

Morgan's own rationale for choosing the Cowpens to make his stand offers an insight into how decisions made under the duress of circumstances are often revisited years later to appear to be acts of pure (and of course, brilliant) volition. Morgan wrote to his friend Captain William Snickers only nine days after the battle that he intended to cross the river and find "a Strong piece of Ground & there decide the Matter but, as matters were Circumstanced, no time was to be lost, I prepared for battle."[9] Years later Morgan would present a very different, and much more heroic, account. He had deliberately chosen to fight with the river to his back to prevent desertion: it would be a glorious do-or-die stand: "As to retreat, it was the very thing I wished to cut off all hope of . . . Had I crossed the river, one half of the militia would immediately have abandoned me."[10] It was a complete contradiction of his after-battle report, which had emphasized a terrain with an escape route in case of defeat: "My situation at the Cowpens enabled me to improve any advantages I might gain, and to provide better for my own security should I be unfortunate."[11] On the evening of 15 January Andrew Pickens and a sizable militia force joined Morgan. It may have influenced his decision to stand and fight, but the truth, despite his later embroidering, was that he had little choice.

Tarleton fancied his chances when he surveyed the Cowpens battlefield: an open meadow (for cattle grazing) about 500 yards deep and the same wide, gently undulating, not much by way of trees and undergrowth; good terrain to maneuver infantry; good space to use cavalry once the enemy was broken. The Mill Gap Road bisected the field north to south. It suited, said Tarleton, "the nature of the troops under . . . [my] command. The situation of the enemy was desperate in case of misfortune."[12] But Morgan was about to prove his mastery of deployment: that science and art of fitting troops to terrain, of understanding the men in his command, their strengths and weaknesses and how best to use them. Morgan, by some osmosis that is difficult to describe (he had no extensive experience of major field command and certainly no formal training in the military craft), was about to demonstrate a genius for battlefield command unmatched on either side at any point during the war.

In the bright but "bitterly cold" early hours of 17 January Morgan laid out a defense in depth that, by drawing Tarleton (who he knew would favor a frontal attack—"down right fighting," as Morgan called it) into a series of linear firefights, would soak up his attacker's resources and energy. It was an instinctive understanding of the dynamics of battle of which Clausewitz would have been proud. First, he made sure his men were fed. (Tarleton's, by comparison, had nothing to eat that day and had been on the march since three that morning.) The main line (sometimes referred to as the light infantry line) was posted on the reverse slope of a slight ridge that traversed the battlefield; the flanks were close to the marshy ground of two creeks that bracketed the Cowpens. The main line had three elements: 120 North Carolina State Troops, Virginia Continentals, Virginia State Troops, and Virginia riflemen all under the command of Captain Edmund Tate on the right wing; in the center were 280 Maryland and Delaware Continentals under the command of Lieutenant Colonel John Eager Howard; and 200 Virginia militia were on the American left under Major Francis Triplett.[13] Howard also had control of the whole line, which totaled 600 men and was tightly aligned along a 200-yard front. Behind the main line, in a shallow gully, Morgan parked William Washington's 3rd Continental Light Dragoons (82 men), together with 45 volunteer horsemen drawn from the militia.

One hundred and fifty yards in front of Howard were 300 of Pickens's North and South Carolina and Georgia militia. (About 20 percent of them had seen previous service as Continentals, and in fact about three-quarters of Morgan's force had combat experience.)[14] The battalions of Colonel Joseph Hayes and Colonel Thomas Brandon were to the left of the road; those of Lieutenant Colonel Benjamin Roebuck and Colonel John Thomas Jr. to the right. A skirmish screen of 120 picked militia riflemen under Joseph McDowell, Samuel Hammond, and Charles Cunningham went forward and stood about 150 yards in front of the militia.

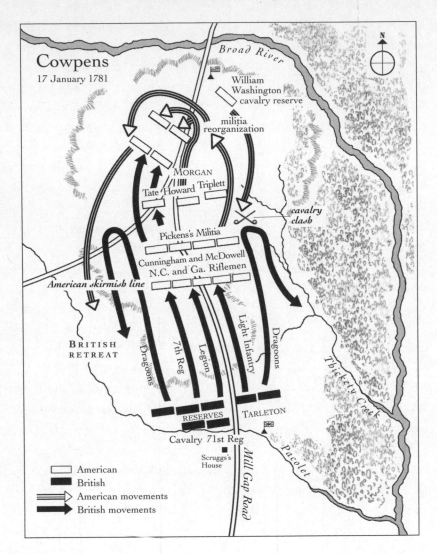

Cowpens
17 January 1781

Broad River

William
Washington
cavalry reserve

militia
reorganization

cavalry
clash

MORGAN
Tate Howard Triplett

Pickens's Militia

Cunningham and McDowell
N.C. and Ga. Riflemen

American skirmish line

BRITISH
RETREAT

Dragoons

7th Reg

Legion

Light Infantry

Dragoons

Thickety Creek

RESERVES

TARLETON

Cavalry 71st Reg

Scruggs's
House

Mill Gap Road

Pacolet

☐ American
■ British
⇒ American movements
➡ British movements

Tarleton's column debouched onto the southern fringe of the
Cowpens close to 6:45 AM. Fifty or so dragoons drew sabers and moved
quickly against the patriot skirmishers, but accurate rifle fire took out
almost a third of them. Tarleton hurriedly began to display his infantry
into line. From right to left, he posted 50 dragoons of the 17th; next to
them were about 150 light infantry made up of the 16th Foot, the light
companies of the 71st Highlanders, and the Prince of Wales American

Regiment (largely Connecticut Loyalists). Then came about 250 of the infantry component of Tarleton's own British Legion, and to their left were 177 men of the 7th Foot, with another 50 or so dragoons anchoring the far left of the line. The two 3-pounders were placed between the light infantry and the Legion and between the Legion and the 7th. (Morgan had no artillery.) To his rear Tarleton had the remaining companies of the 1st Battalion of the 71st (approximately 263 of all ranks) and about 200 British Legion cavalrymen.[15]

Tarleton advanced his line without waiting for it to be completed. Officers of the 7th were still trying to get their men into formation, and it seems rash that the bulk of the 71st—the finest assault troops in the army—were not involved in the initial attack. Morgan described to Nathanael Greene how the British "formed into one Line Raisd a prodjious Yell, and came Running at us as if they intended to eat us up."[16] The militia officers were trying desperately to impose some sort of fire discipline on troops who must have been as tightly wound as men can be. Sixteen-year-old Thomas Young, a mounted volunteer with Washington's force, reported: "Every officer was crying, 'Don't fire!' for it was a hard matter to keep us from it . . . The militia fired first. It was for a time, pop-pop-pop, and then a whole volley; but when the regulars fired, it seemed like one sheet of flame from right to left. Oh! It was beautiful."[17] In fact it was "the most beautiful line I saw."

Morgan intimately knew the capabilities of his militia, and his prebattle directive was designed to minimize their exposure and maximize their impact. They were to give two solid volleys and then retire. Many modern historians have the militia filing off en masse, left to right, across the face of the main line, but Morgan had in fact prepared his Continentals for the militia's retreat by ordering them to open ranks to let the militia pass to the rear. A British account confirms Morgan's innovative tactic: the main line, "observing confusion and retrograding in their front, suddenly faced to the right, and inclined backwards; a manoeuvre by which a space was left in the front line to retreat, without interfering with the ranks of those who were now to oppose the advance."[18] (That Greene would use the same technique, along with others copied from Morgan, at Guilford Courthouse two months later, also suggests this was indeed the method by which the militia extracted themselves.)

Tarleton launched the dragoons of his right, who crashed through the patriot left wing in pursuit of the retreating militia before they in their turn were attacked by Washington's dragoons (who outnumbered them by three to one) and routed. The advancing British line had been badly mauled by skirmishers and the militia, and Tarleton had to dress it before it could again advance. With the length of his line now foreshortened by casualties, Tarleton ordered up his reserves—the 71st Foot and a fifty-man detachment of the Legion dragoons under Captain David Ogilvie—to come into the left wing and attempt to outflank the American right. The battle was being fought in the classic European manner, and both armies exchanged volleys. The British continued to advance.

First the Legion dragoons and then the 71st relentlessly worked their way through McDowell's militia on the American right and began to enfilade the right flank of the Continentals. In response, John Eager Howard ordered a "refusal" of that flank (a maneuver that involved turning the companies on the far right of the line back at right angles so that they were facing the 71st). In the noise and confusion the order was misheard, and to Howard's horror, Andrew Wallace's Virginia Continentals about-faced and fell back, albeit in good order. Along the main line, the retreat spread. It was the critical moment of the battle, and, understandably agitated, Morgan demanded to know from Howard what on earth he was playing at. Howard coolly pointed out that there was no panic and that, fortuitously, the retreat had removed his men from the danger of being outflanked.

For the Highlanders of the 71st, Howard's retreat was simply too inviting, and sensing a rout they charged. Their wild enthusiasm spread farther down the British line, and the 7th joined in "thinking," remembered a Delaware Continental, "that We Were broke [they] set up a great Shout [and] Charged us With their bayonets but in no Order.[19] The Continentals, calmly reloading as they marched back about eighty yards, turned, and delivered a shattering volley at about fifteen yards. Some historians have claimed they "fired from the hip,"[20] but this seems highly unlikely. It would have been physically very difficult (especially swinging around with bayonets attached), unnecessary, inefficient, and contrary to all their training. In any event, it was followed by a bayonet attack as well as the reentrance to the battle of Washington's dragoons and Pickens's militia. Exhausted and surrounded, the 71st surrendered, along with most of the rest of the infantry. Tarleton tried unsuccessfully to persuade the Legion cavalry to rescue the guns, but the battle was over, and with it went 310 British and Loyalist casualties (of whom 110 were killed, over one-third of them officers) and a further 527 captured. (American casualties, according to Morgan, were 12 killed, 60 wounded, but he probably suffered twice that.)[21] Tarleton had lost 86 percent of his force, and it would cripple all of Cornwallis's later efforts. ("The late affair has almost broke my heart," he confided to Lord Rawdon.) In effect, it paved the way to Yorktown.

"Long, Obstinate, and Bloody"

GUILFORD COURTHOUSE,
15 MARCH 1781

Well before disaster had struck him at Cowpens, Cornwallis was determined to defy Henry Clinton's warning not to risk South Carolina by an incursion into North Carolina. On 6 August 1780 Cornwallis wrote to his commander in chief, "It may be doubted by some [that is, Clinton] whether the invasion of North Carolina may be a prudent measure; but I am convinced it is a necessary one, and that if we do not attack that province, we must give up both South Carolina and Georgia, and retire within the walls of Charlestown."[1] It was the eighteenth-century version of flipping the bird.

Stung by the humiliation of Cowpens, shorn of his light infantry, his army reduced to about 2,500, Cornwallis decided to cross his own Rubicon. On 19 January, at Ramsour's Mill, he ordered the baggage of the army to be burned and "sett the example by burning all his wagons, and destroying the great part of his Bagage, which was followed by every Officer of the Army without a murmur," wrote Charles O'Hara to Lord Grafton.[2] He kept only sufficient wagons to transport his medical and ammunition supplies. He even destroyed the rum, which for an eighteenth-century army was drastic indeed. If he was going to travel fast, he had to travel light, and he was going after Morgan and

Greene single-mindedly and wholeheartedly. It was the kind of one-shot strategy that not only appealed to his impetuous pugnacity but also was dictated by the quickly fading prospects of a knockout victory.

For Greene too there was a similar balance between necessity and choice. His army was in no shape or size to offer a knock-down drag-out fight. He had to run, and dreaded it because it demoralized the militia and encouraged the Tories. But in running he saw the possibility of luring Cornwallis over the cliff. Perhaps he could turn Cornwallis's strategy against itself. On 30 January he wrote to Brigadier General Isaac Huger, "I am not without hopes of ruining Lord Cornwallis, if he persists in his mad scheme of pushing through the Country." And in a later letter to Huger Greene predicted, "From Lord Cornwallises pushing disposition, and the contempt he has for our Army, we may precipitate him into some capital misfortune."[3] It would have made Sun Tzu proud.

Like Cornwallis, Greene was also forced to take huge, and for him uncharacteristic, risks in his attempt to outrun the British and get his little army safely over the Dan River. The chase started on 28 January 1781, and as Greene approached the Dan on 10 February he again divided his force. Otho Williams with 700 light troops (including Henry Lee's Legion)—the "flower" of the army, as Greene described it—was to act as a drag to decoy Cornwallis away from the ferry points on the Dan that Greene would need to use to cross to safety. Williams would then double back and join Greene. At one point the Legion was surprised by the British vanguard and came within a whisker of being destroyed, a "loss," said Lee, in the coolest understatement, "[that] would have been severely felt by the American general."[4] On 14 February Williams got his men over Irwin's Ferry, the last at about 11:00 PM, only five or six hours ahead of the British vanguard.

Frustrated, Cornwallis dropped back down to Hillsborough, North Carolina, and then on to Alamance, farther west. After having spent only about one week on the north side of the Dan, Greene recrossed and dogged Cornwallis westward. He needed to "make a show" to encourage the local patriots and by 10 March could report to Thomas Jefferson that "the Militia indeed have flocked in from various quarters," giving him an effective strength of around 4,500. On 12 March Greene headed for Guilford Courthouse. ("We marched yesterday to look for Cornwallis . . . We are now strong enough," reported Major St. George Tucker to his wife.)[5] And in the early hours of 15 March, outnumbered by more than two to one, Cornwallis went looking for Greene.

Although Morgan had been forced to quit the army on 10 February, tormented with sciatica and rheumatism—in fact just plumb worn out—he had bequeathed a game plan for Greene based on the two principles of Cowpens. First, the effective use of militia ("If they fight, you win"). Second, the "collapsing box" defense (drain the British as they are forced to battle through defense-in-depth). All this Greene did, but what he lacked as a field commander was Morgan's understanding of the common soldier, his inspirational presence, and, crucially, Morgan's nerve: the instinct to go for the coup de grâce. Greene was a good strategist and inspired organizer, and "in only one respect can Greene be criticized," according to the historian Christopher Ward. "He lost every battle."

Greene knew well the terrain at Guilford Courthouse. It was about one and a half miles deep west-to-east with the New Garden Road running up through the middle. Its north-to-south width was just over one mile. At the western extremity were some cleared fields; behind them, woodland; and behind that a "vale," about 250 yards deep and 1,000 yards wide, at the eastern end. At the edge of the woodland facing out onto the first clearing Greene put his least experienced militia—North Carolinians. On the right of the road was Brigadier General Thomas Eaton's brigade, about 500 strong. On the left, about the same number of Brigadier General John Butler's. Between them, on the road, were two 6-pounders directed by Captain Anthony Singleton. Eaton's right flank was bolstered by Colonel Charles Lynch's 200 Virginia riflemen, 110 Delaware Continentals under the stalwart Captain Robert Kirkwood, and 86 of Washington's dragoons. Over on the far left flank, Butler had 82 foot and 75 horse of Lee's Legion, plus 200 Virginia riflemen, veterans of Kings Mountain, under Colonel William Campbell.

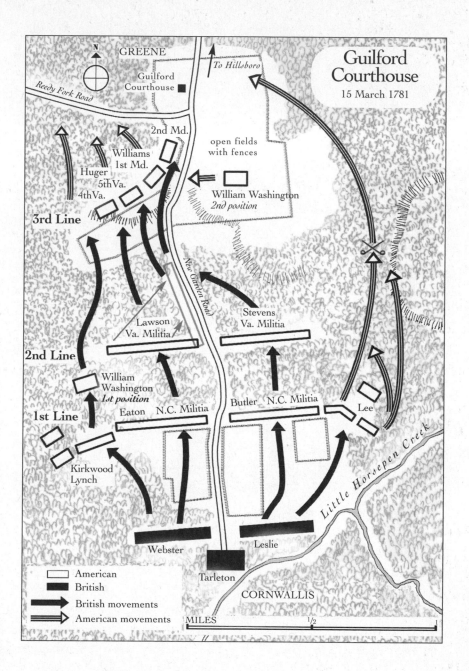

Guilford Courthouse
15 March 1781

GREENE

Guilford Courthouse

Reedy Fork Road

To Hillsboro

open fields with fences

2nd Md.

Williams 1st Md.

Huger 5thVa. 4thVa.

William Washington *2nd position*

3rd Line

New Garden Road

Lawson Va. Militia

Stevens Va. Militia

2nd Line

William Washington **1st position**

1st Line

Eaton   N.C. Militia

Butler   N.C. Militia

Lee

Kirkwood Lynch

Little Horsepen Creek

Webster   Leslie

Tarleton

CORNWALLIS

☐ American
■ British
➡ British movements
⇨ American movements

MILES   1/2   1

The second line, 300 yards back from the first, was in fairly dense woodland and consisted of Virginia militia with a high proportion of reenlisted regulars. To the right of the road were 600 men under Brigadier General Robert Lawson; to the left, a similar number under Brigadier General Edward Stevens.[6]

The third line—the regulars—were up on the eastern edge of the vale, where they had woodland to their back and a relatively clear prospect to their front. On the right flank was Brigadier General Isaac Huger's brigade: 400 of the 5th Virginia under Lieutenant Colonel Samuel Hawes, and another 400 of the 4th Virginia under Lieutenant Colonel John Green. Immediately to their left was Brigadier General Otho Williams's brigade made up of the 1st Maryland under Colonel John Gunby (360 battle-hardened men), and out on the far left of the line 360 men of Colonel Benjamin Ford's 2nd Maryland, who were, for the most part, green troops.[7] Unlike Morgan, Greene posted himself behind his third line, a long way from the front line ("a curious command post for a man of his tactical ability and experience").[8]

When Cornwallis emerged from the woodland onto the western edge of the first clearing, he divided his army into two wings straddling the road. The right wing was commanded by Major General Alexander Leslie and consisted of 212 men of the 2nd Battalion, 71st Foot (the 1st Battalion had been lost at Cowpens), commanded by Lieutenant Colonel Duncan McPherson, their left flank on the road. To their right was the Hessian-Cassel Regiment von Bose (313 men) under Lieutenant Colonel Johann DuBuy (there are variations on the spelling). In support of von Bose and the 71st were the 200 men of the 1st battalion of the Guards under Lieutenant Colonel Norton.

The British left wing was commanded by Lieutenant Colonel James Webster. With its right flank on the road, the 23rd Foot (Royal Welch Fusiliers),[9] comprised 258 men. To their left was Webster's own regiment, the 33rd, with 322 men. In support were 84 jägers, and the two light companies of the Guards, about 100 men. In reserve, Cornwallis kept 420 men of the 2nd Battalion of Guards and Guards' grenadiers under Brigadier General Charles O'Hara, and 174 Legion dragoons under Tarleton (who, earlier that day during a cavalry clash, had received a gunshot wound to his right hand which would eventually entail amputation of his thumb and index finger).[10]

Cornwallis had three 3-pounders stationed on the road, and at about 1:00 PM they started a lively exchange with the two 6-pounders positioned in the center of the North Carolina line. The 1,000-yard British front, consisting of the Regiment von Bose, the 71st, 23rd, and 33rd Foot, began their steady advance across the 400 yards that separated them from the first rank of militia. Greene, as had Morgan at Cowpens, asked for two solid volleys from his North Carolinians and had warned the Virginians behind them not to be panicked when the first line retired. As a result of the militia's opening volley, according to Captain Dugald Stewart, "one half of the Highlanders dropped on that spot."[11] William Montgomery, a North Carolinian, recorded that the British fallen looked like "the scattering stalks of a wheat field, when the harvest man passed over it with his cradle."[12]

Undeterred, the British line moved "in excellent order in a smart run," said Sergeant Roger Lamb, and "arrived within forty yards of the enemy's line [where] it was perceived that their whole line had their arms presented, and resting on a rail fence . . . They were taking aim with the nicest precision."[13] It created on both sides, added Lamb with wonderful understatement, "a most anxious suspense." In fact it was the white-hot core of the eighteenth-century battle experience. In all probability both sides volleyed ("dreadful was the havoc on both sides," observed Lamb), and the British charged with the bayonet. The militia broke and ran. It was, said "Light Horse Harry" Lee, an "unaccountable panic, for not a man of the corps had been killed or even wounded," which is difficult to square with Lamb's "dreadful havoc."

At this point Lynch's Virginian riflemen began enfilading the left flank of the 33rd, which caused the British left wing to skew left in response. Similarly on the right, the collapse of the militia forced Campbell and Lee to retire and pulled after them the von Bose and 1st Guards. It would take them off into their own side-battle, depriving Cornwallis of over a quarter of his fighting strength. These diversions meant that the British center split and Cornwallis moved the 2nd Guards and Guards' grenadiers to fill the void and sent the jägers and

Guards' light infantry to support Webster. Now the British center and left wing entered the dense woodland to take on the Virginia militia.

Here the fighting was decided by desperate and vicious small-unit firefights. The 33rd and 2nd Guards forced Lawson's militia to buckle. The whole of Lawson's line swung back on the road like a door on its hinges, and many used the road as an escape route, provoking William Richardson Davie to retort that Lawson's men "fought as illy as the No[rth] Carolinians."[14] Major St. George Tucker of Lawson's brigade attributed the poor performance to being outflanked. His and John Holcombe's regiment "broke off without firing a single gun and dispersed like a flock of sheep frightened by dogs."[15] Stevens, on the other side of the road, had the shame of Camden to expunge and, recorded Lee, "placed a line of sentinels in his rear to shoot every man that flinched." Whatever the motivation, his men put up such a hard fight against the 71st that the Highlanders were almost totally spent. Stevens was shot in the thigh but would have the satisfaction of later being able to report, "The brigade behaved with the greatest bravery, and stood until I ordered the retreat."[16]

Having broken Lawson, Webster—with the 33rd, Guards' light infantry, jägers, and 23rd following up—broke out onto the western rim of the vale. On the opposite rim the elite of Greene's army calmly stood awaiting them. Webster launched his men down the slope (with "more ardour than prudence") and up the other side, where Continental volleys at thirty yards shattered them (as well as Webster's knee, a wound from which he would die a few days later) and forced their retreat back across the vale.

Now the 2nd Guards and the Guards' grenadiers cleared the wood and found themselves opposite the 2nd Maryland, who, being mainly "raw recruits," ran. As the Guards gave pursuit they were hit in their left flank by the 1st Maryland (who, unlike the 2nd, were battle-hardened) as well as Washington's dragoons who came charging through their rear. The 1st Maryland and the Guards exchanged point-blank volleys ("most terrific, for they fired at the same instant, and they appeared so near that the blaze from the muzzles of their guns seemed to meet," recorded an observer).[17] This was desperate hand-to-hand combat during which the Guards' commanding officer, Colonel James Stuart (standing in for O'Hara, who had been hit in the thigh and chest during the woodland fighting against Lawson's Virginians and Lynch's riflemen), was killed by Captain John Smith of the 1st Maryland. Colonel William Davie left a rare account of personal combat.

> *Colonel Stewart [sic] seeing the mischief Smith was doing, made up to him through the crowd, dust and smoke, and made a violent lunge at him with his smallsword . . . It would have run through his body but for the haste of the Colonel, and happening to set his foot on the arm of a man Smith had just cut down, his unsteady step, his violent lunge, and missing his aim brought him down to one knee on the dead man. The Guards came rushing up very strong. Smith had no alternative but to wheel around and gave Stewart a back-handed blow over, or across the head, on which he fell."[18]*

In his turn Smith was felled by a shot into the back of his head which, amazingly, he survived. Cornwallis, realizing the Guards were on the point of breaking, ordered his artillery officer, Lieutenant John McLeod, to fire canister into the melee. It was a shocking decision, but it worked. The Continentals and dragoons were stopped, and the position was stabilized.

The 23rd and the 71st came up out of the wood, and Cornwallis renewed the action along the entire British line. (Von Bose and the 1st Guards were still embroiled with Campbell and Lee off to the south.) Webster tried a second attack on the now reformed 1st Maryland, Huger's Virginia Continentals, Lynch's riflemen, and Kirkwood's Delawares, but with no more success than the first time. O'Hara, by his own account, led the shredded remnant of the 2nd Guards into the gap created by the routed 2nd Maryland. (How he did this with two serious wounds is a conundrum.) But by this time Greene had ordered a general retreat. It was about 4:00 PM. Over in the woods on the British right the 1st Guards were taking serious punishment from Campbell's riflemen and were lucky to make contact with the von Bose in the nick of time.

For the Continentals the price had been heavy: 57 killed, 111 wounded, 161 missing (most of them captured),[19] with the 1st Maryland taking the brunt. By Otho Williams's estimation the militia lost 22 killed, 73 wounded, and 885 missing (most of whom had left the battlefield). British casualties were 96 killed and 413 wounded (50 of whom died in the night following the battle). The Guards alone took close to 50 percent casualties. Overall it represented a 27 percent loss, whereas total American casualties represented only 6 percent of their total battle force. It was a piece of arithmetic that Greene liked very much. O'Hara, on the other hand, knew how disastrous the "victory" had been: "I wish it had produced one substantial benefit to Great Britain, on the contrary, we feel at the moment the sad and fatal effects our loss on that Day, nearly one half of our best Officers and Soldiers [speaking as the commander of the 2nd Guards] were either killed or wounded, and what remains are so completely worn out by the excessive Fatigues of the campaign . . . [that it] has totally destroyed this army."[20]

22

"Handsomely in a Pudding Bag"

THE CHESAPEAKE CAPES,
5–13 SEPTEMBER 1781;
AND YORKTOWN,
28 SEPTEMBER–19 OCTOBER 1781

I n our beginning is our end." The bluntly insoluble problem that faced Britain at the war's outbreak, symbolized by the siege of Boston, where overwhelming numbers forced the humiliating evacuation of the royal garrison, was as true in the finale. Cornwallis's army, bottled up at Yorktown by an American-French force that outnumbered it two to one, was bombarded into submission, and the would-be jailers of insurrectionists found themselves in the slammer.

During the Seven Years' War the Royal Navy had been the primary instrument of Britain's ascendancy, but now, through a combination of systemic weakness and command failure, it would be the chief pallbearer of Britain's hopes in North America. "With a naval inferiority it is impossible to make war in America," wrote Lafayette to the French minister of war, the comte de Vergennes, on 20 January 1781. The advice did not fall on deaf ears. On 14 August 1781 Washington was informed by the commander of the French expeditionary force in North America, the comte de Rochambeau, that a French fleet under

Admiral the comte de Grasse had set sail from the French West Indies for the Chesapeake.

The British fleet in the West Indies, commanded by Admiral George Rodney, was, of course, fully aware of de Grasse's presence in its waters. In fact, Rodney had had a first-class opportunity to inflict serious damage on de Grasse when the two fleets had met off Tobago on 5 June, but Rodney, who was exhausted and ill with gout and prostate problems, decided against battle even though his second-in-command, Samuel Hood, thought Rodney would almost certainly have had the better of the fight.[1] It was a muffed opportunity that would have the profoundest effect on the course of the American war. Rodney then compounded his misjudgment with another fateful miscalculation.

The West Indies trade was as valuable to France as it was to Britain, and it seemed inconceivable to Rodney that de Grasse would not detach some of his fleet to act as escorts for the merchantmen waiting to sail back to France. He estimated only fourteen French warships would head for America, and consequently instructed Hood to set off in pursuit with a matching number. But de Grasse confounded Rodney by taking his whole fleet—an act of extraordinary political as well as military courage that would put the British at a grave numerical disadvantage. De Grasse sailed for America with twenty-eight ships of the line as well as 3,200 troops commanded by Major General the marquis de Saint-Simon. Rodney, on the other hand, sailed back to Britain as part of the British merchant escort.

Although leaving five days after de Grasse, Hood made the Chesapeake Bay before the Frenchman and, finding it empty, followed his orders to sail on up to New York. De Grasse entered the bay on 30 August[2] and anchored in three lines in Lynnhaven Bay at the southern end of Chesapeake Bay. He immediately disembarked his troops and detached four ships to blockade the James. Five days before his arrival, Admiral the comte de Barras had set sail from Newport with eight ships of the line, four frigates, and nineteen transports carrying the heavy artillery necessary for a full-scale siege. In New York Hood was joined by the five available ships of Admiral Thomas Graves's squadron, and the combined force of nineteen ships (mounting a total of 1,402 guns) sailed for the Chesapeake on 31 August with Graves in overall command. At 8:00 AM on Wednesday, 5 September, the British fleet was spotted sailing down off Cape St. Anne, the northern tip of Chesapeake Bay. De Grasse scrambled into action. Rather than weighing anchor, he ordered the cables slipped (the anchors would be secured to buoys), and the twenty-four available ships of the fleet (totaling 1,788 guns) began to tack their way out of the Chesapeake Bay, even though it meant leaving hundreds of crewmen who had been put ashore to forage. The French emerged, recorded Hood, "in line of battle ahead but by no means regular and connected."[3]

As Graves approached the bay holding the "weather-gage" with a north-northeast wind at his back, de Grasse was going in the opposite direction, leaving the bay. It was imperative that de Grasse should lure Graves away out into the Atlantic. De Barras was out there somewhere and would need to enter the Chesapeake, and an interdiction by the British would be disastrous. Fortunately for the French, Graves obliged by turning his column to the southwest, the last ship turning first, a reversal of the sailing order that left Hood, perhaps the most aggressive admiral of the Royal Navy at the time, in the rear. Now both fleets were heading roughly in the same direction, although on a colliding course. The front of each—the vans—would come together like the point of a *V*. It would bring the vans and centers into combat range but leave the rears too far apart to engage.

Having the weather-gage (the benefit of the wind coming more or less into the back of the sails) was, on the one hand, a great advantage because it allowed for much more maneuverability. On the other, there was a distinct penalty.[4] As the wind came down into the port (left-hand) side it caused the starboard (right-hand) side—the side now facing the French fleet—to keel over periodically, submerging the lower gun-ports which housed the heavier armament.[5] The French, on the other hand, received the wind on their port side, and it tended to lift the lower guns clear of the water. As the wind caused their ships to roll back, they could fire crippling volleys into the rigging of their British counterparts— something of a French tactical trademark.

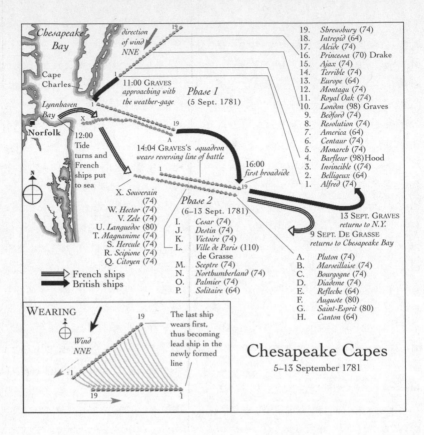

Chesapeake Bay

direction
of wind
NNE

Cape Charles

Lynnhaven Bay

Norfolk

11:00 GRAVES approaching with the weather-gage

Phase 1
(5 Sept. 1781)

12:00 Tide turns and French ships put to sea

14:04 GRAVES'S squadron wears reversing line of battle

16:00 first broadside

N

X. *Souverain* (74)
W. *Hector* (74)
V. *Zele* (74)
U. *Languedoc* (80)
T. *Magnanime* (74)
S. *Hercule* (74)
R. *Scipione* (74)
Q. *Citoyen* (74)

Phase 2
(6–13 Sept. 1781)

I.   *Cesar* (74)
J.   *Destin* (74)
K.   *Victoire* (74)
L.   *Ville de Paris* (110) de Grasse
M. *Sceptre* (74)
N. *Northumberland* (74)
O. *Palmier* (74)
P. *Solitaire* (64)

⟹ French ships
➤ British ships

13 SEPT. GRAVES returns to N.Y.

9 SEPT. DE GRASSE returns to Chesapeake Bay

A. *Pluton* (74)
B. *Marseillaise* (74)
C. *Bourgogne* (74)
D. *Diademe* (74)
E. *Reflecke* (64)
F. *Auguste* (80)
G. *Saint-Esprit* (80)
H. *Canton* (64)

19. *Shrewsbury* (74)
18. *Intrepid* (64)
17. *Alcide* (74)
16. *Princessa* (70) Drake
15. *Ajax* (74)
14. *Terrible* (74)
13. *Europe* (64)
12. *Montagu* (74)
11. *Royal Oak* (74)
10. *London* (98) Graves
9.   *Bedford* (74)
8.   *Resolution* (74)
7.   *America* (64)
6.   *Centaur* (74)
5.   *Monarch* (74)
4.   *Barfleur* (98)Hood
3.   *Invincible* ((74)
2.   *Belligeux* (64)
1.   *Alfred* (74)

WEARING

Wind NNE

The last ship wears first, thus becoming lead ship in the newly formed line

Chesapeake Capes
5–13 September 1781

Whatever the advantage of the wind, Graves failed to use it. First, he ordered the fleet to be "brought to," essentially halting it. The straggling French had time to consolidate, and Hood fumed at the squandered chance to destroy the French van, which had been some mile or two ahead of its center: "It was a glorious opportunity but it was not embraced," complained Hood after the battle.[6] Second, Graves ran signals that, for Hood stuck back in the rear, effectively kept him out of the action. At 3:45 PM the white pennant was hoisted on Graves's flagship, the ninety-eight-gun *London*, indicating "line of battle": the whole line was to follow Indian file with one cable (720 feet) between each two ships. It was a cornerstone signal of the Royal Navy's "Fighting Instructions," the rule book of combat procedures, and had to be obeyed. Given the oblique angle at which the fleets were approaching, it left

Hood far away from the action. At 4:03 PM Graves had raised the blue-and-white checkered flag indicating "close action." The two signals flew together at various times, and it was not until 5:30 PM that Hood saw that the restrictive "line of battle" was no longer displayed and the "close action" pennant flew alone. By then the battle was practically over, and the rear guard of the British line had not fired a shot. The van and the center, on the other hand, had taken a pounding.

Naval cannons of the period ranged from the 42-pounder (rare because of its extreme weight) through the 32-, 24-, 18-, 12-, 9-, 6-, 4-, and 3-pounders down to little half-pound swivel guns. If elevated to ten degrees, a long 32-pounder could send its cast-iron ball almost a mile and half. Besides cannonballs there was grapeshot: nine iron balls clustered (hence *grape*) into a cylindrical canvas bag. From a big gun, grape could range up to 1,300 yards. The naval historian Jack Coggins recorded that in a test firing a 32-pounder loosed off three rounds of grape (a total of twenty-seven balls) at a target 750 feet away. Ten hits were recorded, "one of them with force enough to penetrate four inches of oak."[7] Chain- and bar-shot (like gigantic cufflinks) were particularly destructive against rigging.

As with land cannon, sights were rudimentary. Notches on the breach and the flared end ("swell") of the muzzle offered some small technological aid to the master gunner. But it was his instinctive feel for his gun that counted most. Good gunners understood the gun in relation to the pitch and the roll of the ship. They knew the charge and the timing between ignition and discharge and were expert in adjusting the coin (the wedge at the base of the barrel that controlled elevation). They knew how to control the breeching (the ropes attached to the inside of the hull that held the gun in place against the roll of the ship, checked its recoil after firing, and adjusted the barrel through the horizontal plane), and they knew how to work the gun when men were down. It was knowledge that had been built up over the years, like a patina.

Eight British ships had been in close combat with fifteen French and had taken the great majority of the Royal Navy's casualties: 90 killed and 246 wounded. (The French took 220 casualties.) Although this kind of parallel exchange could cause significant casualties, it was nothing compared with ships that broke through the line at right angles, firing broadsides into the stern and bow of the enemy vessels they passed. For example, when Nelson's flagship, *Victory*, broke the line at Trafalgar in 1805, it fired a double-shotted broadside into the stern of the French *Bucentaure* (*Victory*'s starboard guns firing "as they bear," that is, in sequence as she slid passed), filling the *Bucentaure* with smashing, ricocheting balls and a murderous whirlwind of wooden splinters that killed and wounded over 400 French sailors in a single devastating volley. In the same battle, the French seventy-four-gun *Redoutable* took 522 casualties out of a crew of 600—a staggering 87 percent.[8] Thirteen-year-old Samuel Leech, a powder monkey on a British frigate during the war of 1812, left this searing description of close-quarter carnage.

> *The cries of the wounded now rang through all parts of the ship. These were carried to the cockpit [a small area deep in the bowels of the ship] as fast as they fell, while those more fortunate men, who were killed outright, were immediately thrown overboard. As I was stationed a short distance from the main hatchway, I could catch a glance at all who were carried below.... I saw two ... lads fall nearly together. One of them was struck in the leg by a large shot; he had to suffer amputation above the wound. The other had grape or canister shot sent through his ankle.... He had his foot cut off, and was thus made lame for life.... Two of the boys stationed on the quarter deck were killed.... A man who saw one of them killed, afterwards told me that his powder caught fire and burned the flesh almost off his face. In this pitiable situation, the agonized boy lifted up both hands, as if imploring relief, when a passing shot instantly cut him in two.... A man named Aldrich had his hands cut off by a shot, and almost at the same moment he received another shot, which tore open his bowels in a terrible manner. As he fell, two or three men caught him in their arms, and, as he could not live, threw him overboard.[9]*

For the next four days the two fleets stood off, warily assessing the strategic possibilities. Could Graves, as Hood thought, have beaten de

Grasse to the punch and taken up station within Chesapeake Bay? De Grasse was certainly worried by the possibility: "I was very much afraid that the British might try to get to the Chesapeake . . . ahead of us."[10] But by doing so Graves might also have condemned his fleet to virtual imprisonment, a loss that might have cost Britain not only America but also the West Indies and India too. Graves could not take the chance. De Grasse moved back into the bay on 10 September, prompting Hood to send his superior a message positively dripping with sarcastic contempt: "I flatter myself you will forgive the liberty I take in asking you whether you have any knowledge of where the French fleet is? . . . if he should enter the bay . . . will he not succeed in giving most effectual succour to the rebels?"[11] Part of the "succour" de Grasse was happy to see was de Barras safely at anchor, his siege train delivered up to Rochambeau. Graves, meanwhile, sloped off back to New York.

Having effectively wrecked Sir Henry Clinton's southern strategy, Cornwallis abandoned South Carolina. British victories at Hobkirk's Hill on 25 April, Ninety Six on 19 June, and Eutaw Springs on 8 September 1781 were nothing more than the final flares of a dying fire, successes in name only. In fact, Nathanael Greene, on hearing of Washington's cornering of Cornwallis at Yorktown, would be able to write to Henry Knox, "We have been beating the bush and the General has come to catch the bird."[12] Cornwallis dragged the sorry remnant of his army from Wilmington, North Carolina, into Virginia, arriving at Petersburg on 20 May 1781. On 10 April he had written to his friend Major General William Phillips, who was already in Virginia with an expeditionary force (but would die before Cornwallis could join him), an extraordinary letter that virtually ignored Clinton's role as commander in chief: "Now, my dear friend, what is our plan? Without one we cannot succeed, and I assure you that I am quite tired of marching about the country in quest of adventures. If we mean an offensive war in America, we must abandon New York and bring our whole force into Virginia."[13] On 26 May he sent to Clinton a repudiation of his chief's plan to create a port-stronghold through which the army in Virginia could receive the support of the Royal Navy: "One maxim appears to me to be absolutely necessary for the safe and honourable conduct of this war, which is, that we should have as few posts as possible."[14]

Cornwallis may have sounded confident, but the truth was that Britain had come to the end of its strategic rope. Clinton, clearly harassed to exhaustion, spun himself dizzy trying to cope with threats perceived and real. Perhaps the Franco-American force would attack him in New York as seemed likely from intercepts of a meeting at Wethersfield between Washington and Rochambeau. As a safeguard Clinton instructed Cornwallis on 26 June to send back to New York six regiments of infantry, all his cavalry, and a good portion of his artillery, only to countermand the order when the threat shifted south to Virginia. Cornwallis, for his part, was orchestrating raids by Tarleton and Simcoe in western Virginia that may have been locally alarming but posed no real threat. His limited forays seemed only to lead him up a strategic cul-de-sac. He had nowhere to go.

On 6 July Cornwallis, falling back southward to cross the James River, tried to ambush an American force of about 5,000 under Lafayette at Green Spring Farm. Realizing that a strike against his army during a vulnerable river crossing would be tempting, Cornwallis sequestered most of his men (about 3,000 out of total of 4,600) in woods on the north bank while posting a decoy force out front in an attempt to lure Lafayette into a full-scale attack on what seemed to be only a rear guard waiting its turn to cross. Although Anthony Wayne, with his vanguard of about 1,370, went for the bait, Lafayette held off committing his full strength and so managed to limit his losses. Cornwallis was left like a washed-up cardsharp whose last ace-up-the-sleeve refuses to work its usual magic.

No doubt grinding his teeth, Cornwallis, complying with Clinton's instructions, proceeded to the southern tip of the Yorktown Peninsula, between the York and the James rivers, to establish a base of operations; after some investigation he settled on Yorktown, a once substantial port that had fallen in importance. It sat on a steep bluff looking out onto the York River. About a mile across from it was Gloucester Point, the terminal of a long peninsula that led back up into Maryland. On 8

September he could confidently report to Clinton, "I am now working hard at the redoubts of the place. The army is not very sickly. Provisions for six weeks."[15] The Virginian Brigadier General George Weedon saw it differently: "We have got him handsomely in a pudding bag."[16] Cornwallis would last exactly one day short of six weeks.

If the British were in a dark hole in the early months of 1781, the Americans were not much better off. January was not an auspicious opening of the year. On New Year's Day the Pennsylvania Line ("reduced" says Dr. Thacher, "almost to despair")[17] camped near Morristown, mutinied, killed two officers, and set out for Philadelphia to seek redress from Congress, only to be headed off at Trenton by a delegation from Congress who conceded their demands. On 20 January the New Jersey Line at Pompton followed suit but were treated much more harshly (two were executed). In the same month Washington announced, "There is not a single farthing in the military chest." It was, he said, "an awful crisis."[18] To the French it looked as though their wager might not pay off. The patriot nag on whom they had placed their bet was emaciated and seemed to be running in the wrong direction.

Washington, however, for all his anxiety, saw the strategic situation clearly and hopefully. The pieces were all poised to fall into place, if only the timing could be orchestrated. He wrote to de Grasse from Williamsburg on 25 September, begging him to stay on in the Chesapeake Bay: "The enterprise against York under the protection of your Ships is as certain as any military operation can be rendered by a decisive superiority of strength and means; that it is a fact reducible to calculation."[19] He was entitled to his optimism. The Franco-American army stood at close to 17,000; Cornwallis had about 8,000 men.[20]

Cornwallis was beginning to sweat. On 23 September Clinton received two letters from him (both written on the sixteenth). The first calmly rejected the risk of a "desperate" breakout for the following reason: "Admiral Digby is hourly expected and promise[s] every exertion to help me."[21] The other, written in the fuller knowledge of de Grasse's and de Barras's juncture, threw him into despair: "This place is in no state of defense. If you cannot relieve me very soon, you must be prepared for the worst."[22] Clinton replied that a rescue mission was being assembled in New York and that "there is every hope [it] shall start from hence the 5th of October."[23]

On 28 September French and American forces moved down from Williamsburg, effectively sealing Cornwallis off. The siege had begun. Two days later Cornwallis abandoned his outer defenses, for which he was much criticized both at the time (Clinton thought it extraordinary that "such works in such a position" should have been given up "without a conflict") and subsequently. Actually, given the odds against him, it was sensible, not to say inevitable, that Cornwallis should contract to a smaller perimeter.

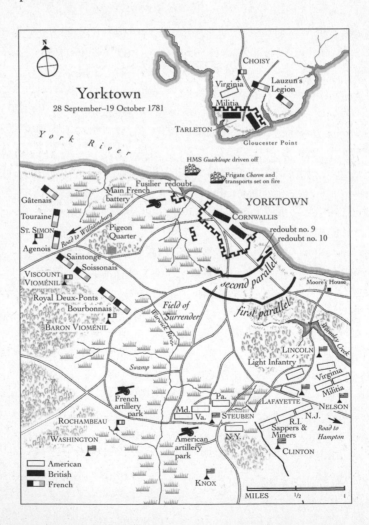

Yorktown
28 September–19 October 1781

It was as Washington had predicted to de Grasse. Without naval support Cornwallis was doomed. The rigorous timetable of the siege ground on relentlessly. The French, particularly, were expert at siege warfare; after all it had been a Frenchman, Sébastien Le Prestre de Vauban (1633–1707), who had codified the science in the latter part of the seventeenth century. On 6 October the allies opened the first parallel (trench) at the prescribed distance of 600–800 yards (beyond the reach of small arms). On the eighth the heavy-caliber siege guns opened up, and the ability of the British to respond was suppressed by overwhelming fire; two days later the second parallel was begun.

The night of the fourteenth saw a synchronized attack against two key British redoubts, No. 9 and No. 10, on the southern front. The comte de Deux-Ponts led 400 men of the Deux-Ponts and Gâtinais regiments[24] against redoubt No. 9. The fighting was vicious, and after the French had taken the strongpoint over one-quarter of their men were dead or wounded.[25] Colonel Alexander Hamilton led the American attack on redoubt No. 10; because he employed ax-wielding pioneers to hack through the abatis, the infantrymen were able to get into the defenses much more quickly and without the high casualties suffered by the French.

With the two redoubts now incorporated into the second parallel, the artillery cordon tightened and the Anglo-German army in Yorktown had to absorb a fearful hammering: "We dare not show a gun to their old batteries," wrote Cornwallis to Clinton. "Experience has shown that our fresh earthen works do not resist their powerful artillery, so that we shall soon be exposed to an assault in ruined works, in a bad position, and with weakened numbers. The safety of the place is, therefore, so precarious that I cannot recommend that the fleet and the army should run great risque in endeavouring to save us."[26]

Death in the British enclave was profligate and gruesome. Major Charles Cochrane, who had arrived as a messenger from New York on 10 October, asked to be allowed to fire a gun and, "anxious to see its effect, looked over to observe it, when his head was carried off by a cannon ball." Hundreds of dead horses lay on the beach and washed around in the shallows of the bay. Johann Conrad Döhla, a private in the

Ansbach-Bayreuth regiment, described in his diary how "the bombs and cannon balls hit many inhabitants and negroes of the city, and marines, sailors, and soldiers. One saw men lying nearly everywhere who were mortally wounded and whose heads, arms, and legs had been shot off." By the night of the sixteenth the American and French batteries were so close to the British line they "could have thrown a stone into it," recorded Döhla.[27]

There was a valiant but forlorn-hope counterattack on 16 October into the second parallel by about 350 Guards under Lieutenant Colonel Robert Abercrombie. A few guns were spiked but were soon back in action. That night Cornwallis tried to ferry his army over the channel to Gloucester Point, from where he thought he might make a breakout north back up into Maryland, maybe all the way back to Pennsylvania or even New York where he could link up with Clinton. It would have been a desperate undertaking even if all the cards fell his way. But a terrific storm came down that night, and Cornwallis was left with a busted flush. Those troops that did get across, for some reason not easy to understand, were recalled. Perhaps it was a matter of honor. As Cornwallis explained to John Simcoe, "the whole of the army" would "share one fate."[28] In any event, it was highly unlikely that an escape could have been effected through Gloucester. Döhla recorded that when the English light infantry returned from Gloucester they "said it was impossible to break out there, because all the surrounding country was strongly occupied and fortified by the enemy. Also a cordon had been drawn around the entire region by several squadrons of French Hussars, so that not the least thing could enter or leave."[29]

On the morning of the seventeenth Cornwallis went to the most forward point in the center of the line. After observing the enemy positions he made up his mind immediately. That afternoon a British drummer boy and an officer emerged from the British lines beating for parley. By the nineteenth surrender terms had been agreed on, and General O'Hara, substituting for the sick Cornwallis, performed the lugubrious duty of offering Cornwallis's sword. The grand occasion was fumbled, even slightly farcical. By some accounts[30] O'Hara tried to offer the sword to Rochambeau and was sent on to Washington, who, in turn, sent him on like a penitent schoolboy to General Lincoln, who, as Washington's second-in-command, was O'Hara's counterpart in rank. Of course, Washington, always punctilious about these things, was following military etiquette, but Lincoln, who had been humiliated at the fall of Charleston in May 1780 by being denied the honors of war (to march out with colors unfurled, playing a march from the victor's repertoire), must have taken some satisfaction in seeing those same ignominies laid on Cornwallis. The British came out with colors unfurled, the fifes and drums playing "a slow march." (If it was "The World Turned Upside Down," as legend has it, there is no hard evidence.)

The defeated marched between the ranks of French and American troops to a "great heath," where they grounded their weapons. Captain Samuel Graham of the hard-fighting 71st Foot described how "some cursed, some went so far as to shed tears, while one man, a corporal, who stood near me, embraced his forelock and then threw it to the ground, exclaiming, 'May you never get so good a master again!'"[31]

The day the British and German soldiery trudged into captivity (the officers enjoyed the privilege of their class—parole), the now utterly redundant "rescue" fleet put out from Sandy Point, and two men sat down to write letters. Cornwallis's, to Clinton, began, "I have the mortification to inform your Excellency..." Washington's, to Congress, began, "Sir, I have the Honor to inform Congress..."

NOTES

INTRODUCTION

1. Christopher Duffy, *The Military Experience in the Age of Reason*, p. 206.
2. George F. Scheer and Hugh Rankin, *Rebels and Redcoats*, p. 447.
3. Christopher Ward, *The War of the Revolution*, p. 92.
4. The similarities, both large and small, are intriguing. First, they were both monarchs (or, in George W. Bush's case, monarchical). In fact George III would have envied President Bush's power. He would certainly have appreciated the environment created by the presidential courtiers that so reassuringly insulated the president from embarrassing contact with hoi polloi. Drifting, for example, in the regal plush of Air Force One, high above devastated New Orleans in 2005, the president could affect a jocular detachment of the "let-them-eat-cake" variety: a sort of Marie Antoinette in flyboy getup, if that can be imagined. Both Georges held sincere religious conviction. Neither could be accused of intellectualism. They knew who they were, by God, and were damned well pleased with it. With the muscular self-confidence of those untroubled by introspection they believed in sticking to their guns. If they had been admirals they would have gone down with the ship, standing stiffly to attention and saluting smartly. Yet in some way they lacked arrogance; both had a childlike egotism that was genuinely confused, hurt even, by criticism. They were both widely ridiculed ("Farmer George" and Alfred E. Neuman, respectively) and because of this ridicule, their passionate attachment to hereditary position and power was underestimated (or "misunderestimated," in Bush's case). They both truly enjoyed being associated with the military and loved dressing the part, although neither had been anywhere close to a battle. Finally, they were both happiest when amiably bumbling around on their estates—an uncomplicated joy in a vexingly complicated world.

5. John Shy, quoted in Piers Macksey, *The War for America*, p. xxi.

6. R. Arthur Bowler, *Logistics and the Failure of the British Army in America, 1775–1783*, p. 239.

7. Charles Royster, *A Revolutionary People at War*, p. 116.

8. "Lastly, there was always the fear of missile attack. Most infantry knew of the damage inflicted even to armored men by well-trained slingers, archers, and javelin throwers. There was no desire to stay still. . . . Running the last 200 yards of no-man's land, as the hoplites at Marathon showed, limited such exposure to attack until the general cover of the protective melee could be reached." Victor Davis Hanson, *The Western Way of War: Infantry Battle in Classical Greece*, p. 140.

9. Hanson, *The Western Way of War*, p. 112, citing the nineteenth-century historian H. Frohlich's analysis of wounds in the *Iliad*.

10. Jonathan Gregory Rossie, *The Politics of Command in the American Revolution*, p. 80.

I. "A CHOAKY MOUTHFUL": THE AMERICAN SOLDIER

1. C. F. Adams, *The Works of John Adams*, vol. 3, p. 4.

2. Charles Lesser, *The Sinews of Independence*, p. xii.

3. Don Higginbotham, *The War of American Independence*, p. 10.

4. James Kirby Martin and Mark Edward Lender, *A Respectable Army*, p. 16.

5. John Shy, *Toward Lexington*, p. 13.

6. Charles Patrick Neimeyer, *America Goes to War*, p. 11.

7. Don Higginbotham, *George Washington and the American Military Tradition*, p. 26.

8. Ibid., p. 25.

9. Gordon S. Wood, *The Radicalism of the American Revolution*, p. 57.

10. Robert A. Gross, *The Minutemen and Their World*, p. 71.

11. Gordon S. Wood, *The Radicalism of the American Revolution*, p. 118.

12. Jonathan Gregory Rossie, *The Politics of Command in the American Revolution*, p. 27. Wolfe to Lord George Sackville, 7 August 1758; quoted in Shy, *Toward Lexington*, p. 416.

13. Rossie, *The Politics of Command*, p. 27.

14. Daniel J. Boorstin, *The Americans*, p. 366.

15. Charles Knowles Bolton, *The Private Soldier Under Washington*, p. 131.

16. Gordon S. Wood, *The Radicalism of the American Revolution*, p. 27.

17. See Henry Steele Commager and Richard B. Morris, eds., *The Spirit of Seventy-Six*, p. 808.

18. Higginbotham, *The War of American Independence*, p. 7.

19. Rossie, *The Politics of Command*, p. 63.

20. Benjamin Rush to John Adams, 1 October 1777, quoted in Charles Royster, *A Revolutionary People at War*, p. 115.

21. Higginbotham, *The War of American Indepencence*, p. 111.

22. Ibid., p. 21.

23. Commager and Morris, *The Spirit of Seventy-Six*, p. 482.

24. Piers Macksey, *The War for America, 1775–1783*, p. 79.

25. Rossie, *The Politics of Command*, p. 27.

26. Boorstin, *The Americans*, p. 369.

27. John C. Dann, ed., *The Revolution Remembered*, p. 218. It may well have been this experience that led Greene to keep what remained of the North Carolina militia in reserve during the battle of Hobkirk's Hill, reversing what had been his preferred practice of using militia as a first-line buffer, with his Continentals behind them to stiffen a few backbones.

28. Higginbotham, *The War of American Independence*, p. 5.

29. Sir John Fortescue, *History of the British Army* (1899–1930), vol. 4, p. 185, quoted in Richard Holmes, *Redcoat*, p. 100.

30. Higginbotham, *George Washington and the American Tradition*, p. 23.

31. John Rhodehamel, ed., *The American Revolution*, p. 752.

32. Christopher Ward, *The War of the Revolution*, p. 50.

33. Richard D. Brown, ed., *Major Problems in the Era of the American Revolution, 1760–1791*, p. 196.

34. Martin and Lender, *A Respectable Army*, p. 46.

35. Bolton, *The Private Soldier*, p. 232.

36. Rossie, *The Politics of Command*, p. 19.

37. William Moultrie, *Memoirs of the Revolution* (1802), quoted in Royster, *A Revolutionary People*, p. 322.

38. Rhodehamel, *The American Revolution*, p. 619.

39. John Buchanan, *The Road to Guilford Courthouse*, p. 140.

40. Ibid., p. 125.

41. George F. Scheer and Hugh F. Rankin, *Rebels and Redcoats*, p. 176.

42. Rhodehamel, *The American Revolution*, p. 223.

43. Scheer and Rankin, *Rebels and Redcoats*, p. 408.

44. Boorstin, *The Americans*, p. 369.

45. Ward, *The War of the Revolution*, p. 756.

46. Scheer and Rankin, *Rebels and Redcoats*, p. 445.

47. Ibid., p. 447.

48. Bolton, *The Private Soldier*, p. 43.

49. Rhodehamel, *The American Revolution*, p. 205.

50. Buchanan, *The Road to Guilford Courthouse*, p. 123.

51. Ward, *The War of the Revolution*, p. 383.

52. Neimeyer, *America Goes to War*, p. 127.

53. Bolton, *The Private Soldier*, p. 30.

54. George Washington to Joseph Reed, 28 November 1775. Quoted in Erna Risch, *Supplying Washington's Army*, p. 6.

55. Higginbotham, *George Washington and the American Military Tradition*, p. 89.

56. Walter Edgar, *Partisans and Redcoats*, p. 43.

57. Joseph Plumb Martin, *A Narrative of a Revolutionary Soldier*, p. 19.

58. John Adams to Joseph Reed, 7 July 1776; quoted in Neimeyer, *America Goes to War*, p. 112.

59. John Shy, "Hearts and Minds: the Case of 'Long Bill' Scott," in Brown, *Major Problems*, p. 211.

60. Higginbotham, *The War of American Independence*, p. 392.

61. Rhodehamel, *The American Revolution*, p. 173.

62. Higginbotham, *The War of American Independence*, p. 394.

63. Edgar, *Partisans and Redcoats*, p. 43.

64. Allen Bowman, *The Morale of the American Revolutionary Army*, p. 13.

65. Neimeyer, *America Goes to War*, p. 26.

66. Royster, *A Revolutionary People*, p. 69.

67. Not without opposition. The county committee of Cumberland County wailed, "All Apprentices and servants are the Property of their masters and mistresses, and every mode of depriving such masters and mistresses . . . is a Violation of the rights of mankind" (Higginbotham, *The War of American Independence*, p. 394).

68. Ibid., p. 314.

69. Royster, *A Revolutionary People*, p. 133.

70. Ward, *The War of the Revolution*, p. 614.

71. Bowman. *The Morale of the American Revolutionary Army*, p. 20.

72. *Elijah Fisher's Journal,* 10 February 1779 (1880 edition), pp. 14–22.

73. This quote and most of the information on wages and cost of living come from Jackson Turner Main, *The Social Structure of Revolutionary America*.

74. Neimeyer, *America Goes to War*, pp. 16, 19, 21.

75. Ibid., pp. 20, 35, 37.

76. Ibid., p. 14.

77. Marion Balderston and David Syrett, eds., *The Lost War*, p. 33.

78. Neimeyer, *America Goes to War*, pp. 28, 40.

79. A. Goodwin, ed., *The New Cambridge Modern History*, vol. 8, *The American and French Revolutions, 1763–93* (1965), p. 514.

80. Neimeyer, *America Goes to War*, p. 48.

81. For comparison: The population of the United States in 2005 was 296 million, of whom about 40 million (or 13 percent) were African-American (22 percent in 1775); the population of Britain during the war was about 11 million.

82. Higginbotham, *The War of American Independence*, p. 389. The British historian Piers Macksey puts the numbers somewhat higher: "If one supposed that a quarter of the white population [he assumes the overall white population was 1.9 million] could bear arms there would be about 450,000 [actually, 475,000]. From this total

Loyalists, religious pacifists, and seamen would have to be deducted." *The War for America*, p. 29.

83. Howard H. Peckham, *The Toll of Independence*, p. 133.

84. Higginbotham, *The War of American Independence*, p. 389. Higginbotham does not give a printed source for the Peckham number.

85. Jay Luvaas, ed. and trans., *Frederick the Great on the Art of War*, p. 75.

86. The returns for March 1777 have been lost. The numbers in this section draw almost exclusively on Lesser, *The Sinews of Independence*, p. xxxi.

87. Lee Kennett, *The French Forces in America, 1780–1783*, p. 59.

88. Ibid., p. 20.

89. Ibid., p. 23.

90. Ibid., p. 24.

91. Ibid., p. 28.

92. Ibid., p. 39.

93. Richard M. Ketchum, *Victory at Yorktown*, p. 168.

2. LOBSTERBACKS: THE BRITISH SOLDIER

1. Higginbotham, *The War of American Independence*, p. 16.

2. Benson Bobrick, *Angel in the Whirlwind*, p. 327.

3. Stephen Brumwell, *Redcoats*, p. 2.

4. Richard Holmes, *Redcoat*, p. 139.

5. Roy Porter, *English Society in the 18th Century*, p. 315.

6. Ibid., p. 130.

7. Ibid., p. 334.

8. Sylvia R. Frey, *The British Soldier in America*, p. 15.

9. Brumwell, *Redcoats*, p. 56.

10. Macksey, *The War for America*, p. 5.

11. Edward E. Curtis, *The British Army in the American Revolution*, p. 76.

12. Balderston and Syrett, *The Lost War*, p. 62.

13. Neimeyer, *America Goes to War*, p. 126.

14. Gunther E. Rothenberg, *The Art of Warfare in the Age of Napoleon*, p. 13.

15. Sir Charles Jenkinson to Captain Bennett, 26 February 1780; in Curtis, *The British Army*, p. 62.

16. Ibid., p. 62.

17. Frey, *The British Soldier*, p. 6.

18. Ibid., p. 23–26.

19. Ibid., pp. 12–18.

20. Holmes, *Redcoat*, p. 54.

21. Brumwell, *Redcoats*, p. 73.

22. Luvaas, *Frederick the Great*, p. 314.

23. Correlli Barnett, *Britain and Her Army, 1509–1970*, p. 213.

24. Curtis, *The British Army*, p. 1.
25. Twenty-nine thousand in Britain and 7,700 in Ireland. Macksey, *The War for America*, p. 39.
26. Macksey, "Appendix: Troops in America (including Canada)," in *The War for America*.
27. Rothenberg, *The Art of Warfare*, p. 61.
28. "Except for a minor war against the Caribes, the British Army had seen no action for 12 to 13 years; relatively few soldiers were over 30 years of age." Anthony D. Darling, *Red Coat and Brown Bess*, p. 8.
29. Balderston and Syrett, *The Lost War*, p. 33.
30. Rhodehamel, *The American Revolution*, p. 68.
31. Captain John Bowater to the earl of Denbigh, 22 May 1777, New York; in Balderston and Syrett, *The Lost War*, p. 126.
32. Philip R. N. Katcher, *King George's Army, 1775–1783*, p. 135.
33. Rhodehamel, *The American Revolution*, p. 11.
34. David Chandler, *The Art of Warfare in the Age of Marlborough*, p. 69.
35. Curtis (*The British Army*, p. 3), for example, states that the 1st Foot Guards were the Grenadier Guards.
36. Robert K. Wright Jr., *The Continental Army*, pp. 46–49.
37. Ibid., p. 87.
38. Ibid., p. 108.
39. Holmes, *Redcoat*, p. 48.
40. Ibid.
41. Edward J. Lowell, *The Hessians*, p. 18.
42. Elisha Bostwick of the 7th Connecticut describing Hessians captured at Trenton; in Rhodehamel, *The American Revolution*, p. 512.
43. Neimeyer, *America Goes to War*, p. 54.
44. Rhodehamel, *The American Revolution*, p. 196.
45. Balderston and Syrett, *The Lost War*, p. 126.
46. Barnet Schecter, *The Battle for New York*, p. 129.
47. John J. Gallagher, *The Battle of Brooklyn, 1776*, p. 93.
48. Commager and Morris, *The Spirit of Seventy-Six*, p. 437.
49. Neimeyer, *America Goes to War*, pp. 55–56. Note: the figures quoted are only for Hesse-Cassel men, the largest contingent of Britain's German auxiliaries.
50. Ernst Kipping, *The Hessian View of America 1776–1783*, p. 10.
51. Higginbotham, *The War of American Independence*, p. 134.
52. Goodwin, *The New Cambridge Modern History*, vol. 8, p. 489.
53. Macksey, *The War for America*, p. 36.
54. Rhodehamel, *The American Revolution*, p. 241.
55. William H. Nelson, "The American Tory," in Richard D. Brown, *Major Problems*, p. 284.
56. Louis Birnbaum, *Red Dawn at Lexington*, p. 365.

57. Rhodehamel, *The American Revolution*, p. 427.

58. Paul H. Smith, *Loyalists and Redcoats*, p. 122.

59. Ibid., pp. 45–46.

60. Ibid., p. 49.

61. Ibid., p. 73.

62. Ibid., p. 76.

63. Buchanan, *The Road to Guilford Courthouse*, p. 204.

64. Balderston and Syrett, *The Lost War*, p. 131.

65. Wallace Brown, *The Good Americans*, p. 117.

66. Quoted ibid., p. 135.

67. Quoted in Smith, *Loyalists and Redcoats*, p. 146.

68. Buchanan, *The Road to Guilford Courthouse*, p. 131.

69. Quoted in Scheer and Rankin, *Rebels and Redcoats*, p. 109.

70. Commager and Morris, *The Spirit of Seventy-Six*, p. 661.

71. Germain to William Knox, 31 December 1776, in Smith, *Loyalists and Redcoats*, p. 43.

72. Rhodehamel, *The American Revolution*, pp. 755–56.

73. "He treats a Loyalist like his friend, embarked on the same cause, and if the meanest of them has business with him, he attends them himself and makes them eat and drink with him." *Rivington's Gazette*, 28 February 1781; Wallace Brown, *The Good Americans*, p. 115.

74. Smith, *Loyalists and Redcoats*, p. 159.

3. "MEN OF CHARACTER": THE OFFICER CLASS

1. Richard D. Brown, *Major Problems*, p. 195.

2. Bolton, *The Private Soldier Under Washington*, p. 24.

3. Sheer and Rankin, *Rebels and Redcoats*, p. 82.

4. Ibid., p. 128.

5. Commager and Morris, *The Spirit of Seventy-Six*, p. 479.

6. Higginbothan, *George Washington and the American Military Tradition*, p. 53.

7. Gordon S. Wood, *The Radicalism of the American Revolution*, p. 118.

8. Dann, *The Revolution Remembered*, p. 116.

9. Higginbotham, *The War of American Independence*, p. 400.

10. Main, *The Social Structure*, p. 214.

11. Ibid.

12. Martin and Lender, *A Respectable Army*, p. 106.

13. Wright, *The Continental Army*, p. 81.

14. Holmes, *Redcoat*, p. 162.

15. Richard D. Brown, *Major Problems*, p. 208.

16. Frey, *The British Soldier*, p. 135.

17. Holmes, *Redcoat*, p. 164.
18. J. A. Houlding, *Fit for Service*, p. 109.
19. Ibid.
20. Commager and Morris, *The Spirit of Seventy-Six*, p. 1233.
21. Rhodehamel, *The American Revolution*, p. 607.
22. Quoted in Godwin, *The New Cambridge Modern History*, vol. 8, p. 204.
23. Scheer and Rankin, *Rebels and Redcoats*, p. 457.
24. Royster, *A Revolutionary People*, p. 199.
25. Ibid., p. 91.
26. Curtis, *The British Army*, pp. 27–28.
27. Higginbotham, *The War of American Independence*, p. 207.
28. Royster, *A Revolutionary People*, p. 210.
29. Lieutenant Samuel Shaw, quoted ibid., p. 210.
30. Holmes, *Redcoat*, p. 287.
31. Brumwell, *Redcoats*, p. 90. The incident occurred in Nova Scotia in 1757.
32. Royster, *A Revolutionary People*, p. 201.
33. John Ellis, *Armies in Revolution*, p. 53.
34. Risch, *Supplying Washington's Army*, p. 253.
35. Higginbotham, *The War of American Independence*, pp. 400–401.
36. Ibid., p. 401.
37. Royster, *A Revolutionary People*, p. 343.
38. Gouverneur Morris to John Jay, 1 January 1783; quoted ibid., p. 340.

4. WHAT MADE MEN FIGHT

1. Goodwin, *The New Cambridge Modern History*, vol. 8, p. 519.
2. Charles Patrick Neimeyer, *America Goes to War*, p. 8.
3. Ward, *The War of the Revolution*, p. 549.
4. Ibid., p. 295.
5. Martin, *Narrative*, p. 241.
6. Holmes, *Redcoat*, p. 179.
7. Martin, *Narrative*, p. 243.
8. Royster, *A Revolutionary People at War*, p. 19.
9. Ibid., p. 172.
10. Bolton, *The Private Soldier*, p. 161.
11. Royster, *A Revolutionary People*, p. 251.
12. Ibid., p. 226.
13. Bolton, *The Private Soldier*, p. 161.
14. Rhodehamel, *The American Revolution*, p. 171.
15. Frey, *The British Soldier in America*, p. 115.
16. Porter, *English Society*, p. 279.

17. Frey, *The British Soldier in America*, p. 117.
18. Ibid.
19. Ibid.
20. Higginbotham, *The War of American Independence*, p. 104.
21. Royster, *A Revolutionary People*, p. 17.
22. Commager and Morris, *The Spirit of Seventy-Six*, p. 625.
23. Rhodehamel, *The American Revolution*, p. 332.
24. Commager and Morris, *The Spirit of Seventy-Six*, p. 600.
25. Rhodehamel, *The American Revolution*, p. 348.
26. Richard M. Ketchum, *Victory at Yorktown*, p. 253.
27. Holmes, *Redcoat*, p. 392.
28. Porter, *English Society*, p. 340.
29. Frey, *The British Soldier*, p. 117.
30. Luvaas, *Frederick the Great on the Art of War*, p. 78.
31. Frey, *The British Soldier*, p. 123.
32. Holmes, *Redcoat*, p. 213.
33. Richard M. Ketchum, *Saratoga*, p. 427.
34. Holmes, *Redcoat*, p. 214.
35. Frederick William, Baron von Steuben, *Regulations for the Order and Discipline of the Troops of the United States*, p. 114.
36. Frey, *The British Soldier*, p. 128.
37. Luvaas, *Frederick the Great*, p. 110.
38. John Keegan and Richard Holmes, *Soldiers*, p. 54.
39. Holmes, *Redcoat*, p. 145.
40. Dann, *The Revolution Remembered*, p. 366.
41. Scheer and Rankin, *Rebels and Redcoats*, p. 223.
42. Frey, *The British Soldier*, p. 112.
43. P. E. Kopperman, "'The Cheapest Pay': Alcohol Abuse in the Eighteenth-Century British Army," *Journal of Military History* 40, no. 3 (July 1996): 449. Cited in Brumwell, *Redcoats*, p. 105.
44. Holmes, *Redcoat*, p. 47.
45. Scheer and Rankin, *Rebels and Redcoats*, p. 60.
46. Rhodehamel, *The American Revolution*, p. 581.
47. Reported by Colonel Otho Williams, in ibid., p. 708.
48. Scheer and Rankin, *Rebels and Redcoats*, p. 462.
49. Ibid., p. 464.
50. Ward, *The War of the Revolution*, p. 92.
51. Luvaas, *Frederick the Great,* p. 77.
52. Curtis, *The British Army*, p. 28.
53. Holmes, *Redcoat*, p. 322.
54. Ibid., p. 323.
55. Scheer and Rankin, *Rebels and Redcoats*, p. 369.

56. Dann, *The Revolution Remembered*, p. 376.
57. Porter, *English Society*, p. 141.
58. Gordon S. Wood, *The Radicalism of the American Revolution*, p. 16.
59. Frey, *The British Soldier*, pp. 87–90.
60. Dann, *The Revolution Remembered*, p. 410.
61. Bolton, *The Private Soldier*, p. 173.
62. Neimeyer, *America Goes to War*, p. 154.
63. Ebenezer Denny's journal, 1–15 May 1781; in Rhodehamel, *The American Revolution*, p. 679.
64. Washington to Lee, 9 July 1779; quoted in Scheer and Rankin, *Rebels and Redcoats*, p. 359.
65. Ibid., p. 359.

5. FEEDING THE BEAST

1. Commager and Morris, *The Spirit of Seventy-Six*, p. 807.
2. Ibid., p. 808.
3. Risch, *Supplying Washington's Army*, p. 208.
4. Ibid., p. 75.
5. Ibid., p. 82.
6. Ibid., p. 190.
7. Oscar Reiss, *Medicine and the American Revolution*, p. 7.
8. Risch, *Supplying Washington's Army*, p. 193.
9. Curtis, *The British Army*, p. 89.
10. Commager and Morris, *The Spirit of Seventy-Six*, p. 837.
11. Curtis, *The British Army*, p. 94.
12. Martin, *Narrative*, p. 22.
13. Risch, *Supplying Washington's Army*, p. 198.
14. Curtis, *The British Army*, p. 95.
15. Frey, *The British Soldier*, p. 33.
16. Reiss, *Medicine*, p. 9.
17. Risch, *Supplying Washington's Army*, p. 218.
18. Rhodehamel, *The American Revolution*, p. 401.
19. Martin, *Narrative*, p. 90.
20. Ward, *War of the Revolution*, p. 613.
21. Commager and Morris, *The Spirit of Seventy-Six*, p. 200. The unfortunate canine was Captain Dearborn's pet.
22. Scheer and Rankin, *Rebels and Redcoats*, p. 119.
23. Ward, *The War of the Revolution*, p. 825.
24. Scheer and Rankin, *Rebels and Redcoats*, p. 417.
25. Harold Peterson, *The Book of the Continental Soldier*, p. 149.

26. George C. Neumann and Frank Kravic, *Collector's Illustrated Encyclopedia of the American Revolution*, p. 94.

27. Macksey, *The War for America*, p. 66. David Chandler in *The Art of Warfare in the Age of Marlborough* makes the point that the rise in the size of European armies in the eighteenth century made staggering demands on commissaries. Most armies aimed to issue each man with about two pounds of bread each day. An army of 60,000 would need approximately 100 million pounds of bread for the six-month campaign season. Only a few areas of Europe could sustain that level of consumption, and this is one of the major factors in explaining the concentration of campaigning in northern France, the Spanish Netherlands, Westphalia, parts of the Rhineland, and Lombardy—some of the most productive areas of Europe.

28. Curtis, *The British Army*, p. 103.

29. Ibid. p. 97.

30. Ibid., p. 130.

31. Ibid., p. 125.

6. THE THINGS THEY CARRIED: WEAPONS, EQUIPMENT, AND CLOTHING

1. Harold L. Peterson, *The Book of the Continental Soldier,* p. 12.

2. Risch, *Supplying Washington's Army*, p. 350.

3. Ibid., p. 348.

4. J. A. Houlding, *Fit for Service*, p. 41.

5. Peterson, *Arms and Armor*, p. 183. One of the myths Peterson dispels is that American gunsmiths were too frightened to "sign" their products in case of a British victory. In fact, many Committee of Safety muskets were identified by their makers.

6. Dann, *The Revolution Remembered*, pp. 124–5.

7. Commager and Morris, *The Spirit of Seventy-Six*, p. 208.

8. Anthony D. Darling in *Red Coat and Brown Bess* cites H. L. Blackmore, *British Military Firearms, 1650–1850*, p. 45.

9. Darling cites H. L. Blackmore.

10. Lawrence E. Babits, *A Devil of a Whipping*, pp. 13 and 165. Babits achieved these results on two test firings, each on windless days with unfouled bores and new flints.

11. Gallagher, *The Battle of Brooklyn, 1776*, p. 43fn.

12. Babits, *A Devil of a Whipping*, pp. 103–4.

13. Major General B. P. Hughes, *Firepower*, p. 26.

14. Dowling, *Red Coat and Brown Bess*, p. 11.

15. Hughes, *Firepower,* p. 27.

16. Rothenberg, *The Art of Warfare*, p. 65.

17. Chandler, *The Art of Warfare*, p. 131.

18. Lieutenant Colonel Dave Grossman, *On Killing*, p. 12.

19. Houlding, *Fit for Service*, p. 167.
20. Luvaas, *Frederick the Great on the Art of War*, p. 78.
21. Robert L. O'Connell, *Of Arms and Men*, p. 158.
22. Author's correspondence with Clay Smith, gunsmith at Colonial Williamsburg.
23. Scheer and Rankin, *Rebels and Redcoats*, p. 67.
24. Gallagher, *The Battle of Brooklyn*, p. 125.
25. Peterson, *The Book of the Continental Soldier*, p. 204.
26. Author's correspondence with Clay Smith.
27. Gallagher, *The Battle of Brooklyn*, p. 118.
28. Babits, *A Devil of a Whipping*, p. 18.
29. John Buchanan, *The Road to Guilford Courthouse*, p. 217.
30. Babits, *A Devil of a Whipping*, p. 20.
31. Peterson, *Arms and Armor*, p. 200.
32. John W. Wright, "The Rifle in the American Revolution," *The American Historical Review* 2, no. 29, pp. 292–299.
33. Peter Force, *American Archives* (1839–53), 5th series, vol. 2, p. 1247.
34. Ward, *The War of the Revolution*, p. 108.
35. Dann, *The Revolution Remembered*, p. 409.
36. Ferguson's design owed much to John Wilmore's mechanism of the late seventeenth century as well as that of Isaac de la Chaumette, a Huguenot living and working in Britain who patented his system in 1704. Ferguson was also indebted to John Warsop's 1720 breech plug design. "Ferguson's key improvements were completely piercing the breech from top to bottom; one turn to open or close . . . The breech plug could not be accidentally removed or dropped in heat of combat." (Lance Klein, "This Barbarous Weapon," *Muzzle Blasts* 5, no. 1.
37. Peterson, *Arms and Armor*, p. 219.
38. Klein, "This Barbarous Weapon."
39. Glenys Crocker, *The Gunpowder Industry*, p. 5.
40. Orlando W. Stephenson, "The Supply of Gunpowder in 1776," *The American Historical Review* 30, no. 2, p. 274.
41. Risch, *Supplying Washington's Army*, p. 341.
42. Commager and Morris, *The Spirit of Seventy-Six*, p. 776.
43. Bolton, *The Private Soldier*, p. 118.
44. Commager and Morris, *The Spirit of Seventy-Six*, p. 618.
45. Peterson, *The Book of the Continental Soldier*, p. 64.
46. Houlding, *Fit for Service*, p. 148.
47. Commager and Morris, *The Spirit of Seventy-Six*, p. 1083.
48. Peterson, *The Book of the Continental Soldier*, p. 61.
49. Barnett, *Britain and Her Army*, p. 128.
50. Luvaas, *Frederick the Great*, p. 78 regarding firepower; the quote about the bayonet is on p. 147.
51. Felix Reichman, "The Pennsylvania Rifle," p. 6

52. Earl J. Hess, *The Union Soldier in Battle: Enduring the Ordeal of Combat*, p. 51.

53. Commager and Morris, *The Spirit of Seventy-Six*, p. 576.

54. Grossman, *On Killing*, p. 123.

55. Rhodehamel, *The American Revolution*, p. 269.

56. Ibid., p. 353.

57. "These redcoats were masters of the bayonet, and they employed it with grim effectiveness" (Higginbotham, *The War of American Independence*, p. 63). "A lack of precision made the Brown Bess less useful as a gun than as a stock for a bayonet. Indeed, the 'white weapon' was much the most effective British arm throughout the war." (Ward, *The War of the Revolution*, p. 29).

58. Houlding, *Fit for Service*, p. 261, fn10.

59. Rhodehamel, *The American Revolution*, p. 341.

60. Hughes, *Firepower*, p. 11.

61. Chandler, *The Art of Warfare*, p. 83.

62. Scheer and Rankin, *Rebels and Redcoats*, p. 432.

63. Peterson, *The Book of the Continental Soldier*, p. 102.

64. Epaphras Hoyt, *A Treatise on the Military Art* (1798), quoted in Peterson, *Arms and Armor*, p. 263.

65. Commager and Morris, *The Spirit of Seventy-Six*, p. 419.

66. Higginbotham, *The War of American Independence*, p. 305.

67. Wright, *The Continental Army*, p. 115.

68. Bowman, *The Morale of the American Revolutionary Army*, p. 20.

69. Risch, *Supplying Washington's Army*, p. 285.

70. Ibid., p. 298.

71. Ward, *The War of the Revolution*, p. 551.

72. Victor Neuberg, *Gone for a Soldier*, p. 99.

73. Risch, *Supplying Washington's Army*, p. 285.

74. Commager and Morris, *The Spirit of Seventy-Six*, p. 125.

75. Rhodehamel, *The American Revolution*, p. 106.

76. Commager and Morris, *The Spirit of Seventy-Six*, p. 1197.

77. Rhodehamel, *The American Revolution*, p. 321.

78. Ward, *The War of the Revolution*, p. 779. Is there something a little fishy about this anecdote? Ward cites as his source Charles Stedman's *The History . . . of the American War*, which was published in London in 1794. Stedman was a Loyalist historian with a particular ax to grind. The action (which Stedman claims resulted in 300 Loyalist deaths, a massive number by the standards of the time) is not recorded in Howard H. Peckham's meticulously researched listing of every engagement and its resulting casualties, *The Toll of Independence*.

79. Reiss, *Medicine*, p. 200.

80. Commager and Morris, *The Spirit of Seventy-Six*, p. 837.

81. Frey, *The British Soldier*, p. 35.

82. Reiss, *Medicine*, p. 11.

83. John Lancaster, *American Heritage: History of the American Revolution*, p. 107 (2003 illustrated edition).
84. Richard M. Ketchum, *Decisive Day*, p. 154.
85. This list comes from an exhibit at Bunker Hill National Historic Site.
86. For example, the full kit, including musket, of a soldier of the French expeditionary force to America of 1780 weighed just over sixty pounds (Kennett, *The French Forces in America*, p. 24).
87. Philip J. Haythornthwaite, *Weapons and Equipment of the Napoleonic Wars*, p. 127.

7. THE BIG GUNS: ARTILLERY

1. Commager and Morris, *The Spirit of Seventy-Six*, p. 1107.
2. Scheer and Rankin, *Rebels and Redcoats*, pp. 53–54.
3. Dann, *The Revolution Remembered*, p. 237.
4. Hughes, *Firepower*, p. 32.
5. Scheer and Rankin, *Rebels and Redcoats*, p. 56.
6. Martin, *Narrative*, p. 113.
7. Ibid., p. 206.
8. Commager and Morris, *The Spirit of Seventy-Six*, p. 1105.
9. The Madras Artillery of the Honourable East India Company tests carried out between 1810 and 1817, cited in Hughes, *Firepower*, p. 40.
10. Haythornthwaite, *Weapons and Equipment of the Napoleonic Wars*, p. 60.
11. Holmes, *Redcoat*, p. 244.
12. Ibid., p. 244.
13. Rothenberg, *The Art of Warfare*, p. 77.
14. Dr. James Thacher, *Military Journal of the American Revolution, 1775–1783*, p. 284.
15. Haythornthwaite, *Weapons and Equipment of the Napoleonic Wars*, p. 76.
16. Risch, *Supplying Washington's Army*, pp. 357–59.

8. THE SANGUINARY BUSINESS: WOUNDS, DISEASE, AND MEDICAL CARE

1. Matthew Naythons, M.D., *The Face of Mercy: A Photographic History of Medicine at War*, p. 232.
2. A test carried out for this book by Dr. Martin Fackler, Letterman Army Institute of Research, Division of Military Trauma Research, Presidio of San Francisco, 18 March 2004.
3. Holmes, *Redcoat*, p. 253.
4. Thacher, *Military Journal*, p. 113.
5. *The Napoleonic War Journal of Captain Thomas Henry Browne, 1807–1816*, quoted in Holmes, *Redcoat*, p. 254.
6. Commager and Morris, *The Spirit of Seventy-Six*, pp. 1111–12.

7. Frey, *The British Soldier*, p. 104.
8. Dann, *The Revolution Remembered*, p. 103.
9. Commager and Morris, *The Spirit of Seventy-Six*, p. 602.
10. Frey, *The British Soldier*, p. 110.
11. Buchanan, *The Road to Guilford Courthouse*, p. 236.
12. Scheer and Rankin, *Rebels and Redcoats*, p. 277.
13. Luvaas, *Frederick the Great*, p. 147.
14. Dann, *The Revolution Remembered*, p. 231.
15. Reiss, *Medicine*, p. 40.
16. Ibid., p. 387.
17. Frey, *The British Soldier*, p. 49.
18. Commager and Morris, *The Spirit of Seventy-Six*, p. 831.
19. Risch, *Supplying Washington's Army*, p. 387.
20. Reiss, *Medicine*, p. 37.
21. Buchanan, *The Road to Guilford Courthouse*, p. 204.
22. Rhodehamel, *The American Revolution*, p. 367.
23. Scheer and Rankin, *Rebels and Redcoats*, p. 240.
24. Dann, *The Revolution Remembered*, p. 296.
25. Mary C. Gillett, *The Army Medical Department, 1775–1818*, p. 18.
26. Ibid.
27. Holmes, *Redcoat*, p. 251.
28. Bolton, *The Private Soldier*, p. 178.
29. Commager and Morris, *The Spirit of Seventy Six*, p. 820.
30. Frey, *The British Soldier in America*, p. 50.
31. Gillett, *The Army Medical Department*, p. 11.
32. John C. Fitzpatrick, ed., *The Writings of George Washington from the Original Manuscript Sources* 3:309–10.
33. Reiss, *Medicine*, p. 104.
34. Risch, *Supplying Washington's Army*, p. 379.
35. Ibid., p. 127.
36. R. Arthur Bowles, *Logistics and the Failure of the British Army in America, 1775–1783*, p. 8.

9. "TRULLS AND DOXIES": WOMEN IN THE ARMIES

1. Don Higginbotham, *The War of American Independence*, p. 397.
2. Benjamin Thompson, Boston, 4 November 1775; in Commager and Morris, *The Spirit of Seventy-Six*, pp. 153–54.
3. John U. Rees, "'The Proportion of Women which Ought to be Allowed . . .': An Overview of Continental Army Female Camp Followers," *Continental Soldier* 8, no. 3 (Spring 1995): 51–58. Rees disputes Linda Grant De Pauw's assertion that at times

women may have been 20 percent of the total Continental army strength (Linda Grant De Pauw, "Women in Combat—The Revolutionary War Experience," *Armed Forces and Society* 7, no. 2 (Winter 1981): 209–26.

4. Rees, "The Proportion of Women," p. 55.
5. Ray Raphael, *A People's History of the American Revolution*, p. 163.
6. John U. Rees, "The Number of Rations Issued to the Women in Camp," *Brigade Dispatch* 28, no. 1 (Spring 1998): 2–10; 28, no. 2 (Summer 1998): 2–12, 13.
7. "Revolutionary Services of Captain John Markland," *Pennsylvania Magazine of History and Biography* 9 (1885): 105. Quoted in John U. Rees, *Minerva, Quarterly Report on Women and the Military* 4, no. 2 (Summer 1996).
8. Rees, 'The Proportion of Women.' p. 58.
9. Alfred F. Young points out: "Molly, of course, was a diminutive for Mary and Margaret, but in England it had long been slang for an effeminate man (whence: 'Don't mollycoddle the boy') . . . Several different women were known as 'Molly Pitcher.'" (*Masquerade: The Life and Times of Deborah Sampson, Continental Soldier*, p. 106.)
10. Martin, *Narrative*, p. 115.
11. Thomas Anburey, *Travels Through the Interior Parts of America* (1969 edition), vol. 1, pp. 436–37.
12. Scheer and Rankin, *Rebels and Redcoats*, p. 37.
13. Although Alfred F. Young points out the homosexual connotation of the word *Molly* (see note 9, above), *shirt-lifter* was British working-class slang for a homosexual, certainly in use in London to this day.
14. Young, *Masquerade*, p. 114.
15. Raphael, *A People's History*, p. 157.
16. Martin, *Narrative*, p. 170.
17. Frey, *The British Soldier*, p. 20.
18. Rhodehamel, *The American Revolution*, p. 319.
19. Frey, *The British Soldier*, p. 20.
20. Holmes, *Redcoat*, p. 364.

10. CUFF AND SALEM, DICK AND JEHU: BLACKS IN THE WAR

1. Thomas Fleming, foreword to *The American Heritage History of The American Revolution* (2003 edition), p. 4.
2. Sidney Kaplan, *The Black Presence in the American Revolution*, p. 32.
3. Bolton, *The Private Soldier*, p. 20.
4. Neimeyer, *America Goes to War*, p. 72.
5. Rossie, *The Politics of Command*, p. 165.
6. Sylvia R. Frey, *Water from the Rock*, p. 78.
7. Bolton, *The Private Soldier*, p. 20.
8. Neimeyer, *America Goes to War*, p. 73.

9. Benjamin Quarles, *The Negro in the American Revolution*, p. 10.
10. Rhodehamel, *The American Revolution*, p. 10.
11. Scheer and Rankin, *Rebels and Redcoats*, p. 88.
12. Quarles, *The Negro*, p. 54.
13. Neimeyer, *America Goes to War*, p. 18.
14. Quarles, *The Negro*, p. 58.
15. Ibid., p. 71.
16. Wright, *The Continental Army*, p. 125.
17. Dann, *The Revolution Remembered*, p. 28.
18. Main, *The Social Structure*, p. 198.
19. Quarles, *The Negro*, p. 19.
20. Frey, *Water from the Rock*, p. 142.
21. Frey, *The British Soldier*, p. 18.
22. Ibid., p. 19.
23. Quarles, *The Negro*, p. 146.
24. Ibid., p. 148.
25. Frey, *Water from the Rock*, p. 76.
26. Rhodehamel, *The American Revolution*, pp. 410–12.
27. Ibid., pp. 523–25.
28. Ibid., pp. 523–24.
29. Ibid., p. 526.
30. Neimeyer, *America Goes to War*, p. 79.
31. Quarles, *The Negro*, p. 68.
32. Ibid., p. 108.

11. THE "PROPER SUBJECTS OF OUR RESENTMENT": INDIANS

1. Alan Axelrod, *Chronicle of the Indian Wars*, (1993), p. 113.
2. Ibid., p. 41.
3. It became the "Six Nations" after the assimilation of the Tuscarora in 1722; ibid.
4. Ibid., p. 42.
5. Brumwell, *Redcoats*, p. 209.
6. John Adams to Horatio Gates, 27 April 1776; quoted in Neimeyer, *America Goes to War*, p. 91.
7. Commager and Morris, *The Spirit of Seventy-Six*, p. 1000.
8. Gary Nash, "The Forgotten Experience: Indians, Blacks and the American Revolution," in Richard D. Brown, *Major Problems*, p. 281.
9. Colin G. Calloway, The American Revolution in Indian Country, p. 3.
10. Patrick M. Malone, *The Skulking Way of War*, p. 98.
11. Barbara Graymont, *The Iroquois in the American Revolution*, p. 155.
12. Ibid., p. 18.
13. Frey, *Water from the Rock*, p. 54.

14. Rhodehamel, *The American Revolution*, p. 487. Neimeyer states (*America Goes to War*, p. 100) that Joseph Reed, when president of the Pennsylvania legislature, set a bounty of $2,500 for every Indian scalp—surely an impossibly high figure. Calloway (*The American Revolution*, p. 49) says that South Carolina paid $75 for male Indian scalps while Pennsylvania offered $1,000.

15. Axelrod, *Chronicle*, p. 121.

16. Ward, *The War of the Revolution*, p. 860. It is interesting that Ward merely describes these executions as Rogers having "dealt sternly" with the five victims.

17. Scheer and Rankin, *Rebels and Redcoats*, p. 351.

18. Axelrod, *Chronicle*, p. 112.

19. Calloway, *The American Revolution*, p. 97.

20. Ibid., p. 98.

21. Ibid., p. 104.

22. Rossie, *The Politics of Command*, p. 80.

23. Calloway, *The American Revolution*, p. 122.

24. Ibid., p. 167.

25. Thomas Jefferson to John Page, 5 August 1776; quoted in Neimeyer, *America Goes to War*, p. 162.

26. Nash, "The Forgotten Experience," in Richard D. Brown, *Major Problems*, p. 283.

27. Calloway, *The American Revolution*, p. 281.

28. Graymont, *The Iroquois*, p. 262.

12. AMBUSH: LEXINGTON AND CONCORD, 19 APRIL 1775

1. Smith led grenadier companies from the 10th, 4th, 18th, 38th, 47th, 52nd, 59th, 43rd, 23rd, and 5th Foot (about 385 rank and file in all). John Pitcairn led the light companies of the 10th, 4th, 23rd, 43rd, 59th, 52nd, 47th, 38th, and 5th Foot (about 350 rank and file). See Brendan Morrissey, *Boston, 1775*, p. 41.

2. Higginbotham, *The War of American Independence*, p. 61.

3. The British asserted that one of their men, wounded on the far side of North Bridge, had been scalped. The American version was that a patriot boy, startled by the sudden movement of the wounded redcoat, took off the top of his head with an ax. For one side it was an atrocity; for the other, "an unfortunate accident."

4. Michael Pearson, *The Revolutionary War*, p. 74.

5. Louis Birnbaum, *Red Dawn at Lexington*, p. 179.

6. Mark M. Boatner (*Encyclopedia of the American Revolution*, p. 631) suggests enigmatically "whether, at this stage of the Revolution, it would have been good from a political point of view for the Americans to have annihilated these 1,800 British regulars." Considering that the militia lacked even basic military leadership during the retreat, it seems unlikely that anyone was considering, far less communicating,

the sophisticated political implications of something the militia must have devoutly wished.

7. Pearson, *The Revolutionary War*, p. 82.
8. Richard M. Ketchum, *Decisive Day: The Battle for Bunker Hill*, p. 18.

13. "A COMPLICATION OF HORROR . . . ": BUNKER'S HILL, 17 JUNE 1775

1. Gage's colleague Admiral Samuel Graves had advocated a radical scorched-earth strategy: burn Roxbury and Charlestown, which Gage rejected as "too rash and sanguinary."
2. Ketchum, *Decisive Day*, p. 75.
3. Commager and Morris, *The Spirit of Seventy-Six*, p. 125. Prescott was not the only one confused by the names. Charles Stedman's map of 1794, published in his *History of the Origins of the American War*, reverses the names of the hills, and one modern American historian comes up with the astounding claim that Breed's Hill was in fact Bunker's Hill but was later renamed.
4. Ibid., p. 123.
5. Ketchum, accepting the Clinton version, describes Gates "heedlessly refusing to stir from his lethargy or interrupt his rest, to investigate, to prepare . . . to act." (*Decisive Day*, p. 115).
6. Rhodehamel, *The American Revolution*, p. 46.
7. Ketchum (*Decisive Day*) asserts that British troops carrying 125 pounds of kit somehow "leaped out [of the landing boats] and jogged up the hill" (p. 138). Those who have carried 125 pounds would be deeply skeptical, as they would be of the claim that these troops carried "about the same weight as . . . a good-sized deer." (p. 154).
8. There seems to be some confusion about the time Stark was summoned to go to Prescott's aid. Ketchum states in *Decisive Day* (p. 130) that Putnam persuaded Ward to order 200 of Stark's force to the peninsula in the "early morning" soon after the *Lively* opened fire at dawn. At 10:30 AM Ward ordered the rest of Stark's regiment together with Reed's to Breed's Hill. However, Stark claimed (p. 254) he got his orders at 2:00 PM. Considering that the first British attack against Stark's position occurred at around 3:00 PM, it seems unlikely that Stark could have marched his men from Medford, got to the beach, and built his wall all in one hour.
9. Rawdon to his uncle, the earl of Huntingdon, 20 June 1775; in Commager and Morris, *The Spirit of Seventy-Six*, p. 130.
10. Ketchum, *Decisive Day*, p. 191.
11. Ibid., p. 169.
12. Commager and Morris, *The Spirit of Seventy-Six*, p. 135.
13. Ibid., p. 136.

14. A VAUNTING AMBITION:
QUEBEC, 31 DECEMBER 1775

1. Pearson, *The Revolutionary War*, p. 127.
2. Dr. Isaac Senter, 18 November 1775; in Rhodehamel, *The American Revolution*, p. 100.
3. George M. Wrong, *Canada and the American Revolution* (1935), p. 302. Montgomery had captured at least two 24-pounders at St. Johns but deployed nothing larger than 12-pounders at Quebec—no better than peashooters against the western walls of the city.
4. Rhodehamel, *The Revolutionary War*, p. 103.
5. Wrong, *Canada*, p. 287.
6. Rhodehamel, *The Revolutionary War*, p. 106.
7. Commager and Morris, *The Spirit of Seventy-Six*, p. 189.
8. Macksey, *The War for America*, p. 79, makes the expiration of enlistments the key factor, but it does not figure in firsthand accounts. Three captains apparently balked at joining the assault, but their men were keen to go. (Perhaps Montgomery's offer of plunder helped motivate them.)
9. Pearson, *The Revolutionary War*, p. 137.
10. Commager and Morris, *The Spirit of Seventy-Six*, p. 208.
11. Ibid., p. 207.
12. Brendan Morrissey, *Quebec, 1775*, p. 63.
13. Commager and Morris, *The Spirit of Seventy-Six*, p. 208.
14. Pearson, *The Revolutionary War*, p. 133.
15. W. J. Wood, *Battles of the Revolutionary War, 1775–1781*, p. 52.
16. Don Higginbotham, *Daniel Morgan: Revolutionary Rifleman*, p. 49.
17. Ibid., p. 46.
18. Morrissey, *Quebec, 1775*, p. 61.

15. "WE EXPECT BLOODY WORK": BROOKLYN, 22–29 AUGUST 1776

1. Barnet Schecter, *The Battle for New York*, p. 98.
2. Ibid., p. 84.
3. Ibid., p. 100.
4. John J. Gallagher, *The Battle of Brooklyn, 1776*, p. 92.
5. Ibid., p. 82.
6. Pearson, *The Revolutionary War*, p. 165.
7. Schecter, *The Battle for New York*, p. 135.
8. Ibid., p. 134.
9. Commager and Morris, *The Spirit of Seventy-Six*, p. 440.
10. It is one of those delicious bits of historical trivia that Grant, at the end of his career, became the governor of Stirling Castle in Scotland.

11. This is a little perplexing. Given that Grant was bombarding Stirling, and the Hessians at Flatbush were using cannon, how would they have heard Howe's signal guns two miles away?

12. Gallagher, *The Battle of Brooklyn*, p. 119.

13. Ibid., p. 120.

14. Paul David Nelson, *William Alexander, Lord Stirling*, p. 88.

15. Schecter, *The Battle for New York*, p. 83.

16. Gallagher, *The Battle of Brooklyn*, p. 130. Schecter says Stirling led 250 Marylanders, a number that is difficult to reconcile with the casualty figures.

17. Pearson, *The Revolutionary War*, p. 176.

18. Ibid., p. 177.

19. Gallagher, *The Battle of Brooklyn*, p. 138.

20. Henry Onderdonk, *Revolutionary Incidents in Suffolk and Kings Counties* (1884), p. 165 (1970 reprint). Cited in Gallagher, *The Battle of Brooklyn*, p. 139.

21. Ibid., p. 77.

22. Ibid., p. 147.

23. Schecter, *The Battle for New York*, p. 167.

24. Peckham's *The Toll of Independence* has 200 American KIA and 897 captured and wounded, more or less in line with other estimates, but puts British and Hessian casualties (KIA and WIA) at 932, far higher than any other authority.

25. Ward, *The War of the Revolution*, p. 236.

26. Ibid., p. 236.

27. Schecter, *The Battle for New York*, p. 170.

28. Ibid., p. 181.

29. Ibid., p. 185.

30. Martin, *Narrative*, p. 37.

31. Commager and Morris, *The Spirit of Seventy-Six*, p. 467.

32. "There is little evidence . . . that Mary Murray's actions were either deliberately or actually delayed General Howe even in the unlikely case they were intended to. Like Howe, she was almost certainly unaware of Putnam's column and its whereabouts. Furthermore, Howe's decision to halt the army at Inclenberg until all of his troops had landed was written into Clinton's orders before the invasion." Schecter, *The Battle for New York*, p. 190.

33. Ibid., p. 201.

34. Ibid., p. 281.

35. Ward, *The War of the Revolution*, p. 270.

36. Ibid., p. 269.

37. Richard M. Ketchum, *The Winter Soldiers*, p. 112.

16. FIRE AND ICE: TRENTON I, 25–26 DECEMBER 1776; TRENTON II, 30 DECEMBER 1776; AND PRINCETON, 3 JANUARY 1777

1. Ewald's *Diary*, quoted in David Hackett Fischer, *Washington's Crossing*, p. 125.
2. Ibid., p. 125.
3. Ketchum, *Winter Soldiers*, p. 198.
4. Fischer, *Washington's Crossing*, p. 135.
5. Justice Thomas Jones, quoted in Scheer and Rankin, *Rebels and Redcoats*, p. 222.
6. Fitzpatrick, *The Writings of George Washington*, p. 355.
7. Fischer, *Washington's Crossing*, p. 182.
8. Christopher Hibbert, *Redcoats and Rebels*, p. 148.
9. Fischer, *Washington's Crossing*, p. 189.
10. Schecter, *The Battle for New York*, p. 266.
11. As Fischer points out in *Washington's Crossing*, the Durham boats (huge flat-bottomed transports originally used for hauling ore and iron to and from the Durham Iron Works in Pennsylvania) and ferries used by Washington's army had no seats. The men stood, even though Emmanuel Leutze's 1851 painting *Washington Crossing the Delaware* has been ridiculed by many historians for showing Washington standing at the helm.
12. Ibid., p. 219.
13. Ibid., p. 239.
14. Ibid., p. 240.
15. Ibid., p. 259.
16. "Since nothing in his letters or orders reveals any clear-cut plan of action, it is almost impossible to say what the general had in mind to do" (Ketchum, *Winter Soldiers*, p. 281).
17. Occasionally one can glimpse the conventions of the medieval host, even the tribal band, in these eighteenth-century battles. Washington arrayed his men "at arms' length apart, so as to make a numerous and formidable appearance." Cornwallis, likewise, used display to intimidate, forming long lines of battalions in full view but out of range. Shouting to intimidate is ancient (and modern). At a critical moment, the Americans "raised a shout, and such a shout," a patriot soldier wrote, "I never since heard; by what signal or command I know not . . . they shouted as one man" (Fischer, *Washington's Crossing*, p. 306).
18. Ibid., p. 303.
19. Ketchum, *Winter Soldiers*, p. 285.
20. Fischer, *Washington's Crossing*, p. 300.
21. Ibid., p. 305.
22. Ibid., p. 335.
23. Ibid., p. 341.

17. THE PHILADELPHIA CAMPAIGN: BRANDYWINE, 11 SEPTEMBER 1777;
GERMANTOWN, 4 OCTOBER 1777;
AND MONMOUTH COURTHOUSE, 28 JUNE 1778.

1. The notion that Washington could have nipped up north, quickly thrashed Burgoyne, and then come south to deal with Howe does not take into account the political pressure on Washington not to cede Philadelphia without a fight, as he would have done if he had gone north. Although it has the convincing logic of a railway timetable, the theory does not take account of the tactical and logistical problems that would almost certainly have derailed it.

2. "Since Trenton had renewed his [Howe's] conviction that the Continental army must be destroyed before the rebellion would break" (Macksey, *The War for America*, p. 125).

3. Ward, *The War of the Revolution*, p. 329.

4. Ibid., p. 329.

5. Ibid., p. 350.

6. Ibid., p. 331.

7. David G. Martin, *The Philadelphia Campaign*, p. 37.

8. Higginbotham, *The War of American Independence*, p. 183.

9. Ward, *The War of the Revolution*, p. 332.

10. Other explanations abound, but none seem completely convincing. For example, Howe told Lord Carlisle, a year after the event, that he went to the Chesapeake after learning that the American's main arms depot had been moved from Philadelphia to northern Virginia, but did not mention it at the time. Yet another theory suggests that Howe wanted to prevent Washington from going west of the Susquehanna, beyond reach yet threatening the British rear. Perhaps; but if Howe was concerned that Washington might strike north against Burgoyne or Clinton in New York, would it not have been a relief to see him exiled and bottled up in Virginia?

11. Wood, *Battles of the Revolutionary War*, p. 96.

12. Martin, *The Philadelphia Campaign*, p. 61

13. Ibid., p. 63.

14. Commager and Morris, *The Spirit of Seventy-Six*, p. 614.

15. Scheer and Rankin, *Rebels and Redcoats*, p. 238.

16. Ibid., p. 239.

17. Wood, *Battles of the Revolutionary War*, p. 112.

18. Cited in Boatner, *Encyclopedia*, p. 109.

19. Scheer and Rankin, *Rebels and Redcoats,* p. 239.

20. Ward, *The War of the Revolution,* p. 356.

21. Boatner, *Encyclopedia*, p. 829.

22. Peckham, *The Toll of Independence*, p. 41, suggests 200 patriots killed and 100 wounded. Ward, *The War of the Revolution*, p. 469, points out that a letter

purportedly written by a Hessian soldier boasting of bayoneting patriots "like so many pigs . . . until the blood ran out of the touch-hole of my musket," and putting the American killed at "three hundred" was a forgery, as there were no German troops at Paoli.

23. Ward, *The War of the Revolution,* p. 469, fn 21.
24. Martin, *Narrative*, p. 102.
25. Ibid., p. 104.
26. Ibid., p. 369.
27. Ibid.
28. Boatner, *Encyclopedia*, p. 430.
29. Ibid., p. 717.
30. Harlow Giles Unger, *Lafayette*, p. 78.
31. Ward, *The War of the Revolution*, p. 578.
32. Ibid., p. 577.
33. Boatner, *Encyclopedia*, p. 721; Ward, *The War of the Revolution*, p. 579.
34. Unger, *Lafayette*, p. 79.

18. THE SARATOGA CAMPAIGN: FREEMAN'S FARM, 19 SEPTEMBER 1777; AND BEMIS HEIGHTS, 7 OCTOBER 1777

1. James Lunt, *John Burgoyne of Saratoga*, p. 69.
2. Ketchum, *Saratoga*, p. 79.
3. Others—Gage, Howe, and Lord Dartmouth—had also proposed the idea, but the detail was Burgoyne's.
4. Ketchum, *Saratoga*, p. 84.
5. Commager and Morris, *The Spirit of Seventy-Six*, p. 542.
6. Ketchum, *Saratoga*, p. 81.
7. Commager and Morris, *The Spirit of Seventy-Six*, p. 543.
8. Ward, *The War of the Revolution*, p. 544.
9. Ketchum, *Saratoga*, p. 134.
10. Lunt, *John Burgoyne*, p. 173.
11. Ibid., p. 74.
12. Edward B. de Fonblanque, *Political and Military Episodes in the Latter Half of the Eighteenth Century, derived from the Life and Correspondence of the Right Honorable John Burgoyne* (1876). Cited in Lunt, *John Burgoyne*, p. 153.
13. Ibid., p. 150.
14. Ketchum, *Saratoga*, p. 158.
15. Ibid., p. 171.
16. Thacher, *Military Journal*, p. 85.
17. Commager and Morris, *The Spirit of Seventy-Six*, p. 557.
18. Lunt, *John Burgoyne*, p. 179.

19. Ibid.

20. Ketchum, *Saratoga*, p. 203.

21. Christopher Ward makes the dramatic point that the casualties at Hubbardton in relation to the men engaged made it "as bloody as Waterloo" (*The War of the Revolution*, p. 414). The Americans had about 1,200 men engaged and lost 350 (29 percent), including those captured. At Waterloo the French had 65,000 engaged and lost 25,000 (38 percent). Percentages can be revealing, but also misleading when one looks at the absolute numbers involved.

22. Commager and Morris, *The Spirit of Seventy-Six*, p. 567. Most histories describe him as a simpleton and therefore particularly acceptable to the Indians who recognized some kind of spiritual gift. Although this version smacks of condescension, being a simpleton and owning significant amounts of property are not necessarily mutually exclusive.

23. Scheer and Rankin, *Rebels and Redcoats*, p. 271.

24. Lunt, *John Burgoyne*, p. 207.

25. Rhodehamel, *The American Revolution*, p. 314.

26. Commager and Morris, *The Spirit of Seventy-Six*, p. 570.

27. Many histories state that the main reason for the Bennington expedition was to find horses for the Brunswick dragoons, usually depicted as cartoonish buffoons forced to traipse along in massive ten-league cavalry boots, trailing huge cavalry sabers—like something straight out of *The Nutcracker*. In fact they were trained to operate as effective dismounted troops and could certainly not have negotiated the journey from Canada encumbered in their full cavalry uniform: "The infamous 12-pound boots and 10-pound swords (in reality neither weighed much more than four) . . . were carried in the regimental baggage wagons, and they wore either gaiters or the striped overalls adopted by most of the German contingent" (Brendan Morrissey, *Saratoga, 1777* p. 53).

28. Commager and Morris, *The Spirit of Seventy-Six*, p. 575.

29. Morrissey, *Saratoga*, p. 53 fn 5.

30. Scheer and Rankin, *Rebels and Redcoats*, p. 264.

31. Pearson, *The Revolutionary War*, p. 266.

32. Commager and Morris, *The Spirit of Seventy-Six*, p. 577.

33. Boatner goes so far to say that "his selection and fortification of the Saratoga battlefield made possible the American victory" (*Encyclopedia*, p. 590).

34. If the only object of Burgoyne's mission was to join Howe to add to Howe's troop strength, Burgoyne could have gone by sea and landed at New York or Rhode Island or Philadelphia. The whole point was to demolish American opposition on the way to a *juncture*. George III understood early in 1777 that a British army needed to be on the ground between Canada and the patriots: "The idea of carrying the army by sea to Sir William Howe would certainly require the leaving a much larger part of it in Canada [a larger force would have to be left in Canada as a defense, thus reducing Burgoyne's force], as in that case the rebel army would

divide that province from the immense one under Sir W. Howe. I greatly dislike that idea" (Lunt, *John Burgoyne*, p. 124).

35. Lunt, *John Burgoyne*, p. 217.
36. Ketchum, *Saratoga*, p. 354.
37. Ibid., p. 362.
38. Commager and Morris, *The Spirit of Seventy-Six*, p. 584.
39. Lunt, *John Burgoyne*, p. 225.
40. Ibid., p. 227.
41. Ibid.
42. Lunt, *John Burgoyne*, p. 227.
43. Pearson, *The Revolutionary War*, p. 276.
44. Rhodehamel, *The American Revolution*, p. 320.
45. Lunt, *John Burgoyne,* p. 228.
46. Rhodehamel, *The American Revolution*, p. 326–27.
47. Ketchum, *Saratoga*, p. 399.
48. Commager and Morris, *The Spirit of Seventy-Six*, p. 599.
49. Ketchum, *Saratoga*, p. 417.
50. Commager and Morris, *The Spirit of Seventy-Six*, p. 587.

19. THE LAURELS OF VICTORY, THE WILLOWS OF DEFEAT: CAMDEN, 16 AUGUST 1780

1. Buchanan, *The Road to Guilford Courthouse*, p. 337.
2. Ibid., p. 67.
3. Ibid., p. 69.
4. Holding fire until the last minute was risky, given the chance of misfires. At the battle of Belgrade (1717) two imperial battalions waited until the Turks were only thirty yards from them before firing. The volley was ineffective, and the imperialists were overwhelmed.
5. This sort of behavior was characteristic of warfare in the South. In July 1780 Colonel Morgan Bryan's Tory cavalry unit of sixty or so was wiped out by William Richardson Davie at Hanging Rock. ("Banastre Tarleton could have done no better. The feat of Davie . . . was a textbook model of a partisan operation" [Buchanan, *The Road to Guilford Courthouse*, p. 133].) On 28 December 1780 William Washington's dragoons caught a 250-strong Georgia Tory unit under Colonel Francis Walters at Hammond's Store and killed and wounded 150 of them (60 percent).
6. Sergeant Major Seymour of the Delaware Regiment, quoted in Buchanan, *The Road to Guilford Courthouse*, p. 161.
7. Buchanan, *The Road to Guilford Courthouse*, p. 163, citing Thomas Pinckney, a Gates aide.
8. Rhodehamel, *The American Revolution*, p. 582.

9. Ibid., p. 583. Buchanan, *The Road to Guilford Courthouse*, p. 163, says the guns "were placed at various points along the line."
10. Buchanan, *The Road to Guilford Courthouse*, p. 163.
11. Boatner, *Encyclopedia*, p. 169.
12. Otho Williams, quoted in Rhodehamel, *The American Revolution*, p. 583.
13. Dann, *The Revolution Remembered*, p. 195.
14. Boatner, *Encyclopedia*, p. 169, gives the estimate of 188 killed; Peckham, *The Toll of Independence*, p. 74, says the number was 250.

20. THE HUNTERS HUNTED: KINGS MOUNTAIN, 7 OCTOBER 1780; AND COWPENS, 17 JANUARY 1781

1. Like an eighteenth-century John Wayne in *The Green Berets*.
2. Wood, *Battles of the Revolutionary War*, p. 196.
3. Buchanan, *The Road to Guildford Courthouse*, p. 230.
4. Ibid., p. 231.
5. Ibid., p. 232.
6. Boatner puts the patriot force at 1,790; Ward at 1,400; Buchanan at 900–1,000.
7. Henry Lumpkin, *From Savannah to Yorktown*, p. 120.
8. Lawrence E. Babits, *A Devil of a Whipping: The Battle of Cowpens*, p. 176.
9. Buchanan, *The Road to Guilford Courthouse*, p. 328.
10. Ibid., p. 328.
11. Commager and Morris, *The Spirit of Seventy-Six*, p. 1158.
12. Banastre Tarleton, *A History of the Campaigns of 1780 and 1781 in the Southern Provinces of North America* (1786), p. 221. Quoted in Edwin C. Bearss, *Battle of Cowpens*, p. 29.
13. Babits, *A Devil of a Whipping*, p. 76.
14. Ibid., p. 42.
15. Buchanan, *The Road to Guilford Courthouse*, p. 321, takes the British deployment from Tarleton's history.
16. Ibid., p. 186.
17. Buchanan, *The Road to Guilford Courthouse*, p. 322.
18. Babits, *A Devil of a Whipping*, p. 95.
19. Ibid., p. 113.
20. For example, Ward, *The War of the Revolution*, p. 761; and Boatner, *Encyclopedia*, p. 297.
21. Babits, *A Devil of a Whipping*, p. 152.

21. "LONG, OBSTINATE, AND BLOODY": GUILFORD COURTHOUSE,
15 MARCH 1781

1. Buchanan, *The Road to Guilford Courthouse*, p. 306.
2. Ibid., p. 340.
3. Ibid., p. 342.
4. Ibid., p. 357.
5. Ibid., p. 369.
6. Ward, Lumpkin, and Boatner (Boatner is indispensable and indefatigable, but usually follows Ward in these sorts of details) have it the other way round, but John Hairr in *Guilford Courthouse* cites Henry Lee: "They were formed in a deep wood; the right flank of Lawson resting on the great [New Garden] road" (p. 76).
7. Boatner, following Ward, insists emphatically that it was the 5th Maryland (Lumpkin also goes along). But he is emphatically wrong. Lawrence E. Babits has put the matter to rest in "The 'Fifth' Maryland at Guilford Courthouse: An Exercise in Historical Accuracy" (February 1988) at http://www.battleofcamden.org/fifthmr.pdf.
8. Hairr, *Guilford Courthouse*, p. 102.
9. *Welch* is not a misspelling, although many historians have an irresistible urge to "correct" it to the more logical *Welsh*. The regiment was raised in Ludlow, Shropshire, on 16 March 1689 and has been in continuous service ever since. Before going to America it had served in most of the major battles of the eighteenth century: Blenheim, Malplaquet, Dettigen, and Minden. http://www.rwfnet.co.uk.
10. I am indebted to Angus Konstam's *Guilford Courthouse, 1781* for the British dispositions.
11. Ibid., p. 63.
12. Buchanan, *The Road to Guilford Courthouse*, p. 375.
13. Commager and Morris, *The Spirit of Seventy-Six*, p. 1164.
14. Ibid., p. 377.
15. Hairr, *Guilford Courthouse*, p. 106.
16. Konstam, *Guilford Courthouse*, p. 69.
17. Ibid., p. 81.
18. Ibid., pp. 83–84.
19. Hairr, *Guilford Courthouse*, p. 125.
20. Buchanan, *The Road to Guilford Courthouse*, pp. 381–82.

22. "HANDSOMELY IN A PUDDING BAG": THE CHESAPEAKE CAPES, 5–13
SEPTEMBER 1781; AND YORKTOWN 28 SEPTEMBER–19 OCTOBER 1781

1. Macksey, *The War for America*, p. 418.
2. There are a variety of dates in various sources. For example, Boatner (*Encyclopedia*,

p. 1237) says, "I have accepted Freeman's statement [Douglas Southall Freeman, *George Washington*, 7 vols, 1948–57] that De Grasse arrived on 26 Aug., only a day ahead of Hood. Other authorities have a later date." The phrase "only a day ahead of Hood" is puzzling. I have used the dates in Jack Coggins, *Ships and Seamen of the American Revolution*, p. 199; Morrissey, *Yorktown*, p. 53; and W. J. Wood, *Battles of the Revolutionary War*, p. 267.

3. Pearson, *The Revolutionary War*, p. 382.

4. The great historian of American naval warfare, Alfred T. Mahan, succinctly described the weather-gage as "the wind [that] allowed [a ship] to steer for her opponent, and did not let the latter head straight for her."

5. Thirty-two-pounders, either long for more accuracy and range, or short to save weight: the long 32-pounder with its nine-and-a-half-foot barrel weighed in at about 5,500 pounds; the short at eight feet, 4,900 pounds.

6. Pearson, *The Revolutionary War*, p. 383.

7. Coggins, *Ships and Seamen*, p. 153.

8. Ibid., p. 157.

9. Adam Nicolson, *Seize the Fire: Heroism, Duty, and the Battle of Trafalgar*, p. 211.

10. Wood, *Battles of the Revolutionary War*, p. 284.

11. Ibid.

12. Schear and Rankin, *Rebels and Redcoats*, p. 493.

13. Commager and Morris, *The Spirit of Seventy-Six*, p. 1201.

14. Ibid., p. 1202.

15. Pearson, *The Revolutionary War*, p. 387.

16. Commager and Morris, *The Spirit of Seventy-Six*, p. 1218.

17. Martin and Lender, *A Respectable Army*, p. 162.

18. Douglas Southall Freeman, *Washington* (1968 abridged ed.), p. 457.

19. Commager and Morris, *The Spirit of Seventy-Six*, p. 1225.

20. Morrissey, *Yorktown*, p. 34.

21. Pearson, *The Revolutionary War*, p. 388.

22. Higginbotham, *The War of American Independence*, p. 382.

23. Pearson, *The Revolutionary War*, p. 389.

24. Ketchum says that it was 400 Deux-Ponts "plus the Gâtinais regiment" (p. 230). *Yorktown*, The Gâtinais had two battalions totaling 1,000 men, and it seems improbable that the whole regiment was committed; it is more likely that combined elements from the two regiments made up the 400 men in the assault group. However, Private Johann Döhla of the Ansbach-Bayreuth regiment records, "Supposedly three thousand men, French and American, took part in this storming operation." (Bruce E. Burgoyne, *A Hessian Diary of the American Revolution*, p. 170.)

25. Ketchum, *Yorktown*, p. 232.

26. Ibid., p. 397.

27. Burgoyne, *A Hessian Diary*, p. 168.

28. Pearson, *The Revolutionary War*, p. 398.

29. Burgoyne, *A Hessian Diary*, p. 172.

30. Primarily in the memoirs of Count Mathieu Dumas. Thacher, who was there, says that O'Hara went first to Washington. Döhla does not mention Rochambeau.

31. Ketchum, *Yorktown*, p. 253.

SELECT BIBLIOGRAPHY

BOOKS

Alden, John R. *The South in the Revolution, 1763–1789*. 1957.

Anderson, Troyer S. *Command of the Howe Brothers during the American Revolution*. 1936.

Atwood, Rodney. *The Hessians: Mercenaries from Hessen-Kassel in the American Revolution*. 1980.

Axelrod, Alan. *Chronicle of the Indian Wars from Colonial Times to Wounded Knee*. 1993.

Babits, Lawrence E. *A Devil of a Whipping: The Battle of Cowpens*. 1998.

Bailey, D.W. *British Military Longarms, 1715–1815*. 1971.

Bailyn, Bernard. *The Faces of Revolution: Personalities and Themes in the Struggle for Independence*. 1990.

Balderston, Marion, and David Syrett, eds. *The Lost War: Letters from British Officers during the American Revolution*. 1975.

Barnett, Corelli. *Britain and Her Army, 1509–1970: A Military, Political and Social Survey*. 1970, 1974 (paperback edition).

Bearss, Edwin C. *Battle of Cowpens: A Documented Narrative and Troop Movement Maps*. 1967.

Berg, Fred Anderson. *Encyclopedia of Continental Army Units: Battalions, Regiments, and Independent Corps*. 1972.

Bicheno, Hugh. *Rebels and Redcoats: The American Revolutionary War*. 2003.

Billias, George Athan. *George Washington's Generals and Opponents: Their Exploits and Leadership*. 1994 (paperback edition).

Birnbaum, Louis. *Red Dawn at Lexington*. 1986.

Black, Jeremy. *Warfare in the Eighteenth Century*. 1999.

Blackmore, H. L. *British Military Firearms, 1650–1850*. 1961.

Blaufarb, Rafe. *The French Army, 1750–1820*. 2002.

Bliven, Bruce. *Under the Guns: New York, 1775–1776.* 1972.

Blumenthal, Walter Hart. *Women Camp Followers of the American Revolution.* 1994.

Boatner, Mark M. *Encyclopedia of the American Revolution.* 1966.

Bobrick, Benson. *Angel in the Whirlwind: The Triumph of the American Revolution.* 1997.

Bodle, Wayne K. *The Valley Forge Winter: Civilians and Soldiers in War.* 2002.

Bolton, Charles Knowles. *The Private Soldier Under Washington.* 1902.

Boorstin, Daniel J. *The Americans: The Colonial Experience.* 1958.

Bowles, R. Arthur. *Logistics and Failure of the British Army in America, 1775–1783.* 1975.

Bowman, Allen. *The Morale of the American Revolutionary Army.* 1943.

Bowman, Larry G. *Captive Americans: Prisoners during the American Revolution.* 1976.

Bray, Robert C., and Paul E. Bushnell, eds. *Diary of a Common Soldier in the American Revolution, 1775–1783. An Annotated Edition of the Military Journal of Jereiah Greenman.* 1978.

Brogan, Hugh. *The Pelican History of the United States of America.* 1985.

Brown, Richard D., ed. *Major Problems in the Era of the American Revolution, 1760–1791.* 1992.

Brown, Wallace. *The Good Americans: The Loyalists in the American Revolution.* 1969.

Brumwell, Stephen. *Redcoats: The British Soldier and the War in the Americas, 1755–1763.* 2002.

Buchanan, John. *The Road to Guilford Courthouse: The American Revolution in the Carolinas.* 1997.

———. *The Road to Valley Forge: How Washington Built the Army that Won the Revolution.* 2004.

Buel, Richard. *Dear Liberty: Connecticut's Mobilization for the Revolutionary War.* 1980.

Burgoyne, Bruce E., trans. *The Diary of Lieutenant von Bardeleben and other von Donop Documents.* 1998.

———. *A Hessian Diary of the Revolution by Johann Conrad Döhla.* 1990.

Burns, Eric. *The Spirits of America: A Social History of Alcohol.* 2003.

Calhoon, Robert McCluer. *The Loyalists in Revolutionary America, 1760–1781.* 1973.

Callahan, North. *George Washington: Soldier and Man.* 1972.

Calloway, Colin G. *The American Revolution in Indian Country.* 1995.

Carp, E. Wayne. *To Starve the Army at Pleasure: Continental Army Administration and American Political Culture, 1775–1783.* 1990.

Carrington, Henry B. *Battle Maps and Charts of the American Revolution.* 1974.

Caruana, Adrian B. *Grasshoppers and Butterflies: The Light 3 Pounders of Pattison and Townshend.* 1979.

———. *The Light 6-Pdr. Battalion Gun of 1776.* 1977.

Chandler, David. *The Art of Warfare in the Age of Marlborough.* 1976.

———. *Dictionary of the Napoleonic Wars.* 1979.

Chartrand, René. *The French Army in the American War of Independence.* 1991.

———. *The French Soldier in Colonial America.* 1984.

Chidsey, Donald B. *The War in the South: The Carolinas and Georgia in the American Revolution*. 1969.

Coggins, Jack. *Ships and Seamen of the American Revolution*. 1969.

Commager, Henry Steele, and Richard B. Morris, eds. *The Spirit of Seventy-Six: The Story of the Revolution as Told by its Participants*. 1958.

Cook, Don. *The Long Fuse: How England Lost the American Colonies, 1760–1785*. 1996.

Cox, Caroline. *A Proper Sense of Honor: Service and Sacrifice in George Washington's Army*. 2004.

Crocker, Glenys. *The Gunpowder Industry*. 1986.

Crow, Jeffrey J., and Larry E. Tise. *The Southern Experience in the American Revolution*. 1978.

Cummings, William P. *The Fate of a Nation: The American Revolution through Contemporary Eyes*. 1975.

Cunliffe, Marcus. *Soldiers and Civilians: The Martial Spirit in America, 1775–1865*. 1968.

Curtis, Edward E. *The British Army in the American Revolution*. 1926.

Dann, John C., ed. *The Revolution Remembered: Eyewitness Accounts of the War for Independence*. 1980.

Darling, Anthony D. *Red Coat and Brown Bess*. 1971.

Davis, Burke. *America's First Army*. 1962.

———. *The Campaign that Won America: The Story of Yorktown*. 1970.

———. *The Cowpens-Guilford Courthouse Campaign*. 1962.

Desjardin, Thomas. *Through a Howling Wilderness: Benedict Arnold's March to Quebec, 1775*. 2005.

Downey, Fairfax. *Indian Wars of the United States Army, 1776–1865*. 1963.

Duffy, Christopher. *The Military Experience in the Age of Reason*. 1987.

Dupuy, Colonel T. N. *The Evolution of Weapons and Warfare*. 1980.

Edgar, Walter. *Partisans and Redcoats: The Southern Conflict that Turned the Tide of the American Revolution*. 2001.

Egnal, Marc. *A Mighty Empire: The Origins of the American Revolution*. 1988.

Ellet, Elizabeth P. *The Women of the American Revolution*. 1969.

Ellis, John. *Armies in Revolution*. 1974.

Elting, John R. *The Battle of Bunker's Hill*. 1975.

Engelman, R. C., and J. T. Joly. *200 Years of Military Medicine*. 1975.

Fischer, David Hackett. *Paul Revere's Ride*. 1994.

———*Washington's Crossing*. 2004.

Fitzpatrick, John C., ed. *The Writings of George Washington from the Original Manuscript Sources, 1745–1799*. 1931–44.

Fleming, Thomas J. *The Battle of Yorktown*. 1968.

Fortescue, Sir John. *The War of Independence: The British Army in North America, 1775–1783*. 2001. First published as *History of the British Army*, vol. 3, 1911.

Fowler, William M. *Rebels Under Sail: The American Navy During the Revolution*. 1976.

French, Allen. *The First Year of the American Revolution.* 1934.

Frey, Silvia R. *The British Soldier in America: A Social History of Military Life in the Revolutionary Period.* 1981.

———.*Water from the Rock: Black Resistance in the Revolutionary Age.* 1991.

Fuller, Major General J. F. C. *Decisive Battles of the USA.* 1942.

Furneaux, Rupert. *The Battle of Saratoga.* 1983.

Gallagher, John J. *The Battle of Brooklyn, 1776.* 1995.

Galvin, Gen. John R. *The Minutemen: The First Fight: Myths and Realities of the American Revolution.* 1989.

Gibbes, Robert W., ed. *Documentary History of the American Revolution.* 1853.

Gluckman, Colonel Arcadi. *United States Muskets, Rifles, and Carbines.* 1959.

Goodwin, A., ed. *The New Cambridge Modern History.* Vol. 8, *The French and American Revolutions, 1763–93.* 1965.

Gould, Dudley C. *Times of Brother Jonathan: What He Ate, Drank, Wore, Believed In, and Used for Medicine During the War of Independence.* 2001.

Graymont, Barbara. *The Iroquois in the American Revolution.* 1972.

Gross, Robert A. *The Minutemen and Their World.* 1976.

Grossman, Lieutenant Colonel Dave. *On Killing.* 1995.

Gruber, Ira D. *The Howe Brothers and the American Revolution.* 1972.

———, ed. *John Peebles' American War, 1776–82.* 1997.

Hairr, John. *Guilford Courthouse: Nathanael Greene's Victory in Defeat, March 15, 1781.* 2002.

Hamilton, E. P. *The French Army in America.* 1967.

Harris, Sharon, ed. *Women's Early American Historical Narratives.* 2003.

Harvey, Robert. *"A Few Bloody Noses": The Realities and Mythologies of the American Revolution.* 2001.

Hatch, Louis Clinton. *The Administration of the American Revolutionary Army.* 1904.

Hawke, David Freeman. *Everyday Life in Early America.* 1988.

Haythornthwaite, Philip J. *Weapons and Equipment of the Napoleonic Wars.* 1979.

Hess, Earl J. *The Union Soldier in Battle: Enduring the Ordeal of Combat.* 1997.

Hibbert, Christopher. *Redcoats and Rebels: The American Revolution through British Eyes.* 1990.

Higginbotham, Don. *Daniel Morgan: Revolutionary Rifleman.* 1961.

———. *George Washington and the American Military Tradition.* 1985.

———, ed. *George Washington Reconsidered.* 2001.

———. *The War of American Independence: Military Attitudes, Policies, and Practice, 1763–1789.* 1971.

Hoffman, Ronald, and Peter J. Albert, eds., *Arms and Independence: The Military Character of the American Revolution.* 1984.

Holmes, Richard. *Redcoat: The British Soldier in the Age of Horse and Musket.* 2001.

Houlding, J. A. *Fit for Service: The Training of the British Army, 1715–1795.* 1981.

———. *French Arms Drill of the 18th Century.* 1988.

Huddleston, Joe D. *Colonial Riflemen in the American Revolution*. 1978.

Hughes, Major General B. P. *British Smooth-Bore Artillery in the Eighteenth and Nineteenth Centuries*. 1969.

———. *Firepower: Weapons Effectiveness on the Battlefield, 1630–1850*. 1974.

Huston, James A. *Logistics of Liberty: American Services of Supply in the Revolutionary War and After*. 1991.

———. *The Sinews of War: Army Logistics, 1775–1953*. 1966.

Jackson, Donald, and Dorothy Twohig, eds. *Diaries of George Washington*. 1976.

Jackson, Melvin H., and Carel De Beer. *Eighteenth-Century Gunfounding*. 1974.

Jobé, Joseph, ed. *Guns: An Illustrated History of Artillery*. 1971.

Jones, John. *Plain Concise Practical Remarks on the Treatment of Wounds and Fractures*. 1775.

Kail, Jerry, et al., eds. *Who Was Who During the American Revolution*. 1976.

Karsten, Peter, ed. *The Military in America from the Colonial Era to the Present*. 1986.

Katcher, Philip R. N. *King George's Army, 1775–1783: A Handbook of British, American and German Regiments*. 1973.

———. *US Infantry Equipments, 1775–1910*. 1989.

Kaufman, Henry J. *The Pennsylvania-Kentucky Rifle*. 1960.

Keegan, John, and Richard Holmes. *Fields of Battle: The Wars for North America*. 1996.

———. *Soldiers: A History of Men in Battle*. 1985.

Kennett, Lee. *The French Forces in America, 1780–83*. 1977.

Ketchum, Richard M. *Decisive Day: The Battle for Bunker Hill*. 1962.

———. *Saratoga: Turning Point of America's Revolutionary War*. 1997.

———. *Victory at Yorktown: The Campaign That Won the Revolution*. 2004.

———. *The Winter Soldiers: The Battles for Trenton and Princeton*. 1973.

Kipping, Ernst. *The Hessian View of America 1776–1783*. 1971.

Knollenberg, Bernard. *Washington and the Revolution*. 1940.

Konstam, Angus. *Guilford Courthouse, 1781: Lord Cornwallis's Ruinous Victory*. 2002.

Laffin, John. *Surgeons in the Field*. 1970.

Lancaster, Bruce. *The American Revolution*. 2001 (paperback edition).

Leckie, Robert. *George Washington's War: The Saga of the American Revolution*. 1992.

Lee, Robert E., ed. *The Revolutionary War Memoirs of General Henry Lee*. 1869.

Lengel Edward G. *General George Washington: A Military Life*. 2005.

Lesser, Charles H. *The Sinews of Independence: Monthly Strength Reports of the Continental Army*. 1976.

Lively, Robert A. *This Glorious Cause: The Adventures of Two Company Officers in Washington's Army*. 1958.

Lloyd, Ernest M. *A Review of the History of Infantry*. 1908.

Lowell, Edward J. *The Hessians and the Other German Auxiliaries of Great Britain in the Revolutionary War*. 1884.

Lumpkin, Henry. *From Savannah to Yorktown*. 1987.

Lunt, James. *John Burgoyne of Saratoga*. 1976.

Luvaas, Jay, ed. and trans. *Frederick the Great on the Art of War*. 1966.

Macksey, Piers. *The War for America, 1775–1783*. 1964.

Main, Jackson Turner. *The Social Structure of Revolutionary America*. 1965.

Malone, Patrick M. *The Skulking Way of War: Technology and Tactics among the New England Indians*. 1991.

Manucy, Albert. *Artillery through the Ages: A Short Illustrated History of Cannon Emphasizing Types Used in America*. 1949.

Martin, David G. *The Philadelphia Campaign, June 1777–July 1778*. 1993.

Martin, James Kirby, and Mark Edward Lender. *A Respectable Army: The Military Origins of the Republic, 1763–1789*. 1982.

Martin, Joseph Plumb. *A Narrative of a Revolutionary Soldier*. 2001 (paperback edition).

Matloff, Maurice, ed. *The Revolutionary War: A Concise History of America's War for Independence*. 1980.

Merlant, Captain Joachim. *Soldiers and Sailors of France in the America War for Independence*. 1920.

Metzger, Charles H. *The Prisoner in the American Revolution*. 1971.

Middlekauff, Robert. *The Glorious Cause: The American Revolution, 1763–1789*. 1982.

Miller, Nathan. *Sea of Glory: A Naval History of the American Revolution*. 1974.

Milsop, John. *Continental Infantryman of the American Revolution*. 2004.

Mintz, Max M. *The Generals of Saratoga: John Burgoyne and Horatio Gates*. 1990.

Mitchell, Joseph B. *Discipline and Bayonets: The Armies and Leaders in the War of the American Revolution*. 1967.

Montross, Lynn. *Rag, Tag and Bobtail*. 1952.

Moore, Warren. *Weapons of the American Revolution*. 1967.

Morrill, Dan L. *The Southern Campaigns of the American Revolution*. 1993.

Morrissey, Brendan. *Boston, 1775: The Shot Heard Around the World*. 1993.

———. *Monmouth Courthouse, 1778: The Last Great Battle in the North*. 2004.

———. *Quebec, 1775: The American Invasion of Canada*. 2003.

———. *Saratoga, 1777: Turning Point of a Revolution*. 2000.

———. *Yorktown, 1781*. 1997.

Namier, Lewis. *England in the Age of the American Revolution*. 1961.

Neimeyer, Charles Patrick. *America Goes to War: A Social History of the Continental Army*. 1996.

Nelson, Paul David. *William Alexander, Lord Stirling*. 1987.

Nester, William R. *The Frontier War for American Independence*. 2004.

Neuberg, Victor. *Gone for a Soldier: A History of Life in the British Ranks from 1642*. 1989.

Neuman, George C. *History of the Weapons of the American Revolution*. 1976.

Neuman, George C., and Frank Kravic. *Collector's Illustrated Encyclopedia of the American Revolution*. 1975.

Nicolson, Adam. *Seize the Fire: Heroism, Duty and the Battle of Trafalgar*, 2005.

O'Connell, Robert L. *Of Arms and Men: A History of War, Weapons, and Aggression*. 1989.

Onderdonk, Henry. *Revolutionary Incidents in Suffolk and Kings Counties*. 1844.

Palmer, Dave Richard. *The Way of the Fox: American Strategy in the War for America, 1775–83*. 1975.

Pearson, Michael, *The Revolutionary War: An Unbiased Account*. 1972.

Peckham, Howard H. *The Toll of Independence: Engagements and Battle Casualties of the American Revolution*. 1974.

————. *The War for Independence: A Military History*. 1958.

Perrett, Bryan. *The Battle Book: Crucial Conflicts in History from 1469 BC to the Present*. 1992.

Peterson, Harold L. *Arms and Armor in Colonial America, 1526–1783*. 1956.

————. *The Book of the Continental Soldier*. 1968.

Pettengill, Ray W. *Letters from America, 1776–1779*. 1924.

Plumb, J. H. *England in the Eighteenth Century, 1714–1815*. 1950.

Porter, Roy. *English Society in the 18th Century*. 1982.

Quarles, Benjamin. *The Negro in the American Revolution*. 1961, 1973 (paperback edition).

Rankin, Hugh F., ed. *Narratives of the American Revolution*. 1976.

Raphael, Ray. *A People's History of the American Revolution*. 2001.

Reid, Stuart, and Marko Zlatich. *Soldiers of the Revolutionary War*. 2002.

Reiss, Oscar. *Medicine and the American Revolution*. 1998.

Rhodehamel, John, ed. *The American Revolution: Writings from the War of Independence*. 2001.

Rice, Howard C., and Ann S. K. Brown, eds. *The American Campaigns of Rochambeau's Army*. 1972.

Risch, Erna. *Supplying Washington's Army*. 1981.

Roberts, Kenneth, ed. *March to Quebec: Journals of the Members of Arnold's Expedition*. 1938.

Rogers, Colonel H. C. B. *Weapons of the British Soldier*. 1972.

Ross, Steven T. *From Flintlock to Rifle: Infantry Tactics, 1740–1866*. 1996.

Rossie, Jonathan Gregory. *The Politics of Command in the American Revolution*. 1975.

Rothenberg, Gunther E. *The Art of Warfare in the Age of Napoleon*. 1977.

Royster, Charles. *A Revolutionary People at War: The Continental Army and American Character, 1775–1783*. 1979.

Ryan, Dennis P. *A Salute to Courage: The American Revolution as Seen through the Wartime Writings of Officers of the Continental Army and Navy*. 1979.

Sandel, Edward. *Black Soldiers in the Colonial Militia: Documents from 1639–1780*. 1994.

Sawyer, Charles Winthrop. *Firearms in American Society, 1600–1800*. 1910.

Schecter, Barnet. *The Battle for New York*. 2003.

Scheer, George F., and Hugh F. Rankin. *Rebels and Redcoats: The American Revolution Through the Eyes of Those Who Fought and Lived It*. 1957.

Schwalm, Mark. *The Hessians: Auxiliaries to the British Crown in the American Revolution*. 1984.

Seymour, William. *The Price of Folly: British Blunders in the War of American Independence*. 1995.

Shumway, George. *Rifles of Colonial America*. Vols 1 and 2. 2002 (reprint).

Shy, John. *Toward Lexington: The Role of the British Army in the Coming Revolution*. 1965.

Smith, Paul H. *Loyalists and Redcoats: A Study in British Revolutionary Policy*. 1964.

Sosin, Jack M. *The Revolutionary Frontier, 1763–83*. 1967.

Starkey, Armstrong. *European–Native American Warfare, 1675–1815*. 1998.

Svedja, George J. *Quartering, Disciplining, and Supplying the Army at Morristown, 1779–80*. 1970.

Symonds, Craig L. *A Battlefield Atlas of the American Revolution*. 1986.

Tarleton, Banastre. *A History of the Campaigns of 1780 and 1781 in the Southern Provinces of North America*. 1787.

Thacher, James. *Military Journal of the American Revolution, 1775–1783*. 1862.

Thayer, Theodore. *Washington and Lee at Monmouth: The Making of a Scapegoat*. 1976.

Todd, Frederick P. *Soldiers of the American Army, 1775–1954*. 1954.

Tourtellot, Arthur B. *Lexington and Concord*. 1959.

Treacy, M. F. *Prelude to Yorktown: The Southern Campaigns of Nathanael Greene, 1780–1781*. 1963.

Tuchman, Barbara W. *The First Salute: A View of the American Revolution*. 1988.

Underdal, Stanley J., ed. *Military History of the American Revolution: Proceedings of the 6th Military History Symposium, USAF Academy*. 1974.

Unger, Harlow Giles. *Lafayette*. 2002.

Van Tyne, Claude Halstead. *Loyalists in the American Revolution*. 1902.

Voelz, Peter Michael. *Slave and Soldier: The Military Impact of Blacks in the Colonial Americas*. 1993.

Wallace, Willard M. *Appeal to Arms: A Military History of the American Revolution*. 1975.

Ward, Christopher. *The War of the Revolution*. 1952.

Weighley, Russell F. *The American Way of War: A History of the United States Military Strategy and Policy*. 1973

———. *The Partisan War: The Southern Carolina Campaign of 1780–82*. 1970.

Weller, Jac. *Weapons and Tactics: Hastings to Berlin*. 1966.

Wheeler, Richard. *Voices of 1776*. 1972.

Wickwire, Franklin, and Mary Wickwire. *Cornwallis: The American Adventure*. 1970.

Wilbur, C. Keith. *An Illustrated Sourcebook of Authentic Details About Everyday Life for Revolutionary War Soldiers*. 1993.

———*Revolutionary Medicine, 1700–1800*. 1980.

Willard, Margaret Wheeler, ed. *Letters on the American Revolution, 1774–76*. 1925.

Willcox, William B. *The American Rebellion*. 1954.

———. *Portrait of a General: Sir Henry Clinton in the War of Independence*. 1964.

Williams, Glen F. *Year of the Hangman: George Washington's Campaign Against the Iroquois*, 2005.

Williams, Harry T. *Americans at War: The Development of the American Military System*. 1960.

Wilson, Joseph T., and Dudley Taylor Cornish. *Black Phalanx: African American Soldiers in the War of Independence, the War of 1812, and the Civil War*. 1968.

Wolf, Stephanie Grauman. *As Various As Their Lands: The Everyday Lives of Eighteenth-Century Americans*. 1994.

Wood, Gordon S. *The Radicalism of the American Revolution*. 1991.

Wood, W. J., *Battles of the Revolutionary War, 1775–1781*. 1990.

Wright, Robert K. *The Continental Army*. 1989.

Wrong, George M. *Washington and His Comrades in Arms*. 1921.

Young, Alfred F. *Masquerade: The Life and Times of Deborah Sampson, Continental Soldier*. 2004.

Zlatich, Marko. *General Washington's Army*. Vol. 1, *1775–78*. 1994. Vol. 2, *1779–83*. 1995.

ARTICLES

Anderson, Clive. "The Treatment of Prisoners of War in Britain during the American War of Independence." *Bulletin of the Institute of Historical Research* 28 (1955): 63–83.

Atkinson, Christopher T. "British Forces in North America, 1774–81: Their Distribution and Strength." *Journal of the Society for Army Historical Research* 20 (1941): 208–23.

Blano, R. "Medicine in the Continental Army, 1775–81." *Bulletin of the New York Academy of Medicine* 57, no. 8 (1981): 677–701.

Block, F. "Military Medicine in the Eighteenth Century." *Military Surgeon* 65 (1929): 561–76.

Bradford, Sydney S. "A British War Officer's Revolutionary War Journal, 1776–1778." *Maryland Historical Magazine* 56 (1961): 150–75.

———. "The Common British Soldier—from the Journal of Thomas Sullivan, 49th Regiment of Foot." *Maryland Historical Magazine* 62 (1967): 219–53.

Bright, James R. "The Rifle in Washington's Army." *American Rifleman* 8 (August 1947): 7–10.

Bulger, William T., ed. "Sir Henry Clinton's 'Journal of the Siege of Charleston, 1780.'" *South Carolina Historical Magazine* 66 (1965): 147–75.

Caswell, R. D. "Operating the Flintlock Rifle." *Muzzle Blast Online* 4, no. 4 (August–September 1999).

Cox, Elbert. "Winter Encampments of the American Revolution." *Regional Review* 1, no. 2 (August 1938): 3–7.

Davis, Andrew McFarland. "The Employment of Indian Auxiliaries in the American War." *English Historical Review* 2 (1887): 709–28.

Echeverria, Durand, and Orville T. Murphy. "The American Revolutionary Army: A French Estimate in 1777." *Military Affairs* 27, no. 1 (Spring 1963): 1–7.

Evans, Robert M. "The King of Battle: Ruler of Yorktown." *Field Artillery Journal* 49 (September–October 1981): 30–31.

Gottschalk, Louis. "The Attitude of European Officers in the Revolutionary Armies

Toward General George Washington." *Illinois State Historical Society* 32 (1939): 20–50.

Hopkins, Alfred F. "Equipment of the Soldier during the American Revolution." *Regional Review* 4, no. 3 (March 1940): 19–22.

Jameson, Hugh. "Equipment for the Militia of the Middle States, 1775–1781." *Military Affairs* 3 (1939): 26–38.

Kellogg, Louise Phelps. "Journal of a British Officer during the American Revolution." *Mississippi Valley Historical Review* 7, no. 1 (June 1920): 51–58.

Kenmotsu, Nancy. "Gunflints: A Study." *Historical Archaeology* 24, no. 2 (1990): 92–124.

Klein, Lance. "This Barbarous Weapon." *Muzzle Blast Online* 5, no. 1 (February–March 2000) (http://www.muzzleblasts.com).

Knollenberg, Bernard. "Bunker Hill Reviewed." *Massachusetts Historical Society Proceedings* 72 (1957–60): 84–100.

Lutnick, Solomon, M. "The Defeat at Yorktown: A View from the British Press." *Virginia Magazine of History and Biography* 72 (1964): 471–78.

Mahan, John K. "Anglo-American Methods of Indian Warfare, 1676–1794." *Mississippi Valley Historical Review* 45, no. 2 (September 1958): 254–75.

Mann, JoAnn. "Black Americans in the War for Independence." *Soldiers* 30 (January 1975): 30–35.

Nelson, Paul David. "British Conduct of the American Revolutionary War: A Review of Interpretations." *Journal of American History* 65, no. 3 (December 1978): 623–53.

Pickard, Allen H. "Tell General Sullivan to Use the Bayonet." *Airman* 6 (July 1962): 44–48.

Rees, John U. "The Multitude of Women: An Examination of the Numbers of Female Camp Followers with the Continental Army." *Brigade Dispatch* 23, no. 4 (Autumn 1992): 5–17; 24, no. 1 (Winter 1993): 6–16; 24, no. 2: 2–16.

Reichman, Felix. "The Pennsylvania Rifle: A Social Interpretation of Changing Military Techniques." *Pennsylvania Magazine of History and Biography* 69 (1945): 3–14.

Riley, Edward M. "St. George Tucker's Journal of the Siege of Yorktown, 1781." *William and Mary Quarterly*, 3rd ser., vol. 5, no. 3 (July 1948): 375–95.

Robson, Eric. "Purchase and Promotion in the British Army of the Eighteenth Century." *History* 36 (1951): 57–72.

Russell, Peter E. "Redcoats in the Wilderness: British Officers and Irregular Warfare in Europe and America, 1740 to 1760." *William and Mary Quarterly*, 3rd ser., vol. 35, no. 4 (October 1978): 629–52.

Stephenson, Orlando W. "The Supply of Gunpowder in 1776." *American Historical Review* 30, no. 2 (January 1925): 271–81.

Thomas, David Y. "How Washington Dealt with Discontent." *South Atlantic Quarterly* 32 (1933): 63–73.

Tucker, Spencer C. "Cannon Founders of the American Revolution." *National Defense* 60 (1975): 33–37.

Weller, Jac. "The Artillery of the American Revolution." *Military Collector and Historian*, pt. 1, vol. 8, no. 3 (Fall 1956): 61–65; pt. 2, vol. 8, no. 4 (Winter 1956): 97–101.

———. "Irregular but Effective: Partisan Weapons Tactics in the American Revolution, Southern Theatre." *Military Affairs* 21, no. 3 (Autumn 1957): 118–31.

White, Herbert H. "British Prisoners of War in Hartford during the Revolution." *Connecticut Historical Society Bulletin* 19, no. 3, 65–81.

Wright, John W. "The Corps of Light Infantry in the Continental Army." *American Historical Review* 31 (1926): 454–61.

———. "The Rifle in the American Revolution." *American Historical Review* 2, no. 29 (January 1924): 292–99.

Young, Rogers W. "Kings Mountain: A Hunting Rifle Victory." *Regional Review* 3, no. 6 (December 1939): 25–29.

INDEX

battle of); burning of, strategy, 230; fortifications of, 231; patriot defense and evacuation of, 244–48; Washington's army on, 244. *See also* Brooklyn, battle of; Kips Bay, battle of; New York

Marion, Brig. Gen. Francis, 16

Martin, Pvt. Joseph Plumb, 81–82, 89–92, 110–11, 112, 113, 157, 179–80, 181, 243, 245

Maryland: Continental army regiments, occupations prior to service, 29; Continentals, xviii, 18, 66–67, 317, 336, 338, 340; "Dandy Fifth" at battle of Brooklyn, 237; enlistment of indentured servants, 27; German immigration to, 30; militia, 251; provincial regiments, 56; riflemen of, 132–33; rifle use in, 120; vagrants and felons conscripted, 25–26

Massachusetts: Act of Banishment, 53; blacks in service, 186; bounty offered by, 24; Committee of Safety, 184; Concord militia, 8; Continentals, 180, 247, 294; draft in, 25; 4th and 9th regiments, occupations prior to service, 29; free blacks in, 184; Marblehead Regiment, 242–43, 247; troops at battle of Bunker's Hill, 212; troops in battle of Brooklyn, 242

Mathew, Maj. Gen. Edward, 250

Mawhood, Lt. Col. Charles, 263, 265, 266

Maxwell, Brig. Gen. William, 272, 273

McCloud, Alexander, 54

McCullough, David, xviii–xix

McDonald, Gen. Donald, 54

McDonough, Maj. Thomas, 237

McDougall, Gen. Alexander, 77, 248, 278

McDowell, Col. Joseph, 327, 330

McGary, Hugh, 197

McLeod, Lt. John, 339

McPherson, Lt. Col. Duncan, 336

medical care, 170–73; amputation, 171–72; analgesics, 172; drug shortage, 170; drugs used, 170, 175–76; hospitals, 172–73; inoculation for smallpox, 175–76

medical personnel, 167–70; College of Surgeons, 169; Hospital Department, heads of, 168–69; infrastructure for, both sides, 168; medicine chests of, 170; regimental surgeons and surgeon's mates, 169, 170

Memoirs (H. Lee), 19

Memoirs (Lamb), 94

mercenaries, 48–52; in European armies, 43; German, 44. *See also* Hessians

Mercer, Gen. Hugh, 71, 258, 265

Mifflin, Gen. Thomas, 103, 104

Miles, Col. Samuel, 236–37

militia, 6–19; avoidance of service by wealthy, 7, 14; battle experience of, 44; battle of Brooklyn and quitting of, 243–44, 251; battle of Hobkirk's Hill, 13; battle of Quebec, 12; bounties offered by, 24; Concord, Massachusetts, 8; Connecticut, 9, 223, 243; desertion in, 8; draft from into army, 26; exempt populace, 7; failure of, causes, 17; formal warfare and, 17–18; Georgia, 18, 327; Greene's assessment of, 15; independent mindset of, 8, 11, 16; Kentucky, 197; lack of discipline among, 8; land promised to soldiers, 22, 24; leadership of, 17–19; "leveling" tradition in, 8–10; at Lexington and Concord, 16, 205; Loyalists coerced into, 58–59; Loyalist suppressed by, 19; Maryland, 251; New Hampshire, 216, 217, 298; New Jersey, 25, 251, 279; New York, 9, 223; North Carolina, xvii, 7, 13, 16, 17, 18, 19, 287, 327, 334, 357n27; officer corps, 65–66; officers in,